AF251808

TIME & MATTER

Proceedings of the International Colloquium on the Science of Time

TIME & MATTER

Venice, Italy   11 – 17 August 2002

editors

Ikaros I Bigi
University of Notre Dame du Lac, USA

Martin Faessler
Ludwig-Maximilians-Universität, Germany

World Scientific

NEW JERSEY · LONDON · SINGAPORE · BEIJING · SHANGHAI · HONG KONG · TAIPEI · CHENNAI

Published by

World Scientific Publishing Co. Pte. Ltd.

5 Toh Tuck Link, Singapore 596224

USA office: 27 Warren Street, Suite 401-402, Hackensack, NJ 07601

UK office: 57 Shelton Street, Covent Garden, London WC2H 9HE

British Library Cataloguing-in-Publication Data
A catalogue record for this book is available from the British Library.

TIME AND MATTER
Proceedings of the International Colloquium on the Science of Time

ISBN 981-256-634-1

Printed in Singapore by World Scientific Printers (S) Pte Ltd

CONTENTS

Prologue ix

Section I: Measuring Time

The Measurement of Time with Atomic Clocks 3
E. Peik

Measuring the Frequency of Light 17
T. W. Hänsch

Time and Space Variation of Fundamental Constants:
Motivation and Laboratory Search 26
S. G. Karshenboim

Section II: Causality & Signal Propagation

Causality and Superluminal Light 45
G. M. Shore

Quantum Fluctuations and Signal Velocity in Superluminal
Light Pulse Propagation 67
L. J. Wang

Time and Matter in the Interaction between Gravity and Quantum
Fluids: Are there Macroscopic Quantum Transducers between
Gravitational and Electromagnetic Waves? 85
R. Y. Chiao & W. J. Fitelson

Section III: Coherence & Decoherence

Decoherence Unlimited: From Zeno to Classical Motion 103
E. Joos

Coherence and the Clock 117
L. Stodolsky

Decoherence, Chaos and the Second Law 132
W. H. Zurek

vi

Contributed Papers

What Could We Have Been Missing While Pauli's Theorem Was
in Force?   133
E. A. Galapon

Simultaneity and the Concept of 'Particle'   145
C. E. Dolby

Section IV: CP & T Violation

CP and T Violation in the Kaon System   161
E. Blucher

Time-Reversal Non-Invariance   176
D. Zavrtanik

'Per Aspera ad Astra' — A Short Essay on the Long Quest
for CP Violation   185
I. I. Bigi

Section V: Macroscopic Time Reversal and the Arrow of Time

The Thermodynamic Arrow: Puzzles & Pseudo-Puzzles   209
H. Price

Arrow of Time from Timeless Quantum Gravity   225
C. Kiefer

The Evolution of the Universe   235
J. Garcia-Bellido

Did Time and Its Arrow have a Beginning?   250
G. Veneziano

The Wormhole Hazard   260
S. Krasnikov

Elementary Particles as Black Holes   270
G. 't Hooft

Contributed Paper

Counter-Example where Cosmic Time Keeps Its Original Role in
Quantum Cosmology   271
E. I. Guendelman & A. B. Kaganovich

Section VI: New Paradigms

String Theory, Space and Time   285
C. M. Hull

Quantized Space and Time   303
P. Schupp

List of Participants   319

PROLOGUE

About 60 scholars and students attended the conference "Time and Matter" held August 11th - 17 th, 2002 at the site of Venice International University on San Servolo Island in Venice, Italy. It was called an "International Colloquium on the Science of Time" to point to its goal of being an interdisciplinary meeting. A first indication for the success of this goal was reflected by the list of participants and speakers: they came from countries covering the whole width of Europe – from the United Kingdom in the West to Russia in the East – and from the United States, Israel, the Philippines and Australia. There were experimentalists as well as theorists from many different branches of physics – high energy, nuclear and atomic physics, cosmology and astrophysics, nonlinear dynamics; the philosophy of science was represented as well, as it should be.

It amounts to a triviality to state that physics in the twentieth century has done much more than advanced our knowledge: it has changed our view of the whole world around us – and even of the concept of reality – in a most profound way. Yet at the same time we have become more aware of the limitations in our understanding.

Truly central elements in this evolution are the concepts of *time and matter* which have undergone momentous changes. Several paradigm shifts have occurred in the twentieth century:

- The insight that there is no empirical verification of Newton's ansatz of *absolute* time implied that time and space coordinates transform into each other.

- The realization – dawning at first – of the universe's expansion gave rise to the big bang paradigm as the standard model of cosmology; beyond reformulating the problem of the "arrow of time" it raises the issue of even the *origin* of time.

- The suggestion that in the big bang scenario the baryon number of the universe – the observation that as far as we can tell there exists only matter, but no (primary) antimatter in our universe – might be understandable as a *dynamically generated* and thus *predictable* quantity rather than as an a priori arbitrary *initial value*; it requires a violation of CP symmetry, i.e. of the invariance under the combined transformations of parity (or mirror transformations) and charge conjugation, where the latter denotes the exchange of particles and antiparticles; in the context of quantum field theories this almost automatically implies a violation of *microscopic* time reversal invariance.

- Quantum mechanics with its two classes of time evolution – a smooth one described by the Schrödinger equation and an abrupt one reflecting the interaction caused by a measurement – has pointed to many observable phenomena that have remained counterintuitive and therefore have attracted the epiteth of 'quantum paradoxes'; for example, correlations of the Einstein-Rosen-Podolsky (EPR) type force us to revisit the connection between causality and the speed of light championed by Einstein.

- The challenge to quantize gravity has forced us to scrutinize our basic concept of space and time in its interplay with matter distribution they support; it has lead to the suggestion that there might (or even should) be six or seven extra space dimensions in addition to the familiar three.

- An intense discussion in the theoretical literature has been initiated more than twenty years ago on core issues of quantization of gravity, on the completeness of quantum mechanics and on the concept of locality in space and time by Hawking's profound insight that black holes are not truly 'black', but induce radiation.

The whole spectrum of these problems was addressed at this meeting, typically by leading researchers in the field.

There is an apparent irreversability of time – a daily experience in our lifes – that is quantified through the Second Law of thermodynamics stating that entropy cannot decrease; it is usually perceived as a *macroscopic* phenomenon and understood in a statistical sense as reflecting a practical inequality in achieving conjugate initial conditions. This issue has been re-examined in modern cosmology for the universe as a whole; there does not seem to be a need for revising our perspective on the Second Law in a significant way.

CP violation has been observed first in the decay of neutral kaons in 1964. After almost three decades of dedicated experimentation we have recently established a second source of CP violation in neutral kaon transitions, the so-called 'direct' CP violation; it also was shown under very general theoretical assumptions that CP violation is indeed accompanied by a *microscopic* violation of time reversal invariance. It was just in 2001 that CP violation has finally been observed in another system, namely that of neutral 'beauty' mesons, which represents a spectacular success of the so-called CKM theory. Yet we also understand now that the effects of this theory cannot generate the observed baryon number of the universe.

It is intriguing to note that in these experimental efforts the existence of the aforementioned EPR correlations has been an essential and reliable *tool*. Likewise there are other phenomena like the 'quantum Zeno' effect that do not deserve to be called paradoxes: while they might have essential counterintuitive features, they are unequivocal consequences of quantum mechanics and have been observed; they have even been employed routinely in measurements.

It is certainly true that quantum mechanics introduces novel and complex levels into the notion of time. This has been illustrated by the lively debate about 'tunnelling' times that had been going on for some time, yet now can be considered as settled. Ambiguities are avoided once one gives an *operational* definition of what one calls 'time', i.e. specifies how one measures it; then there arise no problems for causality. It was pointed out at the meeting that a careful analysis of such and similar phenomena might shed new light onto the foundations of quantum mechanics – and at the same time guide us towards some novel applications.

Despite quantum mechanics adding new complexities to the very notion of time it enables us to construct atomic clocks of unprecedented accuracy, for which there does not seem to be a limit to be reached soon. Such highly sophisticated instruments allow us to probe the time structure of the world around us in ways that previously could not even be contemplated. In particular it empowers us to scrutinize how constant in time nature's 'constants' really are. While one can claim at best some indications that the fine structure constant α might have changed its value in the past, both the experimental and theoretical tools are available now, with which we can address this issue in a meaningful and accurate way.

The intense debate on the very concepts of space, time and matter that has been going since the early days of general relativity was reviewed at the conference. Recent novel approaches like string theories and now the so-called M theory have revived the rather intriguing notion of extra dimensions that due to their compactification have escaped detection so far – an idea first suggested by Kaluza and Klein about 80 years ago. Yet it leads also to a profound conundrum: the same general formulism that suggest the existence of extra *space* dimensions can equally lead to extra *time* dimensions! However such an occurrance would lead to profound concerns about how causality could be maintained, which is viewed as a *pre*condition for the ability to describe the world around us in terms of a physical theory. Another recent approach was discussed at the conference, where space and time do not merely provide the stage for quantum phenomena to evolve, but are *themselves* quantized suggesting another radical paradigm shift.

Last, but certainly not least, the challenge to reconcile the existence of Hawking radiation with quantum mechanics as a complete theory has lead to the introduction of novel concepts like 'black hole complementarity' or the 'holographic principle', as discussed by Nobel laureate 't Hooft at the conference. A strong case can be made that a resolution of the *apparent* paradox posed by Hawking radiation will establish the incompleteness of quantum field theory per se and its notions of locality of space and time and point to a new paradigm in our understanding of our universe and of reality.

The conference "Time and Matter" with its unorthodox subject and format was not and could not be based on earlier meetings. It thus represented an experiment, the success of which was far from guaranteed. Yet the clear judgement developed among the participants that it had been very successful in its unusual orientation precisely because of its format of a 'colloquium', where experts from different disciplines present their material in a way that – while challenging – is understandable to other scientists with an open mind, where both theoretical as well as experimental aspects are covered in substance.

An important aspect of the success had not been fully appreciated by the organizers beforehand, namely that it provided students of physics a fascinating and highly stimulating novel perspective onto the field of their study, which strengthens their motivation and even enthusiasm to persist. We are deeply grateful for the generosity of our sponsors that allowed us to invite and support ten students.

Another element of the success did not come as a surprise to the organizers, namely the location. It was the singular beauty of Venice that attracted an elite group of speakers that would have been very difficult to assemble otherwise, and the peace and quiet of San Servolo that allowed participants to mingle and continue their discussions in informal ways. Even beyond that the genius loci of Venice was most fitting: a city between Orient & Occident, where things rarely are as they seem on their surface as described by Thomas Mann in 'Death in Venice', a place famed for its carneval and the masks one wears then.

A clear measure of the success is provided by the desire expressed by many participants how much they look forward to receive the proceedings of the conference – and to have another meeting with a similar (maybe even expanded) agenda and format in the foreseeable future. There are already concrete plans in that direction, namely to have a second conference in 2006 on the East-Adriatic coast nearby.

Unfortunately the publication of these proceedings has been delayed considerably for a variety of reasons. Yet we are convinced that the material presented here has lost neither its great interest nor even its topicality. We have arranged it into six groups:

1. In talking about 'time' an obvious question is how and how well we can measure it, on which assumptions it is based. This is covered in three talks:

    (a) E. Peik: "The Measurement of Time with Atomic Clocks"

    (b) T. Haensch: "Measuring the Frequency of Light"

    (c) S. Karshenboim: "Time and Space Variation of Fundamental Constants: Motivation and Laboratory Search"

2. The connection between causality and superluminal speeds is addressed in general and in the specific contexts of the measurement of tunneling times in quantum mechanics; a first reference to the impact of gravity is given:

    (a) G. Shore: "Causality and Superluminal Light"

    (b) L. Wang: "Quantum Fluctuation and Signal Velocity in Superluminal Light Pulse Propagation"

    (c) R. Chiao & W. Fitelson: "Time and Matter in the Interaction between Gravity and Quantum Fluids: Are there Macroscopic Quantum Transducers between Gravitational and Electromagnetic waves?"

3. Quantum Mechanics with its intriguing interplay of 'coherence and decoherence' leads to several new puzzles like the 'Quantum Zeno Effect' and at the same time resolves some old ones:

    (a) E. Joos: "Decoherence Unlimited:From Zeno to Classical Motion"

    (b) L. Stodolsky: "Coherence and the Clock"

    (c) W. Zurek: "Decoherence, Chaos and the Second Law "

    (d) Contributed Papers:
        E. Galapon: "What Could We Have Been Missing While Paulis Theorem Was in Force?"
        C. Dolby: "Simultaneity and the Concept of 'Particle' "

4. The discovery of CP violation came completely unexpected in 1964, partly since it implied violation of *microscopic* time reversal invariance; both have been observed in the decays of neutral K and B mesons, the latter since 2001:

    (a) E. Blucher: "CP & T Violation in the Kaon System"

(b) D. Zavrtanik: "Time-Reversal Non-Invariance"

(c) C. Jarlskog:"Matter, Antimatter and Time Reversal Asymmetry"

(d) D. Hitlin:"T and CP Violation in the Beauty System"

5. The issue of *macroscopic* violations of time reversal invariance has been discussed for more than a century – often under the label of 'arrow of time'; it raises novel fundamental questions and puzzles in the context of quantum gravity. The 'Big Bang Evolution' of the universe – a cornerstone of the standard model of cosmology – has lead to even more questions on the very nature of time and its arrow, on the notion of a beginning, on the distinction between past and future, on lessons to be inferred from black holes:

(a) H. Price: "The Thermodynamic Arrow: Puzzles & Pseudo-Puzzles"

(b) C. Kiefer: "Arrow of Time from Timeless Quantum Gravity"

(c) D.J. Bellido: "The Evolution of the Universe"

(d) G. Veneziano: "Did Time and its Arrow have a Beginning?"

(e) S. Krasnikov: "The Wormhole Hazard"

(f) G. 't Hooft: "Elementary Particles as Black Holes"

(g) Contributed Paper:
E. Guendelman & A. Kaganovich: "Counter-Example where Cosmic Time Keeps its Original Role in Quantum Cosmology"

6. The challenge of bringing gravity into the quantum world has lead to novel paradigms on the nature of space and time and their dimensions:

(a) C.Hull: "String Theory, Space and Time"

(b) P. Schupp: "Quantized Space and Time"

We have not received the write-up of the talks by G. 't Hooft, W. Zurek, C. Jarlskog and D. Hitlin. To fill this gap for the first two cases we have listed papers by 't Hooft and Zurek covering the subjects of their talks that can be downloaded on the World Wide Web from the Los Alamos servers. The material of Jarlskog's and Hitlin's talks is summarized in the article by Bigi.

We hope these proceedings can provide a self-contained introduction to a fascinating subject in many of its interdisciplinary aspects to dedicated readers, including those who did not attend the conference.

ACKNOWLEDGEMENTS

We are indebted to the following companies, institutions, ministries and departments for financial support of the conference:

- Bundesministerium für Bildung und Forschung, Germany

- Bayerisches Staatsministerium für Wissenschaft, Forschung und Kunst, Germany

- DIE ZEIT, Germany

- Heraeus-Stiftung, Germany

- Oscilloquartz S.A., Switzerland

- Physics Department, University of Notre Dame du Lac, U.S.A.

- Department für Physik, Ludwig-Maximilians-Universität München, Germany

We thank all the individuals who have helped directly to provide financial, organizational and intellectual support, in particular:
Dr. Jean-Pierre Aubry, Meike Dlaboha, Dr. Rainer Esser, Ulrich Fuchs, Prof. Ted Haensch, Prof. Don Howard, Prof. Nicola Khuri, Prof. Tony Hyder, Nicolas Hayek, Dr. h.c. Max Mannheimer, Dr. Dieter Müller, Prof. H. Nicolai, Prof. Lev Okun, Prof. Jack Steinberger, Prof. Val Telegdi, Prof. Herbert Walther, Prof. Bruce Winstein, Prof. Hans D. Zeh, Staatsminister Hans Zehetmair.

Very special thanks go to our late colleague, Prof. Jim Cushing of the University of Notre Dame. He helped us greatly in shaping the program of this conference, and he would have liked to participate in it; alas, fate was not kind enough to let it happen. We want to dedicate these Proceedings to his memory in recognition of the many contributions he has made to issues addressed at the conference.

Munich & Notre Dame, January 2005

Ikaros Bigi and Martin Faessler

SECTION I
MEASURING TIME

The Measurement of Time with Atomic Clocks
  E. Peik

Measuring the Frequency of Light
  T. W. Hänsch

Time and Space Variation of Fundamental Constants: Motivation and Laboratory Search
  S. G. Karshenboim

THE MEASUREMENT OF TIME WITH ATOMIC CLOCKS

EKKEHARD PEIK

Physikalisch-Technische Bundesanstalt, Bundesallee 100, 38116 Braunschweig, Germany,
E-mail: ekkehard.peik@ptb.de

This paper presents a brief review of the history of atomic clocks, explains their
fundamental figures of merit and then describes recent developments that have led
to more accurate clocks using laser-cooled and trapped atoms. Future progress can
be expected from clocks that operate at optical frequencies, instead of microwaves.
Some applications of precise clocks in science and technology are outlined.

1 Introduction

Among all physical quantities, time intervals and frequencies are the ones that can
be measured to the highest precision. This fact becomes evident in the relative uncertainties to which the base units of the *systeme international* (SI) can be realised:
The least accurate is the candela with 10^{-4} uncertainty, followed by the kelvin, the
mol and the ampere with uncertainties in the range 10^{-7} to 10^{-8}. The kilogram
is still defined as the mass of the Pt-Ir prototype that is conserved in the Bureau
International des Poids et Mesures (BIPM). While this definition is intrinsically
exact, kilogram comparisons can be done with 10^{-9} relative uncertainty. The meter is not an independent unit but fixed to the second via the assignment of the
value $c \equiv 299\,792\,458$ m/s for the speed of light in vacuum. Interferometric length
measurements can be done with 10^{-11} relative uncertainty. The realisation of the
second, defined via the ground state hyperfine splitting frequency of the caesium
atom, is several orders of magnitude more accurate: an uncertainty of only 10^{-15}
is achieved by the latest generation of caesium clocks.

Conceptually, the measurement of time is very simple: take a stable periodic
phenomenon with a period that is much shorter than the time interval that has to
be measured. In this case the measurement of time intervals or the establishment
of a local time scale can be reduced to the problem of counting the cycles of this
periodic phenomenon. Periodicities and resonance phenomena are ubiquitous in
nature. Historically, the best long term stability was provided by the rotations of
celestial bodies, and still the rotation of the earth determines our everyday feeling
for time. The great progress in the accuracy of clocks over the last decades, however, came from the use of atomic resonance frequencies. This concept of an atomic
clock has been foreseen by Maxwell as early as 1870 when he stated[1]: *"If, then, we
wish to obtain standards of length, time, and mass which shall be absolutely permanent, we must seek them not in the dimensions, or the motion, or the mass of our
planet, but in the wavelength, the period of vibration, and the absolute mass of these
imperishable and unalterable and perfectly similar molecules."* The postulate that
atoms are "imperishable, unalterable and perfectly similar" is also implicitly contained in Einstein's equivalence principle. It is now challenged in several attempts
to create a unified theory of the fundamental interactions, but has so far passed all
experimental tests where temporal drifts in the frequency ratios of different kinds
of atomic frequency standards have been searched for (cf. Sec. 5).

4

The operation of atomic clocks and the realization of atomic time scales is a worldwide endeavor, with more than 50 metrological institutes contributing to the TAI time scale (*temps atomique international*). Research that was and will be relevant to the improvement of clocks is an important branch of atomic physics and quantum optics and has been rewarded with Noble Prizes in physics in 1989 to H. Dehmelt, W. Paul, and N. Ramsey, and in 1997 to S. Chu, C. Cohen-Tannoudji, and W. Phillips. The most precise clocks have always been used in tests of the fundamental theories of physics like QED and relativity, but have now also found a widespread practical application in global satellite navigation systems like GPS. The following short review can only highlight some of these developments, but the reader with a more profound interest in the field can be referred to some comprehensive monographs[2,3,4] and recent review articles[5,6].

2 Caesium Clocks: History and Fundamentals

The work on atomic clocks developed out of microwave spectroscopy of atomic and molecular beams in the 1940s. Microwave oscillators with good spectral purity became available and the interaction with an atomic beam provided relatively undisturbed resonance profiles, especially when Ramsey's method of separated oscillatory fields was used[7]. The first operational caesium beam atomic frequency standard was build in 1955 by Essen and Parry at the British National Physical Laboratory NPL. The device was used to measure the Cs frequency in terms of the ephemeris second, i.e. the rotation of the earth around the sun, in the period 1955-1958 and the result of this measurement[8] was used in the definition of the SI second as it was adopted in 1967: "*The second is the duration of 9 192 631 770 periods of the radiation corresponding to the transition between the two hyperfine levels of the ground state of the caesium 133 atom*". This definition of the atomic unit of time was thus linked to the former astronomical definition to the best available accuracy of the astronomical measurements and opened the way for time and frequency metrology to benefit from the progress in atomic frequency standards.

The caesium atom was chosen for the primary etalon of time partly because of technical convenience: a beam of caesium atoms is generated from an oven at a temperature of only 100°C, and caesium atoms can be detected relatively easily by surface ionisation on a hot platinum wire. Additionally, there is only one stable isotope (Cs-133) so that all atoms in the beam have the same hyperfine structure. The hyperfine splitting frequency of 9.19 GHz is relatively high compared to that of the other alkali atoms. In the "classical" caesium clock (cf. Fig. 1) the preparation and detection of the hyperfine state is done via deflection of the beam in inhomogeneous magnetic fields. The "clock transition" is the Zeeman component $(F = 4, m_F = 0) \rightarrow (F = 3, m_F = 0)$ whose frequency is independent of the magnetic field to first order in small fields (cf. Fig. 2a). In the Paschen–Back domain, the atoms in the $(F = 3, m_F = 0)$ state are magnetic high-field seekers, whereas atoms in the $(F = 4, m_F = 0)$ state are attracted towards lower magnetic field strength. The magnetic fields can be arranged such that atoms in $(F = 4, m_F = 0)$ are focussed through the interaction zone onto the detector. If the clock transition towards the $(F = 3, m_F = 0)$ level is induced in the Ramsey

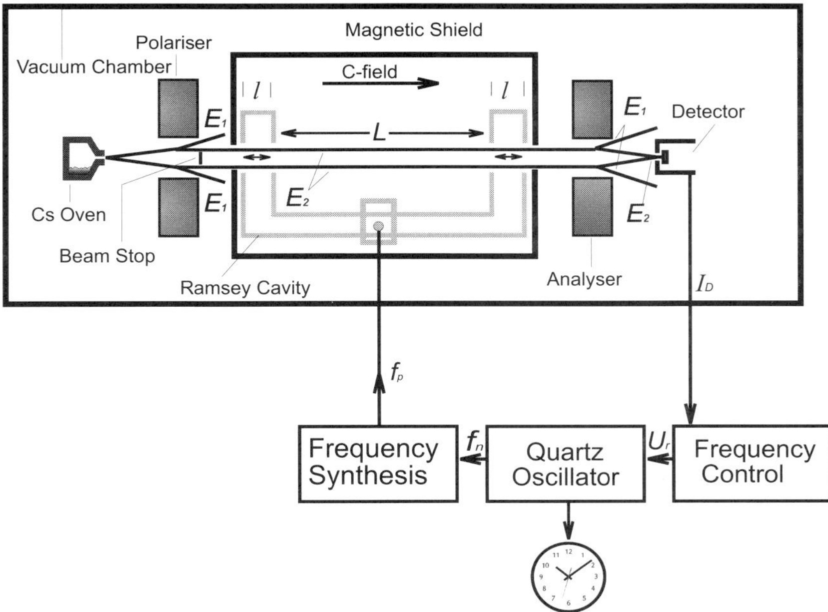

Figure 1. Schematic of a caesium clock with thermal atomic beam and magnetic state selection. Polariser and analyser are hexapole magnets; the beam stop blocks the atoms on the field-free axis where state selection is not effective.

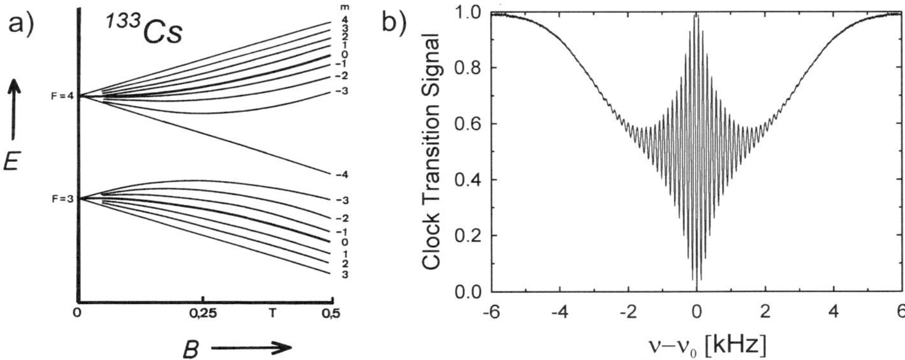

Figure 2. (a) Energy level diagram of the Cs ground state manifold in a magnetic field. The clock transition is the component $(F = 3, m_F = 0) \to (F = 4, m_F = 0)$. (b) Ramsey resonance signal of the primary clock CS2 of PTB.

cavity, a decrease of the atomic flux on the detector will be registered (cf. Fig. 2b). This signal can then be used to steer the frequency of the microwave that is fed into the Ramsey cavity, to keep it in resonance with the atomic transition frequency. Clock output signals at 1 Hz or standard frequencies like 5 MHz are derived from a quartz oscillator in the frequency synthesis chain and can be counted electronically to generate a time scale. A small and homogeneous magnetic field (the so-called C-field) is applied in the microwave interaction region, to separate the other Zeeman components from the clock transition. Because the state-selective magnetic fields deflect the atomic beam, they also act as an effective velocity filter and narrow the velocity distribution of the detected atoms relative to the thermal distribution from the atomic beam oven. This improves the contrast of the Ramsey signal (Fig. 2b) and allows a more accurate correction of the relativistic Doppler effect that has to be taken into account in the evaluation of primary frequency standards. The inverse of the time of flight of the atoms through the microwave interaction zone determines the linewidth of the resonance so that this kind of spectroscopy is said to be Fourier-limited. The maximum length of the interaction region is limited to about 1 m by technical reasons like the homogeneity of the C-field and the symmetry of the microwave distribution to the Ramsey zones, so that the linewidth in such a standard is on the order of 50–100 Hz or 10^{-8} of the transition frequency. Averaging over several days together with careful control and evaluation of systematic frequency shifts allow the center frequency of the resonance to be determined to an uncertainty that represents only a small fraction of the linewidth. In primary clocks of this type, like CS1 and CS2 of PTB, an uncertainty of about $1 \cdot 10^{-14}$ is reached, limited mainly by the cavity phase difference and the quadratic Zeeman effect. These clocks are under practically continuous operation for more than 15 years now and have made a major contribution to the long-term accuracy of TAI[9]. Portable devices of this type of caesium clock with magnetic state selection and about 15 cm length of the interaction zone are also available commercially and can achieve uncertainties in the 10^{-13} range.

The two important parameters that characterize the quality of a frequency standard or a clock are *accuracy* and *stability*. An ideal atomic frequency standard would provide the realization of an unperturbed atomic frequency ν_0 associated with the transition between two atomic states with energy difference $\Delta E = h\nu_0$. In reality, a number of predictable corrections have to be applied to the output frequency ν_{out} of the standard, like e.g. for the relativistic Doppler effect if the velocity of the atoms is known. A certain discrepancy between the corrected ν_{out} and ν_0 has still to be expected because of uncontrollable variations of the operating conditions. This estimated normalized deviation $|\nu_{out} - \nu_0|/\nu_0$ is given as the *accuracy*. For a primary standard, this number is based on an evaluation of the uncertainties associated with various systematic frequency shifts. Comparisons between different clocks of comparable accuracy are used to verify it. The *stability* is a measure of the frequency fluctuations within a certain time interval. In most atomic clocks, an oscillator — like a quartz oscillator for microwaves or a laser for optical frequencies — produces a tunable electromagnetic wave that is used to probe the atomic transition frequency by measuring the transition probability as a function of frequency. This signal is processed and used to steer the oscillator frequency, such that it agrees with ν_0 on average. The atomic spectroscopic signal can be characterised by the linewidth $\Delta\nu$ and by the noise Δp in the measurement of the transition probability. In a stable clock, both these figures have to be as small as possible. Ideally, the linewidth is Fourier-limited, i.e. $\Delta\nu \approx 1/T_c$, where T_c is the cycle time, during which the atoms interact with the radiation. In caesium clocks, T_c is always limited by the time of flight of the atoms through the microwave interaction region. In general, to reach the Fourier-limited regime for long T_c requires both – the linewidth of the oscillator and the natural linewidth of the atomic transition – to be smaller than $1/T_c$. The measurement noise Δp is ideally limited only by the statistical noise of the quantum mechanical state measurement (projection noise), which scales like $1/\sqrt{N}$, where N is the number of atoms. The stability of a standard is usually expressed as the normalized two-sample variance of consecutive frequency measurements over an averaging time τ (Allan variance $\sigma_y(\tau)$). A simple expression (correct to within a factor of order 1) can be given for the Allan variance:

$$\sigma_y(\tau) \approx \frac{\Delta\nu}{\nu_0}\sqrt{\frac{T_c}{N\tau}}. \tag{1}$$

The stability improves with longer averaging times like $1/\sqrt{\tau}$. While a stable clock is not necessarily accurate, accuracy usually requires stability. Otherwise an unpractically long averaging time would be needed to verify the accuracy of the output. For averaging times up to a few hours, the hydrogen maser[7] is more stable than a Cs clock, but the superior accuracy of the latter made it the preferred device for the realization of the unit of time. The simple expression Eq. 1 points out the two most promising strategies for the improvement of the stability of atomic clocks: the first is to reduce the linewidth $\Delta\nu$ by the use of slow or trapped atoms, that allow Fourier-limited spectroscopy with longer interaction times. This approach has enormously benefitted from the application of laser cooling of atoms. The second strategy is to increase the frequency ν_0 and this is the motivation for the development of *optical* clocks, where ν_0 is typically four to five orders of magnitude higher than in the microwave range. These two directions of work are the subjects of the following two sections.

3 Caesium Clock with Cold Atoms: The Caesium Fountain

The kinetic energy of atoms can be reduced very efficiently by exciting them repeatedly to resonance fluorescence with a laser whose photon energy is slightly lower than the excitation energy of the atom at rest. The Doppler effect brings the moving atom to resonance with the detuned laser frequency. On the average, a negative difference appears between the energies of the absorbed and of the emitted photons, and this energy difference is taken from the kinetic energy of the atom. This is the underlying principle of most laser cooling techniques[10]. For a free particle, laser cooling can proceed until the velocity of the atom is comparable to the recoil velocity that an atom at rest acquires after the emission of a single photon: $v_{rec} = \hbar k/m$, where $\hbar k$ is the photon momentum and m the atomic mass. The typical scale of v_{rec} for laser cooling of the heavy alkali atoms on their resonance transitions is a few mm/s. This means that on earth, the atomic velocity will be determined by the gravitational acceleration within a few milliseconds after the cooling is stopped. To have atoms available for longer interaction times, one either has to resort to an environment of reduced gravity, like in space, or throw the atoms upwards on a parabolic flight, in the so-called atomic fountain. The fountain geometry is very convenient for the application of Ramsey's method of separated oscillatory fields: a single microwave cavity is traversed twice by each atom, once on the way upwards and once downwards (cf. Fig. 3). The height of the fountain is typically 1 m, leading to an interaction time of about 1 s and resulting in a factor of 100 improvement in the frequency resolution in comparison to a Cs thermal atomic beam standard. The first fountains with laser cooled atoms were demonstrated in 1989 with sodium[11] and in 1991 with caesium[12].

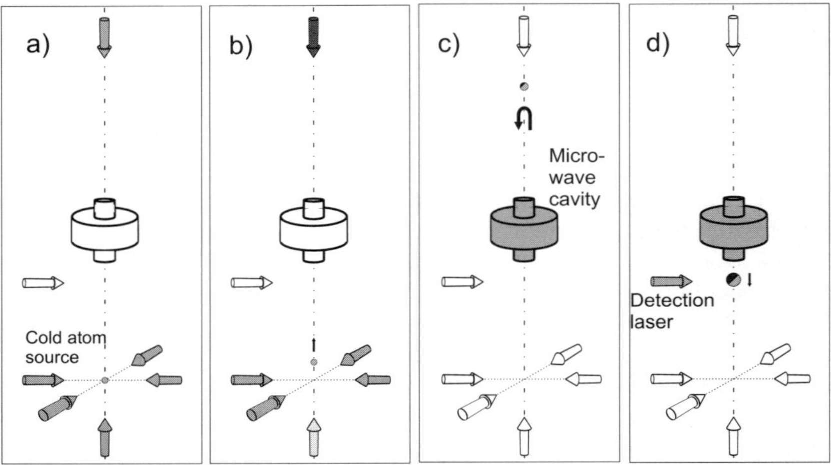

Figure 3. Sequence of operation in a fountain frequency standard: (a) A cloud of cold atoms is loaded. (b) The atoms are launched upwards. (c) In free parabolic flight the atoms pass the microwave cavity twice. (d) State detection via laser excitation and fluorescence detection.

The sequence of operation of a fountain frequency standard is shown in Fig. 3. A cloud of 10^6 to 10^8 atoms is loaded from the background vapor or an atomic beam and cooled in the intersection volume of six laser beams. For caesium atoms, the transition $6\,^2S_{1/2} \rightarrow 6\,^2P_{3/2}$ at 852 nm wavelength is used. Optical pumping ensures that all atoms are in the $F = 4$ hyperfine level of the ground state. The cloud is then launched by slightly detuning the frequency of the two vertically oriented laser beams relative to one another, so that their radiation pressure is balanced in an upwards moving reference frame. Afterwards the atoms pass the microwave cavity two times in free flight. During this time, the lasers are switched off because the presence of resonant photons would cause a strong light shift of the hyperfine resonance. After the second passage through the cavity, a detection laser is switched on and the fluorescence signal is detected first from atoms in the $F = 4$ and then from those in the $F = 3$ state. From the two fluorescence intensities, the normalized transition probability can be determined. Because of the well defined velocity of the atoms, Ramsey fringes with excellent contrast can be obtained (cf. Fig. 4).

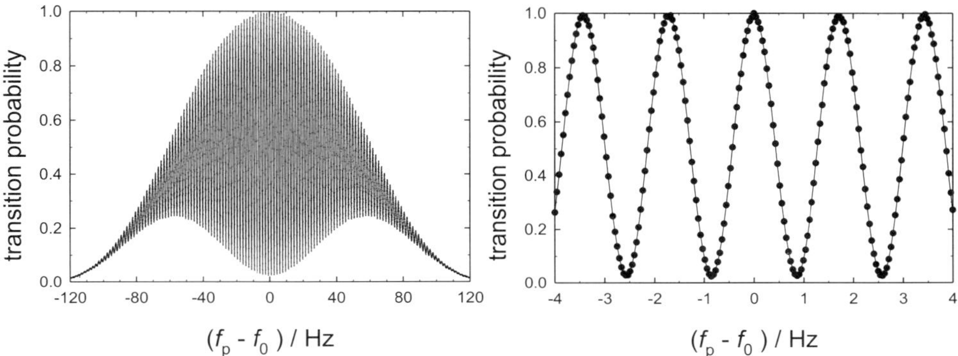

Figure 4. Resonance curve of the clock transition in the caesium fountain CSF1 of PTB (cf. Ref. [14]). Each data point represents the result of one fountain cycle (cf. Fig. 3) for one probe frequency f_p. The right graph shows a high-resolution record of the central Ramsey fringes with 0.88 Hz linewidth.

The national metrology institutes of France (BNM-SYRTE), the USA (NIST) and Germany (PTB) operate primary caesium fountain frequency standards with uncertainties of approximately $1 \cdot 10^{-15}$, about a factor of 10 more accurate than the best Cs beam devices[13]. While effects that limit the accuracy of Cs beam clocks, such as a phase difference between the two Ramsey cavities and the quadratic Zeeman effect, can be strongly reduced by using slow atoms in a fountain, a new systematic effect has appeared: At low temperatures, in the micro- and nanokelvin range, caesium atoms have unusually large collision cross sections and the collisional frequency shift is the dominant contribution to the uncertainty of a Cs fountain frequency standard[13]. In cold ^{87}Rb atoms the collisional shift is smaller by more than a factor of 50, so that a rubidium fountain could be operated with higher atom number, which would lead to a more stable and eventually also more accurate clock[15].

As it was mentioned above, a microgravity environment would allow the operation of a cold atom microwave clock in a regime of still longer interaction time and narrower linewidth. At present, frequency standards with laser-cooled caesium and rubidium atoms are under development for operation on the International Space Station ISS[15].

4 Optical Frequency Standards with Trapped Ions

If an atom is held in a trap, the problem of the limited interaction time can be completely eliminated. One has however to be careful, not to perturb the internal level structure with the trap fields, so that still the resonance frequency of the free atom can be determined. Ions are very convenient in this sense because they carry the electric charge as a "handle", and radiofrequency ion traps (Paul traps[16]) can provide the confinement of an ion around the field-free saddle point of an electric quadrupole potential (cf. Fig. 5). Combined with laser cooling, the ion can be brought to the vibrational ground state of the trap potential[17], where the localisation and residual energy are only determined by the quantum limits. Especially for an optical frequency standard the tight confinement in an ion trap is beneficial, because it is possible to reach the so-called Lamb–Dicke regime where the oscillation amplitude of the particle is much smaller than the wavelength of the radiation that is used to probe it. In the Lamb–Dicke regime — which is obviously much harder to reach for an optical wavelength than for microwaves — the linear Doppler effect and possible shifts due to curvature of the phase front of the radiation can be eliminated. The remaining quadratic Doppler shift is usually smaller than 10^{-18} of the transition frequency for a laser-cooled ion.

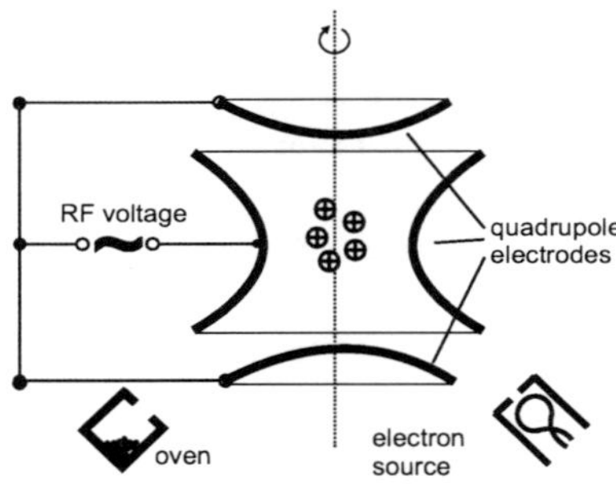

Figure 5. Schematic of a Paul trap: atoms are evaporated from the oven, ionized by electron impact in the trap and stored in an alternating quadrupole field generated by rotationally symmetric ring and endcap electrodes. For single-ion experiments, the diameter of the ring is of the order of 1 mm.

In the 1970s H. Dehmelt published his seminal proposals of the *mono-ion oscillator*[18] and predicted that it should be possible to reach an accuracy of 10^{-18} with an optical clock based on a dipole-forbidden, narrow-linewidth transition in a single, laser-cooled and trapped ion. A single ion, trapped in ultrahigh vacuum, only rarely suffers a collision and interacts with its environment mainly via relatively well controllable electric fields. A number of possible reference transitions with a natural linewidth of the order of 1 Hz are available in different ions, so that a resolution $\Delta\nu/\nu_0 = 10^{-15}$ is possible. In this case, Eq. 1 indicates that even with a single ion ($N = 1$) a stability $\sigma_y(1\,\mathrm{s}) = 10^{-15}$ is possible, about an order of magnitude better than in the best microwave oscillators. Within a few seconds such a clock would allow a more accurate frequency measurement than one of the present Cs fountains after several hours, but to reach a final accuracy of 10^{-18} will still require a rather lengthy averaging of about 10^6 s, or roughly ten days.

To obtain a projection noise limited detection of the transition probability on the forbidden optical transition, Dehmelt proposed the *electron shelving* scheme that allows an observation of quantum jumps (c.f. Fig. 6). The scheme is most conveniently applied if the ion possesses a V-level system, where both a dipole-allowed transition and the forbidden reference transition of the optical clock can be driven with two different laser frequencies from the ground state. The dipole transition is used for laser cooling and the resonance fluorescence that is emitted here can be used for the optical detection of the ion. If the second laser now excites the ion to the metastable upper level of the reference transition, the fluorescence disappears and the ion will only light up again after the metastable state has decayed. Every single excitation of the reference transition suppresses the subsequent scattering of a large number of photons on the cooling transition and can thus be detected with practically hundred percent efficiency. To use this kind of double resonance spectroscopy in an optical clock, the two laser radiations have to be applied alternately in time, because the simultaneous presence of cooling laser radiating would lead to a strong light shift of the reference transition.

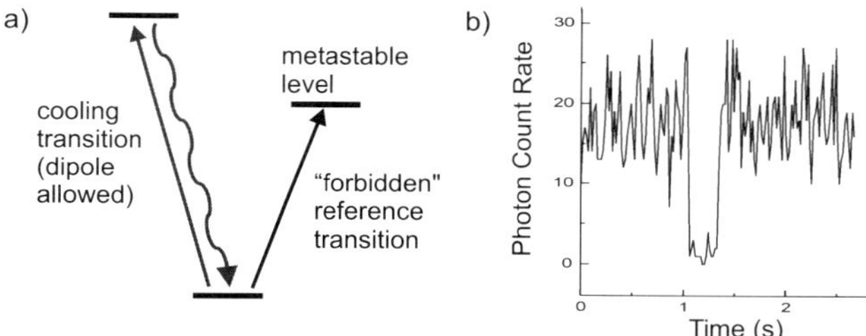

Figure 6. a) Level scheme of an ion used in an optical frequency standard: a strong dipole-allowed cooling transition and the forbidden and narrow reference transition orginate from the ground state. (b) Experimental quantum jump signal of a single ^{115}In$^+$ ion: the fluorescence on the cooling transition shows a dark interval due to an excitation of a metastable state (cf. Ref. [24]).

Presently, a number of groups is pursuing research along the lines of the mono-ion oscillator proposal with different ions[19], like the alkali-like Ba$^+$, Sr$^+$, Ca$^+$, Hg$^+$[20], Yb$^+$[21,22], and the alkaline-earth-like In$^+$[23]. Different types of forbidden transitions are investigated in these ions: for the alkali-like ions, higher order multipole processes lead to the excitation of low-lying metastable ^2D or ^2F levels from the ^2S ground state. As originally favored by Dehmelt, in In$^+$ a transition $^1S_0 \rightarrow {}^3P_0$ exists that connects two levels with $J = 0$ and is made only weakly allowed via the mixing of the electronic and the nuclear spin[23]. These transitions show different sensitivities to possible systematic frequency shifts and also the technical difficulties associated with stable trapping and laser cooling of the various elements are diverse.

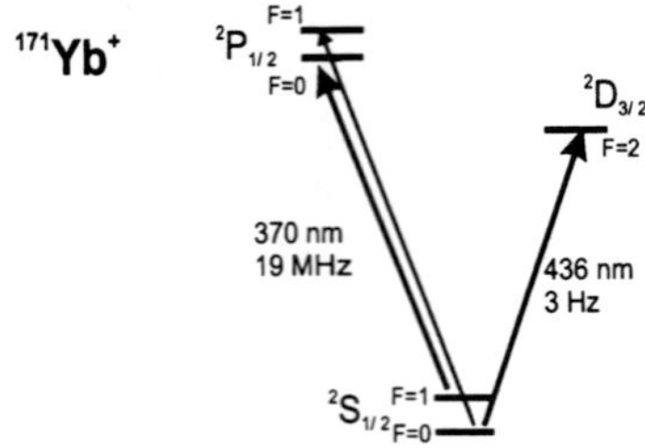

Figure 7. a): Level scheme of the ^{171}Yb$^+$ ion. The $S - P$ transition is used for laser cooling and the $S - D$ quadrupole transition as the reference transition of the frequency standard.

In the following, some recent results from the work on ^{171}Yb$^+$ at PTB will be presented[21,22]. The level scheme of this ion is shown in Fig. 7. The resonance line $^2S_{1/2} \to {}^2P_{1/2}$ with 19 MHz natural linewidth is used for laser cooling and the hyperfine component $(^2S_{1/2}, F = 0) \to (^2D_{3/2}, F = 2)$ serves as the reference transition, with only 3.1 Hz natural linewidth. Like in the Cs clock, this transition between two states with integer total angular momentum quantum number F possesses a component $m_F = 0 \to 0$ whose frequency is independent of the magnetic field to first order. Figure 8 shows three excitation spectra of the $S - D$ electric quadrupole transition of a single ^{171}Yb$^+$ ion. In Fig. 8a, the carrier at detuning 0 is visible, together with two sidebands at ± 0.7 MHz detuning. These sidebands are due to the 0.7 MHz radial oscillation of the ion in the trap. Their relative weakness in comparison to the carrier shows that the Lamb–Dicke regime is reached (the temperature corresponds to about 1 mK). In Fig. 8b the resolved Zeeman structure of the carrier is shown in a weak magnetic field of 1.1 μT. Figure 8c presents a high-resolution spectrum of the $m_F = 0 \to 0$ component under excitation with 30 ms long laser pulses, leading to a Fourier-limited linewidth of 30 Hz, or a resolution $\Delta\nu/\nu$ of $4 \cdot 10^{-14}$. So far, no uncontrolled systematic frequency shifts of the line have been detected at this level. Presently, work is in progress to show by direct optical frequency comparisons of two single-ion standards, that these can really be more accurate than a caesium fountain.

Cold neutral atoms are also investigated for the application in an optical frequency standard[25], most notably the alkaline-earth elements Mg, Ca and Sr with their intercombination lines $^1S_0 \to {}^3P_1$. Like in a Cs fountain, a cloud of 10^8 atoms may be interrogated in each cycle instead of a single ion, leading to an obvious advantage in the stability of such a standard. The high accuracy of the single-ion standard, however, seems difficult to reach here, because the free atoms are not in the Lamb–Dicke regime so that the linear Doppler effect and shifts associated with phase front curvature of the laser beams have to be considered. In addition, a collisional shift may arise from the interaction between the cold atoms. The availability of different kinds of frequency standards is often helpful in metrological applications. In the microwave domain today, the more stable hydrogen maser is frequently used in conjunction with the more accurate Cs clock and a similar distribution of tasks may be possible in the future with neutral-atom and single-ion-based optical

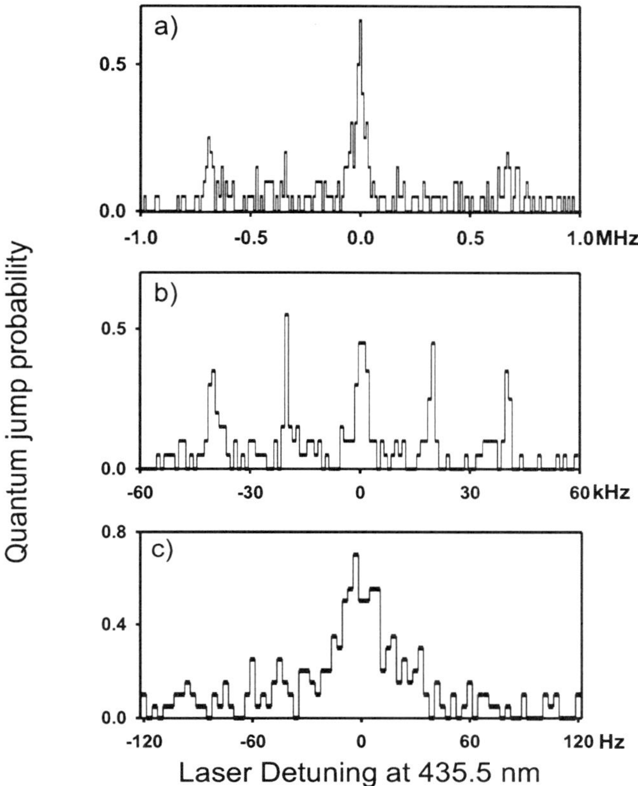

Figure 8. Excitation spectra of the $S - D$ quadrupole transition of a single laser-cooled [171]Yb[+] ion, obtained with the electron shelving method (cf. Ref. [22]). Note the different frequency scales: (a) central carrier resonance and first order motional sidebands; (b) Zeeman pattern of the carrier in a weak magnetic field with the central $\Delta m_F = 0$ component and the adjacent $\Delta m_F = \pm 1, \ \pm 2$ lines; (c) high-resolution scan of the carrier with a Fourier-limited linewidth of 30 Hz. Each data point is the result of 20 cycles of alternating interrogation of reference and cooling transition for each reference laser detuning.

frequency standards.

To realize an optical clock — i.e. a device that displays time — from an optical frequency standard requires means to divide the optical frequency down to the microwave range where cycles can be counted electronically. Ideally, this frequency division should be performed without loss of stability or accuracy. The task is closely related to the precise measurement of absolute optical frequency in terms of the SI second, because this also requires a link between the microwave frequency of a Cs clock and the optical frequency. Until recently, this task was quite difficult, demanding many lasers and microwave oscillators throughout several decades of the electromagnetic spectrum. An important breakthrough was achieved in 1999 with the demonstration that a mode-locked femtosecond laser can be used to span large optical frequency intervals precisely[26,27]. (See the contribution by T. W. Hänsch in this volume). The frequency spectrum of such a laser consists of a broad comb of equidistant modes, where the frequency difference is given by the pulse repetition rate of the laser and is typically in the range of several 100 MHz. Only two radio-frequencies — the repetition rate ν_{rep} and an offset frequency ν_{ceo} — are sufficient to characterise the frequencies of all the optical modes: $\nu_n = n\nu_{rep} + \nu_{ceo}$, $(0 \leq \nu_{ceo} < \nu_{rep})$. Both, ν_{rep} and ν_{ceo} can be adjusted by changing the length and the dispersion of the laser resonator, respectively. So it is possible for example, to operate the femtosecond laser in such a way that its repetition rate is an exact subharmonic of the reference frequency of an optical frequency standard that is locked to one of the comb lines. This technique has been successfully employed to measure absolute optical frequencies of several atoms and ions in direct comparison to primary Cs fountains. The lowest uncertainty of $1 \cdot 10^{-14}$ has been reached for the transitions $^2S_{1/2} \rightarrow \, ^2P_{3/2}$ in $^{171}\mathrm{Yb}^+$ ($688\,358\,979\,309\,312 \pm 6\,\mathrm{Hz}$)[28] at PTB and for $^2S_{1/2} \rightarrow \, ^2P_{5/2}$ in $^{199}\mathrm{Hg}^+$ ($1\,064\,721\,609\,899\,143 \pm 10\,\mathrm{Hz}$)[29] at NIST.

5 Applications in Science and Technology

The quest for still more accurate clocks does not just represent a *l'art pour l'art* attitude of the metrology community. The technical applications of atomic clocks — though not always at the ultimate available accuracy — are widespread, like in the synchronisation of electric power lines and digital telecommunication networks. A quite demanding application is navigation and geodesy with satellites in the GPS, the Russian GLONASS and the planned European GALILEO systems. Here the measurement of the distance between satellite and receiver is performed as a measurement of the propagation time of a microwave signal, so that a nanosecond timing accuracy is required in order to reach a position uncertainty of 1 m. The atomic clocks on board the satellites are adjusted every few hours to be in agreement with a time scale that is generated in ground-based timing stations with a large ensemble of hydrogen masers and caesium clocks. Astronomers are in demand of atomic clocks and time scales for very long baseline interferometry (VLBI) and for the timing of millisecond pulsars. It seems that some millisecond pulsars show a long term stability that is comparable to that of the atomic time scale TAI. Pulsar timing has led to an important test of general relativity: the indirect proof of the emission of gravitational waves[30]. The test of fundamental theories is maybe the most eminent contribution of atomic clocks to science. Apart from pulsar timing, predictions of special and general relativity have been verified in various clock comparison and time transfer experiments, like the precise determination of the gravitational red shift with a hydrogen maser in a rocket by Vessot *et al.*[31]. Highly precise tests of QED have been possible through the accurate measurement of transition frequencies in simple atomic systems — most notably the hydrogen atom — that could be compared with QED calculations.

In closing I would like to come back to Maxwell's statement about the "imperishable, unalterable and perfectly similar molecules" cited in the introduction. The availablity of a number of highly accurate frequency standards that are based on different types of transitions in different elements enables new tests of this postulate to be performed. If the frequency ratio of two different atomic standards would be found to vary in space or time, this would be an indication of a violation of Einstein's equivalence principle and of a temporal or spatial dependency of (at least some of) the fundamental "constants" like α, e, c, *etc.* (See the contribution by S. Karshenboim in this volume). These experiments are complementary to geophysical and astronomical observations, that search for drifts of the constants on cosmological timescales[32,33]. The most stringent limit in a laboratory experiment so far has been obtained from a comparison of a Cs and a Rb fountain over a period of three years[15]. Interpreted in terms of a temporal drift of the fine structure constant, the result was $d\ln(\alpha)/dt = (4.2 \pm 6.9) \cdot 10^{-15}$ per year. Further progress in this field can be expected from the availability of several precise single-ion frequency standards. It has recently been shown that the femtosecond laser optical frequency comb can be used to measure optical frequency ratios — i.e. dimensionless numbers that are not limited by the uncertainty in the realization of the SI second — to an uncertainty of 10^{-18} within averaging times of only 100 s[34]. In this context it might be interesting that recently a very promising candidate for a *nuclear* optical frequency standard has been identified[35] so that also the constancy of the ratio of electromagnetic and strong coupling constants may be scrutinized.

Acknowledgments

I would like to thank my colleagues at PTB's Time Unit laboratory: A. Bauch, T. Schneider, Chr. Tamm and S. Weyers for many stimulating discussions and for providing figures and data for this article.

References

1. J. C. Maxwell, *Address to the Mathematical and Physical Sections of the British Association* (Liverpool, September 15, 1870.) British Association Report, Vol. XL.
2. J. Vanier and C. Audoin: *The Quantum Physics of Atomic Frequency Standards* (Hilger, Bristol, 1989).
3. F. G. Major: *The Quantum Beat: The Physical Principles of Atomic Clocks* (Springer, New York, 1998).
4. C. Audoin and B. Guinot: *The Measurement of Time* (Cambridge University Press, Cambridge, 2001).
5. A. N. Luiten (ed.): *Frequency Measurement and Control – Advanced Techniques and Future Trends* (Springer, Berlin, 2001).
6. A. Bauch and H. R. Telle, Rep. Prog. Phys. **65**, 789 (2002).
7. N. F. Ramsey, Rev. Mod. Phys. **62**, 541 (1990).
8. W. Markowitz, R. G. Hall, L. Essen, and J. V. L. Parry, Phys. Rev. Lett. **1**, 105 (1958).
9. A. Bauch *et al.*, Metrologia **37**, 683 (2000).
10. C. S. Adams and E. Riis, Prog. Quant. Electr. **21**, 1 (1997).
11. M. Kasevich, E. Riis, S. Chu, and R. de Voe, Phys. Rev. Lett. **63**, 612 (1989).
12. A. Clairon, C. Salomon, S. Guellati, and W. Phillips, Europhys. Lett. **16**, 165

(1991).

13. See the contributions by A. Clairon *et al.*, S. R. Jefferts *et al.*, S. Weyers *et al.* in: *Proc. of the 6. Symp. on Frequency Standards and Metrology*, Ed.: P. Gill (World Scientific, Singapore, 2002).

14. S. Weyers, A. Bauch, U. Hübner, R. Schröder, and Chr. Tamm, IEEE Trans. Ultrason. Ferroel. and Freq. Contr. **47**, 432 (2000).

15. C. Salomon *et al.*, in *Atomic Physics 17*, AIP Conference Proc. Vol. 551 (AIP, New York, 2001);
P. Lemonde *et al.* in Ref. [5].

16. W. Paul, Rev. Mod. Phys. **62**, 531 (1990).

17. F. Diedrich, J. C. Bergquist, W. M. Itano, and D. J.Wineland, Phys. Rev. Lett. **62**, 403 (1989).

18. H. Dehmelt, IEEE Trans. Instrum. Meas. **31**, 83 (1982).

19. A. A. Madej and J. E. Bernard, in Ref. [5].

20. R. J. Rafac, B. C. Young, J. A. Beall, W. M. Itano, D. J. Wineland, and J. C. Bergquist, Phys. Rev. Lett. **85**, 2462 (2000).

21. Chr. Tamm, D. Engelke, and V. Bühner, Phys. Rev. A **61**, 053405 (2000).

22. Chr. Tamm, T. Schneider, and E. Peik, in: *Proc. of the 6. Symp. on Frequency Standards and Metrology*, Ed.: P. Gill (World Scientific, Singapore, 2002).

23. Th. Becker *et al.*, Phys. Rev. A. **63**, 051802(R) (2001).

24. E. Peik, G. Hollemann, and H. Walther, Phys. Rev. A **49**, 402 (1994).

25. F. Riehle and J. Helmcke, in Ref. [5].

26. Th. Udem, J. Reichert, R. Holzwarth, and T. W. Hänsch, Opt. Lett. **24**, 881 (1999).

27. S. A. Diddams et al., Phys. Rev. Lett. **84**, 5102 (2000).

28. J. Stenger, Chr. Tamm, N. Haverkamp, S. Weyers, and H. Telle, Opt. Lett. **26** 1589 (2001).

29. Th. Udem et al., Phys. Rev. Lett **86**, 4996 (2001).

30. J. H. Taylor, Rev. Mod. Phys. **66**, 711 (1994).

31. R. F. C. Vessot *et al.*, Phys. Rev. Lett. **45**, 2081 (1980).

32. S. Karshenboim, Can. J. Phys. **78**, 639 (2001).

33. J. P. Uzan, Rev. Mod. Phys. **75** (2003).

34. J. Stenger, H. Schnatz, Chr. Tamm, and H. Telle, Phys. Rev. Lett. **88** 073601 (2002)

35. E. Peik and Chr. Tamm, Europhys. Lett. **61**, 181 (2003).

MEASURING THE FREQUENCY OF LIGHT

T.W. HÄNSCH

Max-Planck-Institut fr Quantenoptik, Hans-Kopfermann-Str. 1, D-85748 Garching, and Sektion Physik, Ludwig-Maximilians-Universitä t, Schellingstr. 4, D-80799 Munich, Germany
E-mail: t.w.haensch@physik.uni-muenchen.de

For three decades, precision spectroscopy of atomic hydrogen has motivated advances in laser spectroscopy and optical frequency metrology. This work has now culminated in the arrival of femtosecond laser optical frequency comb synthesizers which provide revolutionary tools for ultraprecise optical spectroscopy, and which can serve as clockworks for future optical atomic clocks.

1 Introduction

Precise spectroscopy of the simple hydrogen atom has long permitted unique confrontations between experiment and theory. Fig. 1 illustrates how the accuracy of optical spectroscopy of atomic hydrogen has improved over time.

Classical spectroscopists remained limited to about six or seven digits of precision by the large Doppler broadening of hydrogen spectral lines. In 1971, our group at Stanford overcame this barrier by Doppler-free saturation spectroscopy of the red Balmer-a line with a pulsed tunable dye laser. Other groups, notably in New Haven, Oxford, and Paris, joined in to improve the accuracy by three orders of magnitude over the next two decades. Around 1990, a new barrier appeared: the limits of optical wavelength metrology due to unavoidable geometric wavefront errors. Progress beyond a few parts in 10^{10} has been achieved only because we have learned increasingly well how to measure the frequency of light rather than its wavelength. In 2000, the accuracy reached 1.9 parts in 10^{14} [1]. Extrapolating, we should expect an accuracy of a few parts in 10^{15} around 2003. However, further progress is becoming difficult, because we are again approaching a barrier: the limits of how well we know our unit of time, the second. Since 1967 the second has been defined in terms of the 9 GHz ground state hyperfine splitting of atomic cesium. Cesium atomic clocks have been continually refined, as shown by the dashed line in Fig. 1. With the latest generation of laser cooled cesium atomic fountain clocks, one can now reach an accuracy of a few parts in 10^{15}, but the potential for further improvements seems almost exhausted. However, our optical frequency counting techniques make it now feasible to develop optical atomic clocks, based on sharp optical resonances in laser-cooled trapped ions, neutral atoms or molecules. With such clocks future spectroscopic measurements may reach accuracies of parts in 10^{18} and beyond.

In atomic hydrogen, the highest resolution can be achieved on the ultraviolet 1S-2S two-photon resonance with a natural linewidth of only 1 Hz. At Garching, we observe this resonance by collinear excitation of a cold hydrogen atomic beam. The hydrogen atoms are produced by microwave dissociation of molecules and cooled to a temperature of about 6 K by collisions with the walls of a nozzle mounted to a helium cryostat. A collinear standing wave field at 243 nm for Doppler-free

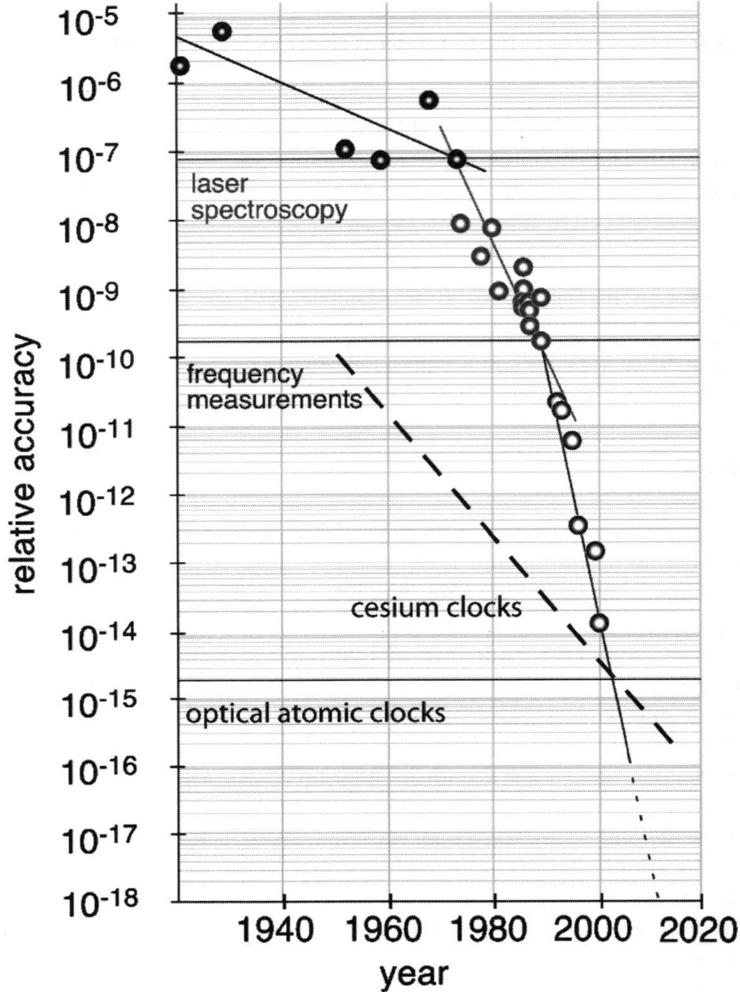

Figure 1. Advances in the relative accuracy of optical spectroscopy of atomic hydrogen

two-photon excitation is produced by coupling the frequency-doubled output of a dye laser into a buildup cavity inside the vacuum chamber. Atoms excited to the 2S metastable state after a travel path of about 10 cm are detected by applying a quenching electric field and counting the emitted vacuum ultraviolet Lyman-α photons. The laser light is periodically blocked by a chopper, and the photon counts are sorted into bins corresponding to different delay times. With slow atoms selected by a delay time of 1.3 ms, the linewidth is now reduced to about 530

Hz at 243 nm corresponding to a resolution of $4.3 \cdot 10^{-13}$. To measure the line position to 1% of this width, we would have to reach an accuracy of 5 parts in 10^{15}. The achievement of such extreme spectral resolution for atomic hydrogen has inspired two international collaborations at CERN (ATHENA and ATRAP) to work towards the production and spectroscopy of slow antihydrogen atoms [2,3,4]. Future precise spectroscopic measurements might thus unveil conceivable differences in the energies or gravitational properties of matter and antimatter.

2 Optical frequency measurements

As recently as 1996, an optical frequency measurement would have required a large and highly complex harmonic laser frequency chain [5]. In 1997, we demonstrated a frequency interval divider chain, which can stay in a convenient region of the spectrum by working with frequency differences rather than with the frequencies themselves. We never built a complete optical frequency counter, but we used a 4-stage divider chain to bridge a 1 THz frequency interval. This approach made it possible to measure the hydrogen 1S2S frequency with a conventional harmonic frequency chain. A transportable CH4-stabilized He-Ne-laser served as an intermediate reference [6].

Since 1999, optical frequency measurements have been enormously simplified with the advent of femtosecond laser optical frequency comb synthesizers [7,8]. In a now common implementation, the pulse train from a Kerr-lens mode-locked Ti:sapphire laser is sent through a microstructured silica fiber, which broadens the spectrum by nonlinear processes so that white light emerges which can be dispersed by a diffraction grating into a rainbow of colors. Remarkably, successive light pulses are so highly phase correlated that the spectrum presents an evenly spaced comb of several hundred thousand sharp spectral lines. The spacing is precisely given by the pulse repetition frequency. The entire comb is displaced by some offset frequency that arises from a slip of the carrier wave relative to the pulse envelope from pulse to pulse. With a frequency comb spanning more than an optical octave, it is straightforward to measure this carrier-envelope offset (CEO) frequency. We only need to produce the second harmonic of the red part of the comb spectrum and observe a beat note with the blue lines of the original comb. Once we can measure the offset frequency, we can control it or even make it go away, so that the frequencies of the comb lines become simply integer harmonics of the pulse repetition rate.

Such a frequency comb provides a direct link between optical frequencies and microwave frequencies. This link can be used in either direction. We can measure or control the pulse repetition rate with a cesium atomic clock and synthesize a dense comb of optical reference frequencies which are directly linked to the primary standard of time. Or we can start with a sharp optical reference line in some cold trapped ion, cold atoms, or slow molecules, and lock a nearby comb line to this optical reference. All the other comb line frequencies are thus rational multiples of the optical reference frequency, and the repetition frequency becomes a precisely known fraction. The comb synthesizer can thus act as a clockwork for future generations of optical atomic clocks. Such clocks will slice time into a hundred thousand times finer intervals than microwave cesium clocks. Standard laboratories in many industrialized nations have begun research programs aimed at the development and perfection of such optical atomic clocks.

Extensive review articles have been written on optical frequency comb synthesizers [8]. Here, I may perhaps add some personal perspective on this development. The idea of using the frequency comb of a mode-locked laser for high resolution spectroscopy is not new. Already in the late seventies, our group at Stanford had demonstrated that a mode-locked picosecond dye laser could produce a frequency comb which we used to measure fine structure intervals in atomic sodium [9]. The origin of the comb spectrum is well explained in a classic textbook [10]. Consider an arbitrary optical waveform circulating inside an optical cavity. During each roundtrip, an attenuated copy escapes through a partly transmitting mirror. A single copy will have a broad and more or less complicated spectrum. However, two identical copies end-to-end will produce interference fringes in the spectrum, somewhat reminiscent of Youngs double slit experiment. Three copies produce a spectrum that resembles the interference pattern of a triple-slit, and an infinite series of copies produces sharp lines which can be identified with the modes of the cavity.

In a real laser, successive pulses will not be identical replicas. Because of dispersion, the carrier wave inside the resonator travels with a phase velocity that differs from the group velocity of the pulse envelope. The resulting carrier-envelope (CE) phase slip may amount to thousands of cycles during a roundtrip, but only the remainder modulo 2π is relevant. As a result, all the comb lines are displaced by a CEO frequency that equals this phase slip per pulse interval. This relationship has already been discussed in detail in the 1978 Stanford Ph.D. thesis of Jim Eckstein. A first interferometric observation of pulse-to-pulse phase shifts was reported in 1996 by the group of F. Krausz in Vienna [11]. In the late 1970s, we did not seriously consider absolute frequency measurements with a laser frequency comb, because the spectrum of the available dye laser was much too narrow. In the early 1990s, the technology of ultrafast lasers advanced dramatically with the discovery of Kerr-lens mode locking. I remember a trade show in 1994, when I was captivated by an exhibit of a (Coherent Mira) mode-locked Ti:sapphire femtosecond laser with regenerative amplifier. The laser beam was focused into a glass slide to produce a white light continuum which a prism dispersed into a rainbow of colors. A striking feature was the laser-like speckle pattern which indicated a high degree of spatial coherence. However, the speckle did not imply anything about the all-important phase correlations between successive pulses, and the pulse repetition frequency of a few hundred kHz remained inconveniently low for frequency comb experiments. Nonetheless, I felt sufficiently intrigued to acquire such a system for our frequency metrology laboratory in 1994.

We did not pursue the femtosecond laser approach seriously right away, because we had come quite far in perfecting our alternative scheme of optical interval division. An accurate measurement of the 1S-2S frequency seemed almost within reach. We also felt that we would need an independent tool to verify any measurement with a femtosecond laser frequency comb, since the frequency metrology community would otherwise distrust any results. The measurements involving optical interval dividers took longer than anticipated. In 1997 we finally published a result for the 1S-2S frequency with an accuracy of 3.4 parts in 10^{13}, a record for optical frequency measurements at the time [6]. From this result, together with other spectroscopic measurements, we could determine a new value of the Rydberg constant and of the 1S ground state Lamb shift. If one believes in QED, the same measurement also yields an accurate value for the mean quadratic charge radius of the proton, and the hydrogen deuterium isotope shift provides the structure radius of the deuteron. We were proud that our table-top experiment exceeded the accuracy of earlier measurements by electron scattering with large accelerators by an order of magnitude.

Exploring alternatives to our frequency interval divider chain, we also began experiments with electro-optical frequency comb generators, kindly provided by M. Kourogi. Such cavity-enhanced phase modulators readily generate combs of modulation sidebands extending over several THz [12].

It seemed now compelling to try to produce even broader frequency combs with a mode-locked femtosecond laser, and Th.Udem, was getting ready to take a closer look at the frequency spectrum of our Mira laser, to be later joined by J. Reichert and R. Holzwarth. At that time, in March of 1997, I visited the European laboratory for nonlinear spectroscopy (LENS) in Florence, Italy, and watched M. Bellini working with an amplified femtosecond laser system. As is common in many ultrafast laboratories, he produced a white light continuum by focusing part of the laser pulse train into a plate of calcium fluoride. I asked what would happen if the laser beam were split into two parts and focused in two separate spots; would the two white light continua interfere? Most people in the laboratory did not expect to see interference fringes, but when we tried the experiment, using a Michelson interferometer to carefully adjust the relative timing of the two pulses, we soon observed stable interference fringes of high contrast for all the colors visible to the eye [13]. The two white light sources had to be mutually phase-coherent. No matter how complicated the process of white light continuum generation might be, the process was reproducible. If the two pulses were separated in time rather than in space .... In the next few days I sat down to write a detailed six page proposal for a universal frequency comb synthesizer which essentially described the now common self-referencing scheme. The idea seemed so close to reality now that I asked M. Weitz and Th. Udem to witness every page of this proposal on April 4, 1997. This vision provided a new motivation for our team to seriously explore the potential of Kerr-lens mode-locked femtosecond lasers for optical frequency comb synthesis.

By that time, hundreds of such lasers were in use in laboratories around the world, but they were mostly used to study ultrafast phenomena. Nobody had ever looked for any comb lines, as far as we could tell. With a repetition frequency of 90 MHz, the comb spectrum of our Mira laser was so densely spaced that no spectrometer in our laboratory could resolve the comb lines. Therefore, we resorted to heterodyne detection, employing a cw diode laser as a local oscillator. The diode laser beam and the pulse train were superimposed with a beam splitter, and a beat signal was detected with an avalanche photodiode after some spectral filtering. After paying attention to the mechanical stability of the femtosecond laser, we did observe stable comb lines. Next, we investigated the spacing of the comb lines.

22

We phase-locked two diode lasers to two arbitrarily chosen comb lines and used an optical interval divider stage to produce a new frequency precisely at the center. A beat note with the nearest comb line confirmed, much to our delight, that the comb lines were pefectly evenly spaced, way out into the wings of the emission spectrum, within a few parts in 10^{17} [14].

It was now certain that the frequency comb of such a mode-locked femtosecond laser could serve as a ruler in frequency space to measure large optical frequency intervals. In a first demonstration of a precision measurement with a femtosecond laser, we determined the frequency interval between the cesium D1 resonance line and the fourth harmonic of a transportable CH4-stabilized 3.39 μm He-Ne-laser, which had been calibrated with a harmonic laser frequency chain at the PTB Braunschweig [15]. The optical cesium frequency allows a determination of the fine structure constant from the atomic recoil energy as measured by atom interferometry in the group of Steve Chu at Stanford.

Afterwards, we focused our efforts on the more ambitious goal to measure an absolute optical frequency relative to a cesium atomic clock in our own laboratory. With a frequency comb spanning an entire octave, we could have simply measured the interval between a laser frequency and its second harmonic. However, in early 1999, we did not yet have such a broad comb. We could produce combs of some 60 THz width by broadening the spectrum of our Mira laser by self-phase modulation in a short piece of ordinary optical fiber. Therefore, we relied on some interval divider stages to produce 1/2 and 4/7 of the frequency of the 486 nm dye laser in the hydrogen spectrometer. We could then bridge the remaining gap with our frequency comb to determine the laser frequency itself. As a reference, we first used a commercial cesium atomic clock, and then a highly accurate transportable cesium fountain clock (PHARAO), built at the LPTF in Paris. In June of 1999, this first absolute optical frequency measurement with a femtosecond laser frequency comb yielded a new value of the hydrogen 1S-2S frequency accurate to $1.4 \cdot 10^{-14}$, surpassing all earlier optical frequency measurements by more than an order of magnitude [1]. Members of the frequency metrology community, such as J.L. Hall in Boulder, who had remained extremely skeptical, soon became ardent evangelists for the new femto-comb approach.

Just before the completion of these measurements, a new tool for the generation of octave spanning frequency combs appeared on the horizon. Researchers at Bell Laboratories demonstrated a microstructured rainbow fiber, which could broaden the spectrum of the nano-joule pulses of a mode-locked femtosecond laser oscillator to a white light continuum. After the white light interference experiments in Florence, I felt rather confident that this magic fiber would preserve the phase coherence of successive pulses and produce useable comb lines with a desirable large frequency spacing. However, our efforts to obtain a sample of this fiber were foiled by the lawyers at Lucent Technologies. Fortunately, we learned that the group of P. St. J. Russel at the University of Bath has long been producing similar photonic crystal (PC) fibers, and in November of 1999, we could finally try such a fiber in our laboratory.

At that time, we had acquired a small Ti:sapphire ring laser (GigaOptics GmbH, GigaJet) producing pulses of about 25 fs at a repetition frequency of 625 MHz.

Launching about 170 mW into a 30 cm long PC fiber, we immediately produced a spectrum spanning more than an octave. In the fiber, a small fiber core is surrounded by air-filled holes which give a large change in the effective refractive index. Part of the light travels as an evanescent wave in air, so that the region of zero group velocity dispersion is shifted down to match the wavelength of the Ti:sapphire laser. The injected laser pulses can thus maintain a high intensity, without quickly spreading by dispersion. The detailed mechanism of spectral broadening is still under investigation, with self-phase modulation and soliton splitting identified as important contributors. However, the process is sufficiently reproducible that useable comb lines can be produced throughout the spectrum. Today we know that similar broad spectra can also be produced with tapered communication fibers, and a growing number of laboratories are demonstrating laser oscillators which produce almost an octave or even a full useable octave without any need for external spectral broadening [16,17].

With such an octave-spanning comb, it is now a simple task to realize a self-referencing frequency comb synthesizer. We only need to select a few thousand comb lines from the red end of the spectrum, which form pulses that are intense enough to be frequency doubled in a nonlinear crystal. The comb lines in the doubled spectrum are displaced from the precise integer harmonics of the pulse repetition frequency by twice the CEO frequency. A collective beat note with corresponding lines in the original comb therefore directly reveals the CEO frequency. The absolute frequency of each comb line is then determined by two radio-frequencies, which can be measured precisely, and an integer mode number, which can be identified uniquely by a wavelength measurement with a common wavemeter. This type of self-referencing frequency synthesizer was first realized by D. Jones et al. in Boulder [18], who obtained a fiber sample from Bell Laboratories a few weeks before we received our fiber from the University of Bath.

In a first stringent test, Ronald Holzwarth has compared such an octave spanning frequency comb synthesizer with the more complex frequency synthesizer used in the 1999 hydrogen frequency measurement [19]. By starting with a common 10 MHz radiofrequency reference and comparing comb lines near 350 THz, he could verify agreement within a few parts in 10^{16}, probably limited by Doppler shifts due to air pressure changes or thermal expansion of the optical tables. In 2002, a group at the PTB Braunschweig demonstrated how a femtosecond laser frequency comb generator can be used as a transfer oscillator to precisely measure optical frequency ratios [20]. As a test case, they measured the frequency ratio between the second harmonic of a Nd:YAG laser and the fundamental frequency, verifying the expected value of 2 with an uncertainty of 7 parts in 10^{19}. More recently, M. Zimmermann in our laboratory has pushed a related experiment to an uncertainty of 6 parts in 10^{21} [21]. So far, we have not identified any systematic error that would limit the precision of spectroscopic measurements or the performance of future optical atomic clocks. Commercial frequency comb synthesizers are now being produced by Menlo Systems GmbH, founded by some of my former students [22].

3 New frequency measurement of hydrogen 1S-2S

A new measurement of the hydrogen 1S-2S frequency is planned for February 2003. M. Fischer and N. Kolachevsky have implemented many improvements in the hydrogen spectrometer. The interaction region is now differentially pumped to reduce line shifts and the loss of slow atoms due to collisions with background gas. The passive reference cavity for stabilization of the dye laser has been much improved with mirrors optically contacted to a monolithic spacer of ULE ceramic. The cavity housing with its temperature stabilization and acoustic shielding is mounted on an active vibration isolation stage. Light from the dye laser is sent through a fiber into the frequency metrology laboratory, where the optical frequency will be compared by M. Zimmermann and Th. Udem to the radio frequency of the transportable Paris PHARAO atomic fountain clock, using an octave-spanning femtosecond laser frequency comb synthesizer.

This experiment is already attracting some attention because it can be considered as a test for a possible slow variation of the fine structure constant. As pointed out by V. Flambaum [23], astronomical observations of spectral lines in the light of distant quasars suggest that the fine structure constant ? had a slightly smaller value during the early age of the Universe. Assuming a linear rate of change, ? might be changing by about 1 part in 10^{15} per year. H. Fritzsch has recently argued, that grand unification would imply that ? cannot change simply by itself [24]. If all known forces are to remain unified at very high energies, other coupling constants must change as well. As a result, the masses and magnetic moments of hadrons should change relative to those of the electron. Fritzsch is pointing out an intersesting magnifying effect: we would expect that radiation emitted in a hyperfine transition should vary in time about 17 times more strongly than light emitted in a normal atomic transition, but in the opposite direction, i.e. the atomic wavelength becomes smaller with time, but the hyperfine wavelength increases. Even if we should not find any evidence for such a change at the expected levels of one or two parts in 10^{14}, it certainly remains an important goal to further improve the accuracy of spectroscopic measurements to establish more stringent limits on possible slow variations of fundamental constants.

4 Conclusions

Our new ability to measure the frequency of light with extreme precision makes it now possible to develop optical atomic clocks of unprecedented accuracy. Such clocks will be enabling tools for a myriad of new precision measurements, ranging from ultraprecise spectroscopy to new tests of special and general relativity. Perhaps the biggest surprise would be if we found no surprise.

References

1. M. Niering *et al.*, *Phys. Rev. Lett.* **84**, 5496 (2000).
2. A. Amoretti *et al.*, *Nature* **419** , 456 (2002).
3. G. Gabrielse *et al.*, *Phys. Rev. Lett.* **89**, 213401 (2002).

4. G. Gabrielse *et al.*, *Phys. Rev. Lett.* **89**, 233401 (2002).

5. H. Schnatz *et al.*, *Phys. Rev. Lett.* **76**, 18 (1996).

6. Th. Udem *et al.*, *Phys. Rev. Lett.* **79**, 2646 (1997).

7. J. Reichert *et al.*, *Opt. Commun.* **172**, 59 (1999).

8. Th. Udem *et al.*, *Nature* **416**, 233 (2002).

9. J.N. Eckstein *et al.*, *Phys. Rev. Lett.* **40**, 847 (1978).

10. A.E. Siegman, Lasers, University Science Books, Mill Valley, 1986.

11. L. Xu *et al.*, *Opt. Lett.* **21**, 2008 (1996).

12. T. Udem *et al.*, *Opt. Lett.* **23**, 1387 (1998).

13. M. Bellini and T.W. Hä nsch, *Opt. Lett.* **25**, 1049 (2000).

14. Th. Udem *et al.*, *Opt. Lett.* **24**, 881 (1999).

15. Th. Udem *et al.*, *Phys. Rev. Lett.* **82**, 3568 (1999).

16. T.M. Fortier, D.J. Jones, and S. Cundiff, submitted to *Opt. Lett.*.

17. www.gigaoptics.de

18. D. Jones *et al.*, *Science* **288**, 635 (2000).

19. R. Holzwarth *et al.*, *Phys. Rev. Lett.* **85**, 2264 (2000).

20. J. Stenger *et al.*, *Phys. Rev. Lett.* **88**, 073601 (2002).

21. M. Zimmermann *et al.*, submitted to *Opt. Lett.*.

22. www.menlosystems.com

23. J.K. Webb *et al.*, *Phys. Rev. Lett.* **89**, 283002 (2002).

24. X. Calmet and H. Fritzsch, *Phys. Lett.* **B540**, 173 (2002).

TIME AND SPACE VARIATION OF FUNDAMENTAL CONSTANTS: MOTIVATION AND LABORATORY SEARCH

SAVELY G. KARSHENBOIM

D. I. Mendeleev Institute for Metrology, 198005 St. Petersburg, Russia
Max-Planck-Institut für Quantenoptik, 85748 Garching, Germany
E-mail: sek@mpq.mpg.de

Fundamental physical constants play important role in modern physics. Studies
of their variation can open an interface to new physics. An overview of different
approaches to a search for such variations is presented as well as possible reasons
for the variations. Special attention is paid to laboratory searches.

1   Introduction

Any interactions of particles and compound objects such as atoms and molecules
are described by some Lagrangian (Hamiltonian) and *constancy* of parameters of
the basic Lagrangian is a cornerstone of modern physics. Electric charge, mass and
magnetic moment of the particle are parameters of the Lagrangian. However, there
are a few simple reasons why we have to expect the nature to be not so simple.

- A theory described by a Lagrangian suggests some properties of the spacetime. It seems that introducing gravitation we arrive to some inconsistency
  of a classical description of the space-time continuum and that means that
  the picture must be more complicated. It is not necessary, however, that the
  complicated nature imply variable constants, but it is possible.

- In particle/nuclear/atomic/molecular physics we deal with the effective Lagrangians. The "true" fundamental Lagrangian is defined at the Planck scale
  for elementary objects (leptons, quarks and gauge bosons) and we can study
  only its "low-energy" limit with a pointlike electron and photon and extended
  hadrons and nuclei.

- One more reason is presence of some amount of matter, which selects a preferred frame and produces some background fields. In usual experiments we
  often have problems with environment and have either to produce some shielding or to subtract the environment contribution. However, we cannot ignore
  the whole Universe and its evolution.

- The expansion of Universe may lead to some specific time and space dependence in atomic transitions which are similar to a variation of "constants".

An illustration can be found in the so-called inflation model of evolution of the
Universe (see e.g. [1]). The Standard Model of evolution suggests a phase transition
in some very early part which dramatically changed properties of electrons and photons. It happens without any changes of the unperturbed parameters of the basic
Lagrangian defined at the Planck scale. A change of the electron mass (from zero to
some nonvanishing value of m_e) arose eventually from cooling of matter caused by

expansion. Meanwhile photon properties were changed via renormalization going from the Planck scale down to our real world (which is very different for a zero and non-zero electron mass).

Considering variation of the fundamental constants we have to clearly recognize two kinds of a search. The first one is related to the most sensitive and easily accessible quantities. In such a case a limitation for the variation is the strongest and easiest to obtain, but sometimes it is not clear what fundamental quantity it is related to. An example is a study of samarium resonance by absorption of a thermal neutron [2]

$$^{149}\mathrm{Sm} + n \to\, ^{150}\mathrm{Sm} + \gamma\,. \tag{1}$$

Estimations led to an extremely low possible variation but it is hard to express it in terms of the fine structure constant or some other fundamental constant (see Sec. 11 for detail).

The other kind of a search is provided by a study of quantities which can be clearly related to the fundamental constants such as optical transitions (see [3] and Sec. 8 for detail).

One may wonder whether it is really important to interpret a variation of some not fundamental value (such as a position of a resonance) in terms of some fundamental quantities. A fact of the variation itself must be a great discovery more important than the exact value of the variation rate of the fine structure constant α or another basic constant. A problem, however, is related to the nature of precision tests and searches. Any of them is realized on the edge of our ability to perform calculations and measurements and any single result on the variation is not sufficient since a number of sources of new systematic effects, which were not important previously at the lower level of accuracy, may appear now. It is crucially important to be able to make a comparison of different results and to check if they are consistent.

In our paper we first try to answer a few basic questions about the constants:

- Are the fundamental constants fundamental?

- Are the fundamental constants constant?

- What hierarchy of the variation rate can be expected for various basic constants?

After a brief overview of most important results we consider advantages and disadvantages of laboratory searches and in particular experiments with optical frequency standards.

2 Are the fundamental constants fundamental?

First of all, we have to note that we are mainly interested in searches for a possible variation of dimensionless quantities. A search of the variation of constants is based on comparison of two measurements of the same quantity separated in time and/or space. For such a comparison, the units could also vary with time and their realization should be consistent for separate measurements. In principle, we can compensate or emulate a variation of a dimensional quantity via a redefinition of the units. To avoid the problem we have to compare dimensionless quantities, which are *unit*-independent. E.g., studying some spectrum we can make a statement on the variation of the fine structure constants α, but not on the variation of speed of light c, Planck constant $\hbar$ or electric charge of the electron e separately.

However, the variation of dimensional quantities can in principle be detected in a different kind of experiment. If we have to recognize which constant is actually varying, we should study effects due to their time- and space- gradients. We do not consider such experiments in this paper.

Precision studies related to astrophysics as well as atomic and nuclear physics deal with characteristics which can be linked to the values of the charge, mass and magnetic moment of an electron, proton and neutron, defined as their properties for real particles (i.e. at $E^2 = p^2c^2 + m^2c^4$) at zero momentum transfer. In the case of nuclear transitions, variation of the pulsar periods *etc* we can hardly interpret any results in terms of the fundamental constants, while in the case of atomic and molecular transitions that can be done (see Sec. 6).

We can combine the constants important for spectroscopy into a small number of basic combinations:

- one dimensional constant (e.g., the Rydberg constant Ry) is needed to describe any frequency;

- a few basic dimensionless constants, such as

    - the fine structure constant α;

    - the electron-to-proton mass ratio m_e/m_p;

    - the proton g factor g_p;

    - the neutron g factor g_n

  are needed to describe any ratio of two frequencies.

As mentioned above, any variation of a dimensional constant cannot be successfully detected: in the case of the astrophysical measurement it will be interpreted as a contribution to the red shift and removed from further analysis, while in the laboratory experiments it will lead to the variation of the second, defined via cesium hyperfine structure. A variation of the value of the Rydberg constant in respect to the cesium hyperfine interval is detectable since it is a dimensionless quantity. However, a physical meaning of such variation cannot be interpreted in terms of the Rydberg constant as a fundamental constant, its possible variation should be due to a variation of the cesium magnetic moment (in units of the Bohr magneton) and the fine structure constant.

Nature of the g factor of the proton and neutron is not well understood and in particular it is not clear if their variations can be considered as independent. Obviously, the g factors are *not* truly fundamental constants, arising as a result of strong interaction in the regime of strong coupling.

Concerning the fine structure constant, we first have to mention that it is a result of renormalization while some more fundamental quantities are defined at the Planck scale.

The origin of the electron and proton mass is different. The electron mass is determined by the details of the Higgs sector of the Standard Model of the electroweak interactions, however, this sector originates from some higher-level theory and a really fundamental constant is rather m_0/M, where m_0 is a "bare" electron mass (i.e. the mass prior to the renormalization which is needed to reach the electron mass m for a *real* electron) and M is a "big" mass related to some combination of the Planck mass and the compactification radius (if we happen to live in a multidimensional world). In the case of proton the situation is different. Most of the proton mass is proportional to Λ_{QCD} (see e.g. [4,5]), which can be expressed in terms of the unperturbed interaction constant and a big mass M'. The latter is some combination of the Planck mass and compactification radius, but it is not the same as M. A small portion of the proton mass and in particular $m_p - m_n$ comes from the mass of current quarks, theory of which is similar to theory of the electron mass. The values of m_0/M and Λ_{QCD}/M can in principle be expressed in terms of the parameters of the basic Lagrangian defined at the Planck scale.

Studies of the gravitational interaction can provide us with a limitation for a variation of m_p/M, however, the limitations are much weaker than those obtained from spectroscopy (see e.g. [6]). Performing spectroscopic measurements we can reach a limitation for a value of m_e/m_p, however, it is rather an accidental value, in contrast to m_0/M and Λ_{QCD}/M, and its interpretation involves a number of very different effects.

3 Are fundamental constants constant?

We have to acknowledge that some variations, or effects which may be interpreted as variations have happened in the past and are present now.

- A Standard Model of the evolution of our Universe has a special period with *inflation of Universe* due to a phase transition which happened at a very early stage of the evolution and significantly changed several properties of particles (see e.g. [1]). In particular, the electron mass and so-called current quark masses (the latter are responsible for a small part of the nucleon mass and in particular for the difference of the proton and neutron mass) were changed. Prior to the phase transition the electron was massless. The proton mass determined by so called Λ_{QCD} was essentially the same. At the present time the renormalization of the electric charge only slightly affects the charge because it has an order of $\alpha/\pi \ln(M/m)$. However, with massless leptons the renormalization has not only ultraviolet divergence but also an infrared one. The phase transition for the electron mass m is also a phase transition for its electric charge e. The transition was caused by cooling of the Universe, and cooling was a result of expansion. The Universe is still expanding and cooling. It should lead to some variation of m and e but significantly below a level of accuracy available for experiments and observations now.

- Expansion of the Universe should modify the Dirac equation for the hydrogen atom and any other atoms and nuclei. However, the expansion itself, without any time and space gradients will just create a red shift common for any atoms and transitions in an area. The second order effect gives an acceleration (note that for a preliminary estimation one can set $H' \sim H^2$). The acceleration will shift energy levels but produce no time variation. And only the H^3 term can give a time dependent correction to the energy levels. It is indeed beyond any experimental possibility.

- In principle, we also have to acknowledge that if the Universe has a finite size, that must produce an infrared cut off which should enter into equations. Since we do not have any real infrared divergence for any observable quantity, the radius of the Universe will enter the expressions for the electric charge and mass of electron in combinations such as $(a_0/R_U)^2$ and the ratio of the Bohr radius and the radius of the Universe is extremely small. With the expansion of the Universe, the radius $R_U(t)$ is time dependent and that will give some small (undetectable) variation of the constants. The real situation is not so simple. First, we do not know if the Universe has a finite size. Second, doing finite time experiments we have to deal with some horizon and that does not depend on a size of the Universe. It is unclear how the cut off due to the horizon problem will interfere with the expansion of the Universe and its radius (if R_U is finite).

The discussed effects are small and not detectable now. It is even not clear whether they may be detected in principle, however, they demonstrate a clear indication that

- a property of fundamental basic particles, like their charge and mass of the electron, should vary with time;

- a property of compound objects, such as atoms and nucleus, should vary with time even if properties of their components are not varying.

The main question is the following: is there any reason for a faster variation, which can be detected with current experimental accuracy? This question has not yet been answered.

4   Time and space variations

Most considerations in literature have been devoted to the time variation. However, an astrophysical search (which has only provided us with possibly positive results) cannot distinguish between space and time variations, since remote astrophysical objects are separated from us both in time and space.

To accept space variation is perhaps essentially the same as to suggest existence of some Goldstone modes. While there is none for the Standard Model of the electroweak interactions, there are some experimental limitations on the Goldstone modes for Grand Unification Theories (see e.g. [7]), but it is difficult to exclude them completely. Another option is some domain structure. In the case of "large" domains with the finite speed of light and horizon any easy conjunction of two domains is unlikely even reducing the total vacuum energy. A domain structure can be formed at the time of inflation when the Universe was expanding so fast that in a very short time two previously causality-connected points could be very far from each other – out of horizon of each other. There is a number of reasons that a domain structure due to a parameter directly related to the vacuum energy cannot exist, since the energy would tend to reach its minimum. But if a construction like the Cabibbo-Kobayashi-Maskawa (CKM) matrix is a result of spontaneously broken symmetry, we could expect some minor fluctuations of CKM parameters, such as the Cabibbo angle, which were approximately, but not exactly, the same at some early time with their evolution being completely independent because of the horizon problem. CKM contributions are due to the weak interactions for hadrons and they slightly shift magnetic moments of proton and neutron at a fractional level of 10^{-5} and that is how such effects could be studied via precision spectroscopy. They are also important for the neutron lifetime and their variation could change the nuclear synthesis phemonena. We also have to underline that the space distribution with an expansion of the horizon and on their way to an equilibrium should provide some time evolution.

5 Scenario and hierarchy

A possibility of time variation of the values of the fundamental constants at a cosmological scale was first suggested quite a long time ago [8,9], but we still have no basic common idea on a possible nature of such a variation. A number of papers were devoted to the estimation of a possible variation of one of the fundamental constants (e.g. the fine structure constant α) while a possible variation of any other properties is neglected. As we stated in [10], one has to consider a variation of all constants with approximately the same rate. However, some hierarchy (with rates different by an order of magnitude or even more) can be presented and it strongly depend on a scenario. There is a number of "model dependent" and "nearly model independent" estimations of the variation of the constants and their hierarchy.

- Any estimation based on running constants in SU(5) or in a similar unification theory is rather "near model independent". In particular, that is related to a statement on a faster variation of m_p/M than α (see e.g. [4,5,11]).

- Any estimation in the Higgs sector of SU(5) and other GUTs [11], SUSY, quantum gravitation, strings *etc* strongly depends on the model.

We would like to clarify what is model-dependent in "near model independent" considerations. It does not strongly depend on model suggestions in particle physics, but one still needs a basic suggestion on why (and how) any variation can happen. There may be a universal cause all the time, or there may be a few "phases" with different causes dominating at different stages *etc*. What could be a basic cause for the dynamics? E.g. the basic suggestion for an SU(5) estimation is that everything can be derived from the Lagrangian [4,5,11] with varying parameters. In other words, for some reason there is dynamics operating within the Lagrangian.

- A supporting example is a multidimensional Universe with compactification of extra dimensions and the compactification radius R as an ultraviolet cut-off $\Lambda = \hbar/Rc$ (see e.g. [4]). Slow variation of R is suggested (e.g. an oscillation at a cosmological time scale). All variations of the constants arise from the basic Lagrangian via the renormalization with a variation of the cut off and a variation in the Higgs sector induced directly by the variation of R.

- On the contrary, it may be suggested that dynamics comes from a quantum nature of space-time and in terms of the Lagrangian that could lead to some new effective terms violating some basic symmetries of the "unperturbed" Lagrangian (indeed as a small perturbation). In such a case no reason due to SU(5) is valid and one has to start with a description of the perturbing terms.

Both options are open.

The "model dependent" estimations involve more unknown factors, they need understanding of both: a unification/extension scheme and a cause for the variation.

We need to mention an option that in principle the fundamental constants might be calculable. That does not contradict their variations, which can be caused by presence of some amount of matter, or by an oscillation of the compactification radius *etc.* In such a case, the truly fundamental constants $\alpha_0 \sim 10^{-2}$ (the bare electric charge), $m_e^0/M_P \sim 10^{-22}$, $\Lambda_{QCD}/M_P \sim 10^{-20}$ are of very different order of magnitude (here M_P is the Planck mass). The constants (α and $(m, \Lambda)/M_P$) of so different order of magnitude can be either coupled logarithmically or not coupled at all. In the case of α and Λ_{QCD}/M there is some understanding of this logarithmic coupling (see e.g. [4,5]) which is mainly model independent (a model dependent part is a relation between M_P and a mass of Grand Unification Theory M which enters relationships between the constants). In the case of m_e^0/M_P model dependence is essential. However, as it is explained above, it is difficult to realize if any approximate relations between the constants are helpful or not. A crucial question is whether the variation supports the relations between the constants or violates them.

6   Atomic and molecular spectroscopy and fundamental constants

There are three most accurate results on a possible variation of the constants achieved recently. One of them is related to the Oklo fossil nuclear reactor [12] and a position of the samarium resonance (1). The result is negative and the assigned variation rate for the fundamental constants varies between 10^{-17} and 10^{-19} yr^{-1} [2,13,14,15]. However, the interpretation is rather unclear because there is no reliable way of studying the position of the resonance in terms of the fundamental constants.

Two other results are related to spectroscopy:

- A study of the absorption spectra of some quasars led to a positive result on a variation of the fine structure constant of a part in 10^{15} per a year at $6\,\sigma$ level [16] (see also earlier papers on a $4\,\sigma$ positive result [17]). Meanwhile,

Table 1. Scaling behavior of atomic transitions. μ is the nuclear magnetic moment. References are given to the papers where the scaling behavior was first discussed. Importance of the relativistic corrections for the hyperfine structure was first understood in [21], while for other transitions it was discussed in [22].

Transition	Energy scaling	Refs.
Gross structure	Ry	[20]
Fine structure	$\alpha^2 Ry$	[20]
Hyperfine structure	$\alpha^2(\mu/\mu_B)Ry$	[20]
Relativistic corrections	Extra α^2	[21,22]

a search for a variation of m_e/m_p showed a variation at a fractional level of $(5 \pm 2) \times 10^{-15}$ yr^{-1} [18].

- A comparison of hyperfine intervals for the ground state in cesium-133 and rubidium-87 shows no variation of the ratio of their frequencies at a level a part in 10^{15} [19]. The ratio of these frequencies is more sensitive to a variation of g_p than α [10].

Because of importance of the spectroscopic data, we briefly discuss the behavior of the frequency of different kinds of transitions as a function of the fundamental constants.

Any transition frequency can be presented in the form

$$f = f_{NR} \times F_{Rel}(\alpha) , \qquad (2)$$

where f_{NR} is the frequency in the leading non-relativistic approximation and $F_{Rel}(\alpha)$ is the relativistic correcting factor. Scaling behavior of the non-relativistic results is summarized in Tables 1 and 2. The relativistic correction is a result of perturbative calculation of some singular terms since the relativistic effects are enhanced at short distances equivalent to the large momentum transfer. In neutral atoms and ions with only a few electrons stripped, the electron is located in the Coulomb field with a low effective charge of the screened nucleus at a long distance (e.g. $Z_{eff} \simeq 1$ for neutral alkali atom). On the contrary, at a short distance the electron interacts rather with the bare nucleus and the effective charge is close to the nuclear charge Z. As a result, the correcting factor behaves as

$$F_{Rel}(\alpha) = 1 + C_2(Z\alpha)^2 + ... , \qquad (3)$$

and at high Z (e.g. for ytterbium and mercury) the correction is not small any more (see e.g. [21,22]).

Different scaling behavior of the non-relativistic transition frequencies allows to perform efficient comparison to search for a possible variation of the fundamental constants. The most accurate astrophysical results were obtained studying transitions of the same type [16,17], but with essentially different values of the nuclear charge Z and thus with different relativistic corrections [22].

Table 2. Scaling behavior of molecular transitions. It is assumed that $m_p = m_n$ and the nuclear mass is $A \times m_p$. References are given to the paper where the scaling behavior was first discussed.

Transition	Energy scaling	Refs.
Electronic structure	Ry	[23]
Vibration structure	$(m_e/m_p)^{1/2} Ry$	[23]
Rotational structure	$(m_e/m_p) Ry$	[23]

Table 3. Nuclear magnetic moments μ, nuclear structure effects and relativistic effects for the atoms involved in precision microwave measurements. The uncertainty of the calculation in [21,24] is estimated by comparing results on cesium and mercury in [21,24] and [22]. The actual values of the nuclear magnetic moments are taken from [25].

Z	Atom	Schmidt value for μ (μ_S/μ_N)	Actual value for μ (μ/μ_S)	Relativistic factor $F_{\mathrm{rel}}(\alpha)$	Sensitivity to α variation $\partial \ln\left(F_{\mathrm{rel}}(\alpha)\right)/\partial \ln \alpha$
1	H	$g_p/2$	1.00	1.00	0.00
4	^{9}Be$^+$	$g_n/2$	0.62	1.00	0.00
37	^{85}Rb	$5/14(8 - g_p)$	1.57	1.15, [21,24]	0.30(6), [21,24]
37	^{87}Rb	$g_p/2 + 1$	0.74	1.15, [21,24]	0.30(6), , [21,24]
55	^{133}Cs	$7/18(10 - g_p)$	1.50	1.39, [22]	0.83, [22]
70	^{171}Yb$^+$	$-g_n/6$	0.77	1.78	1.42(15), [21,24]
80	^{199}Hg$^+$	$-g_n/6$	0.80	2.26, [22]	2.30, [22]

7 Hyperfine structure and nuclear magnetic moments

Looking for a variation of the fundamental constants with the help of the hyperfine structure, one needs to deal with the nuclear magnetic moments. There is no accurate model which allows to present the nuclear magnetic moments in terms of the fundamental constants. The only available model, the Schmidt model, is not really accurate. We summarize in Table 3 the magnetic moments derived from the Schmidt model in comparison with the actual values for the atoms applied for the frequency standards (see also [10]). The Table contains also data on relativistic corrections.

One can see that nuclear effects, responsible for a correction to the Schmidt model, are comparable to the relativistic effects, responsible for atomic corrections. Note the significant corrections to the Schmidt model for cesium-133 and rubidium-85. They are large because of a destructive interference of spin and orbit contributions, an essential cancellation of the leading term enhancing the corrections. The primary frequency standards are based on the hyperfine interval in cesium and the large corrections to the Schmidt value of the nuclear magnetic moment of cesium-133 are annoying for a direct interpretation of any absolute measurement, which is actually a comparison of some transition with the cesium standards.

Table 4. Optical transitions: most accurate results and sensitivity of the optical transitions to a time variation of α.

Z	Atom	Frequency [Hz]	Fractional uncertainty	Sensitivity to α variation $\partial \ln \left(F_{\mathrm{rel}}(\alpha) \right) / \partial \ln \alpha$,[22]
1	H	2 466 061 413 187 103(46), [26]	2×10^{-14}	0.00
20	Ca	455 986 240 494 158(26), [30]	6×10^{-14}	0.03
49	In$^+$	1 267 402 452 899 920(230), [28]	18×10^{-14}	0.21
70	Yb$^+$	688 358 979 309 312(6), [29]	0.9×10^{-14}	1.03
80	Hg$^+$	1 064 721 609 899 143(10), [30]	0.9×10^{-14}	- 3.18

8 Optical transitions

The essential nuclear effects related to the nuclear magnetic moment lead to a problem of a reliable interpretation of the data. Much more reliable results are delivered by studying pure optical transitions [26,27,28,29,30,31] which can be obtained via a direct comparison of two optical frequencies, or indirectly via independent absolute measurements of those frequencies in units determined by the cesium microwave transition. Both kinds of comparison are now available for the frequency metrology after a development of the new frequency chain based on the so-called frequency comb [32]. The most accurate data are summarized in Table 4.

An important feature of the optical transitions related to the gross structure is that they can be described with the help of two constants only: the Rydberg constant and the fine structure constant. As a result, a time variation of any frequency can be presented in the form

$$\frac{\partial \ln f}{\partial t} = A + B \frac{\partial \ln F_{\mathrm{rel}}(\alpha)}{\partial \ln \alpha} , \qquad (4)$$

where

$$A = \frac{\partial \ln (Ry)}{\partial t} \qquad \text{and} \qquad B = \frac{\partial \ln \alpha}{\partial t} . \qquad (5)$$

While a variation of the Rydberg constant as we discussed above could have no simple interpretation in terms of the fundamental constants, a time variation of the fine structure constant would have a direct and simple interpretation. An expected signature of the time variation of α is depicted in Fig. 1. In the near future five accurate results are expected. Three of them are related to "near α-independent" results (hydrogen, calcium, indium) and they should play the role of an anchor. Two others (mercury and ytterbium) are strongly α dependent and the dependence is significantly different (see Table 4).

9 Current laboratory limitations

Current laboratory limitations are summarized in Table 5. A limitation on the time variation of the proton g factor is derived assuming that the nuclear corrections to the Schmidt model are not important. That indeed cannot be considered as a reliable approach. All other limitations are obtained in a more reliable way. As

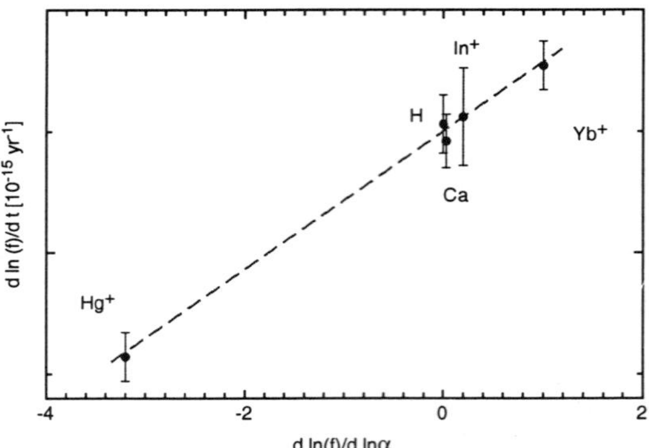

Figure 1. An expected signature of the time variation of the fine structure constant.

Table 5. Current limitations on a possible time variation of the fundamental constants.

Fundamental constant	Limitation for variation rate
α	1×10^{-14} yr^{-1}
m_e/m_p	2×10^{-13} yr^{-1}
$\alpha^2 \mu_p/mu_e$	6×10^{-14} yr^{-1}
$\alpha^{-3} \mu_p/mu_e$	7×10^{-15} yr^{-1}
μ_p/mu_e	2×10^{-14} yr^{-1}
g_p	4×10^{-16} yr^{-1}
$\alpha^2 \mu_n/mu_e$	8×10^{-14} yr^{-1}
μ_n/mu_e	6×10^{-14} yr^{-1}
g_n/g_p	5×10^{-14} yr^{-1}

was pointed out in [10] (see also Table 3), the hyperfine interval in cesium is very difficult for interpretation because of significant nuclear corrections to the Schmidt model. Fortunately, it was demonstrated that there is no variation on the level of one part in 10^{15} per a year for a ratio of cesium-to-rubidium hyperfine structure [19] (as a matter of fact that is the strongest laboratory limitation on a variation of a transition frequency). The hyperfine interval in the ground state of the rubidium-87 in contrast to cesium-133 can be sufficiently well described by its non-relativistic part with use of the Schmidt model for the nuclear magnetic moment [10] (see also Table 3).

10　Precision spectroscopy: tests and reliability

Recent progress in frequency metrology delivered us a number of results, essentially more accurate than any previous data and the expected results can be even more accurate. In such a case we need to be sure that the results are reliable. In this section we briefly discuss possible tests of the accurate frequency measurements.

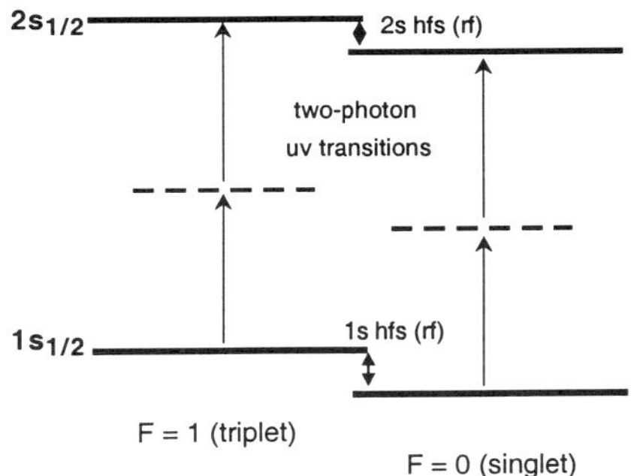

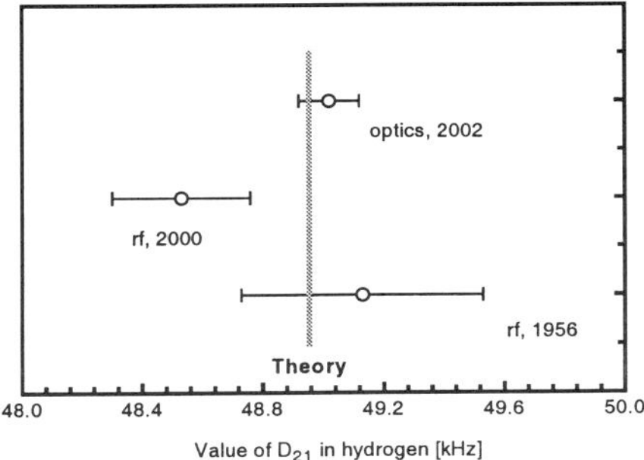

Figure 2. Hyperfine structure in the hydrogen atom: levels scheme of an optical measurement of the $2s$ hfs interval and a comparison of theory to experiment for $D_{21} = 8 \times \nu_{\mathrm{HFS}}(2s) - \nu_{\mathrm{HFS}}(1s)$.

- The cesium hyperfine interval plays a special role in physics because of the definition of the second. It is realized in a number of laboratories and a comparison of different cesium standards is an important metrological work. The comparison shows that we have a sufficient understanding of the accuracy of cesium experiments (see e.g. [33]).

- Study of the $^3P_1 - {}^1S_0$ transition in neutral calcium were performed independently at NIST [30] and PTB [27] and the results are consistent.

- Hyperfine structure of the ground state of ytterbium ion was measured independently at PTB and NML [34]. The results are consistent.

- An important approach to test systematic sources part by part may be a measurement of the isotopic shift and its comparison with theory. If theory is not accurate enough, there is still an option for a precision study. Theory is helpful to fix the form of dependence on the nuclear mass and the nuclear charge radius and the shape of the dependence may be checked via fitting.

- Similar tests can be performed studying the hyperfine structure, like, e.g., a comparison of the $1s - 2s$ transitions in hydrogen for different spin states [35]. Since the hyperfine splitting in the ground state is known with high accuracy [36,10], the comparison of the $1s - 2s$ ultraviolet transitions yields us a value of the hyperfine interval in a metastable $2s$ state. The result is more accurate that one directly derived from a microwave measurement [37] and in good agreement with theory [38]. The transitions under study [35] as well as comparison with theory and early microwave measurements are summarized in Fig. 2.

11   Summary

A comparison of different kinds of search for a possible time and space variations of the fundamental constants is summarized in Table 6. The characteristic level of the limitations is given assuming a linear time dependence. In the case of oscillation

Table 6. Comparison of different kind of search for a possible time and space variation of the fundamental constants.

	Geochemistry	Astrophysics	Laboratory	Laboratory (optics)
Drift or oscillation	$\Delta t \sim 10^9$ yr	$\Delta t \sim 10^9$ yr	$\Delta t \sim 1 - 30$ yr	$\Delta t \sim 1 - 10$ yr
Space variations	$\Delta l \simeq 10^9\ c{\times}$yr	$\Delta l \simeq 10^{10}\ c{\times}$yr	0	0
Level of limitations	10^{-17} yr^{-1}	10^{-15} yr^{-1}	10^{-15} yr^{-1}	10^{-14} yr^{-1}
Present results	Negative	Positive (α)	Negative	Negative
Variation of α	not reliable	accessible	accessible	accessible
Variation of m_e/m_p	not accessible	accessible	accessible	not accessible
Variation of g_p	not accessible	accessible	accessible	not accessible
Variation of g_n	not accessible	not accessible	accessible	not accessible
Strong interactions	not sensitive	not sensitive	sensitive	not sensitive

the limitation from geochemical search and from astrophysical observations should be weakened by a factor $2\Delta t/T$, since the period of oscillation T can be shorter than the time separation Δt. We note that in the case of laboratory limitation the results should depend on a current phase of oscillations. Another problem with interpretation of the astrophysical data is a separation of space and time variations. The different kinds of search offer access to different sets of constants and their reliability depends on whether they are affected by the strong interactions.

Despite a number of advantages and disadvantages of different approaches there is no favorite way. Since we have no background theory, we need to try as many searches as possible and as different as possible.

There are a number of problems which may be of interest and we would like to attract attention to a few of them.

- A comparison of hyperfine intervals in the ground state of ^{85}Rb and ^{87}Rb allows to remove any variation of the fine structure constant due to atomic interactions and the frequency ratio is sensitive only to the proton g factor via the Schmidt model and to strong interactions via corrections to the Schmidt model. Separation of atomic and nuclear physics can be helpful as a test measurement when a number of microwave intervals related to the hyperfine structure will be studied.

- Actual nuclear magnetic moments of ^{199}Hg and ^{171}Yb are very close (the difference is below 5%) and their Schmidt values are the same (see Table 3). If that is a systematic effect, a comparison of the hyperfine intervals in these two ions can give a reliable result on a possible variation of the fine structure constant α. We need better understanding of the magnetic moments of ^{199}Hg and ^{171}Yb.

- Discussing different approaches, we need to mention an idea of [39] to study a 3.5 eV nuclear transition in ^{229}Th which lies in the optical domain. Its comparison with atomic transitions can have indeed no reliable interpretation, but the nuclear transition is very different from atomic transitions and may be sensitive to effects not detectable with other methods.

- Another approach related to the nuclear properties suggested [10] precision studies of the nuclear magnetic moment with extremely small values, which are expected to be very sensitive to detuning of the fundamental constants. Indeed, it is not possible to measure a nuclear magnetic moment accurately enough. However, looking for their variations one can study the hyperfine structure of proper ions. As an example of an extremely small magnetic moment, let us mention $^{198}_{81}$Tl with a magnetic moment below 10^{-3} μ_N ($T_{1/2} = 5.3(5)$ h), $^{153}_{62}$Sm ($\mu = -0.022\,\mu_N$, $T_{1/2} = 46$ h) and $^{192}_{79}$Au ($\mu = -0.009(2)\,\mu_N$, $T_{1/2} = 4.9$ h) [25]. Understanding the nature of such small values is also necessary.

- One more question related to the subject: can we detect the expansion of the Universe in some laboratory experiments? The expansion leads to the red shift of the photons at a level of 10^{-10} yr^{-1}, however, there is no way to study a photon emitted a year ago in laboratory experiments. A chance can appear if we can use objects (planets, spacecrafts) at a distance related to $c \times (1-10)$ min.

A search of a possible variation of the values of the fundamental constant presents a specific field involving both fundamental and applied physics. A search for new physics is based on frequency metrology providing a high motivation. The frequency metrology presents now limitations which are somewhat weaker than those from astrophysics but it has showed significant progress last years and it seems that higher accuracy of the laboratory measurements is just a matter of time and new results will be coming soon.

Acknowledgments

The author is grateful to V. Flambaum, H. Fritzsch, T. W. Hänsch, J. L. Hall, M. Kramer, W. Marciano, M. Murphy, L. B. Okun, E. Peik, T. Udem, D. A. Varshalovich, M. J. Wouters and J. Ye for useful and stimulating discussions. The work has been supported in part by RFBR.

References

References

[1] A. Linde, In *Three Hundred Years of Gravitaion* (Ed. by S. W. Hawking and W. Israel, Cambridge University Press, Cambridge, 1987), p. 604; S. K. Blau and A. H. Guth, *ibid.*, p. 524; G. Börner, *The Early Universe* (Springer-Verlag, 1993).

[2] A. I. Shlyakhter, Nature (London) **264** (1976) 340. See also the preprints: A. I. Shlyakhter, LNPI N. 260 (1976); ATOMKI Report A/1 (1983).

[3] S. G. Karshenboim, In *Laser Physics at the Limits*, ed. by H. Figger, D. Meschede and C. Zimmermann (Springer-Verlag, Berlin, Heidelberg, 2001), p. 165.

[4] W. J. Marciano, Phys. Rev. Lett. **52**, 489 (1984).

[5] X. Calmet and H. Fritzsch, Eur. Phys. J. **C24**, 639 (2002);
H. Fritzsch: E-print hep-ph/0212186.

[6] J.-P. Uzan, E-print hep-ph/0205340.

[7] K. Hagiwara *et al.*, *The Review of Particle Physics*, Phys. Rev. **D66**, 010001 (2002).

[8] P. A. M. Dirac, Nature **139** (1937) 323.

[9] F. J. Dyson, In *Aspects of Quantum Theory* (Cambridge Univ. Press, Cambridge, 1972) p. 213; In *Current Trends in the Theory of Fields* (AIP, New York, 1983), p. 163.

[10] S. G. Karshenboim, Can. J. Phys. **78**, 639 (2000).

[11] P. Langacker, G. Segre, and M. J. Strassler, Phys. Lett. **B528**, 121 (2002)

[12] M. Maurette, Ann. Rev. Nucl. Part. Sci. **26**, 319 (1972);
P. K. Kuroda, *Origin of the Chemical Elements and the Oklo Phenomen* (Springer-Verlag, Berlin, 1982).

[13] J. M. Irvine, Contemp. Phys. **24**, 427 (1983).

[14] T. Damour and F. Dyson, Nucl. Phys. **B480**, 596 (1994).

[15] Y. Fujii, A. Iwamoto, T. Fukahori, T. Ohnuki, M. Nakagawa, H. Hidaka, Y. Oura, and P. Moller, Nucl.Phys. **B573**, 377 (2000).

[16] J. K. Webb, M. T. Murphy, V. V. Flambaum, S. J. Curran, E-print astro-ph/0210531.

[17] J. K. Webb, V. V. Flambaum, C. W. Churchill, M. J. Drinkwater, and J. D. Barrow, Phys. Rev. Lett. **82**, 884 (1999);
J.K. Webb, M.T. Murphy, V.V. Flambaum, V.A. Dzuba, J.D. Barrow, C.W. Churchill, J.X. Prochaska, A.M. Wolfe, Phys. Rev. Lett. **87**, 091301 (2001).

[18] A. Ivanchik, E. Rodriguez, P. Petitjean, and D. Varshalovich, Astron. Lett. **28**, 423 (2002); A. Ivanchik, P. Petitjean, E. Rodriguez, and D. Varshalovich, E-print astro-ph/0210299.

[19] H. Marion, F. Pereira Dos Santos, M. Abgrall, S. Zhang, Y. Sortais, S. Bize, I. Maksimovic, D. Calonico, J. Gruenert, C. Mandache, P. Lemonde, G. Santarelli, Ph. Laurent, A. Clairon, and C. Salomon: physics/0212112.

[20] M. P. Savedoff, Nature **178**, 688 (1956).

[21] J. D. Prestage, R. L. Tjoelker, and L. Maleki, Phys. Rev. Lett. **74**, 3511 (1995).

[22] V. A. Dzuba, V. V. Flambaum, and J. K. Webb, Phys. Rev. **A59**, 230 (1999);
V. A. Dzuba and V. V. Flambaum, Phys. Rev. **A61**, 034502 (2000).

[23] R. I. Thompson, Astrophys. Lett. **16**, 3 (1975).

[24] H. B. G. Casimir, *On the Interaction Between Atomic Nuclei and Electrons* (Freeman, San Francisco, 1963);
C. Schwarz, Phys. Rev. **97**, 380 (1955).

[25] R. B. Firestone, *Table of Isotopes* (John Wiley & Sons, Inc., 1996).

[26] M. Niering, R. Holzwarth, J. Reichert, P. Pokasov, Th. Udem, M. Weitz, T. W. Hänsch, P. Lemonde, G. Santarelli, M. Abgrall, P. Laurent, C. Salomon, and A. Clairon, Phys. Rev. Lett. **84**, 5496 (2000).

[27] G. Wilpers, T. Binnewies, C. Degenhardt, U. Sterr, J. Helmcke, and F. Riehle, Phys. Rev. Lett. **89**, 230801 (2002).

[28] J. von Zanthier; Th. Becker, M. Eichenseer, A. Yu. Nevsky, Ch. Schwedes, E. Peik, H. Walther, R. Holzwarth, J. Reichert, Th. Udem, T. W. Hänsch, P. V. Pokasov, M. N. Skvortsov, and S. N. Bagayev, Opt. Lett. **25** (2000) 1729.

[29] J. Stenger, C. Tamm, N. Haverkamp, S. Weyers, and H. R. Telle, Opt. Lett. **26**, 1589 (2001).

[30] T. Udem, S. A. Diddams, K. R. Vogel, C. W. Oates, E. A. Curtis, W. D. Lee, W. M. Itano, R. E. Drullinger, J. C. Bergquist, and L. Hollberg, Phys. Rev. Lett. **86**, 4996 (2001).

[31] S. Bize, S. A. Diddams, U. Tanaka, C. E. Tanner, W. H. Oskay, R. E. Drullinger, T. E. Parker, T. P. Heavner, S. R. Jefferts, L. Hollberg, W. M. Itano, D. J. Wineland, and J. C. Bergquist, E-print physics/0212109.

[32] T. Udem, J. Reichert, R. Holzwarth, S. Diddams, D. Jones, J. Ye, S. Cundiff, T. Hänsch, and J. Hall, In *Hydrogen atom: Precision physics of simple atomic systems*, ed by S. G. Karshenboim et al., (Springer, Berlin, Heidelberg, 2001), p. 125;
J. Reichert, M. Niering, R. Holzwarth, M. Weitz, Th. Udem, and T. W. Hänsch, Phys. Rev. Lett. **84**, 3232 (2000);
R. Holzwarth, Th. Udem, T. W. Hänsch, J. C. Knight, W. J. Wadsworth, and P. St. J. Russell, Phys. Rev. Lett. **85**, 2264 (2000);
S. A. Diddams, D. J. Jones, J. Ye, S. T. Cundiff, J. L. Hall, J. K. Ranka, R. S. Windeler, R. Holzwarth, Th. Udem, and T. W. Hänsch, Phys. Rev. Lett. **84**, 5102 (2000).

[33] T. Parker, In *Proceedings of the 6th Symposium Frequency Standards and Metrology*, ed. by P. Gill (World Sci., 2002,) p. 89.

[34] R. B. Warrington, P. T. H. Fisk, M. J. Wouters, and M. A. Lawn, In *Proceedings of the 6th Symposium Frequency Standards and Metrology*, ed. by P. Gill (World Sci., 2002), p. 297.

[35] M. Fischer, N. Kolachevsky, S. G. Karshenboim and T.W. Hänsch, Can. J. Phys. **80**, 1225 (2002); further analysis of systematic sources is in progress.

[36] H. Hellwig, R.F.C. Vessot, M. W. Levine, P. W. Zitzewitz, D. W. Allan, and
D. J. Glaze, IEEE Trans. IM **19**, 200 (1970);
P. W. Zitzewitz, E. E. Uzgiris, and N. F. Ramsey, Rev. Sci. Instr. **41**, 81
(1970);
D. Morris, Metrologia **7**, 162 (1971);
L. Essen, R. W. Donaldson, E. G. Hope and M. J. Bangham, Metrologia **9**,
128 (1973);
J. Vanier and R. Larouche, Metrologia **14**, 31 (1976);
Y. M. Cheng, Y. L. Hua, C. B. Chen, J. H. Gao and W. Shen, IEEE Trans.
IM **29**, 316 (1980);
P. Petit, M. Desaintfuscien and C. Audoin, Metrologia **16**, 7 (1980).

[37] N. E. Rothery and E. A. Hessels, Phys. Rev. A**61**, 044501 (2000).

[38] S. G. Karshenboim and V. G. Ivanov, Phys. Lett. B**524**, 259 (2002); Euro.
Phys. J. D**19**, 13 (2002).

[39] E. Peik and Chr. Tamm, Europhys. Lett. **61**, 181 (2003); E. Peik, *contribution
to this book*.

SECTION II
CAUSALITY & SIGNAL PROPAGATION

Causality and Superluminal Light
  G. M. Shore

Quantum Fluctuations and Signal Velocity in Superluminal Light Pulse Propagation
  L. L. Wang

Time and Matter in the Interaction between Gravity and Quantum Fluids:
Are there Macroscopic Quantum Transducers between Gravitational and
Electromagnetic Waves?
  R. Y. Chiao & W. J. Fitelson

CAUSALITY AND SUPERLUMINAL LIGHT

G.M. SHORE

Department of Physics, University of Wales Swansea,
Singleton Park, Swansea SA2 8PP, U.K.
E-mail: g.m.shore@swansea.ac.uk

The causal properties of curved spacetime, which underpin our sense of time in
gravitational theories, are defined by the null cones of the spacetime metric. In
classical general relativity, it is assumed that these coincide with the light cones determined by the physical propagation of light rays. However, the quantum vacuum
acts as a dispersive medium for the propagation of light, since vacuum polarisation in QED induces interactions which effectively violate the strong equivalence
principle (SEP). For low frequencies the phenomenon of gravitational birefringence
occurs and indeed, for some metrics and polarisations, photons may acquire *super-*
luminal phase velocities. In this article, we review some of the remarkable features
of SEP violating superluminal propagation in curved spacetime and discuss recent
progress on the issue of dispersion, explaining why it is the high-frequency limit of
the phase velocity that determines the characteristics of the effective wave equation
and thus the physical causal structure.

1 Introduction

In general relativity, the nature of time and causality are determined by the properties of the null cones of the curved spacetime manifold. A physical realisation
of these geometric null cones is provided in classical electrodynamics by the light
cones traced by the propagation of light rays. As the title of this conference implies,
'Time and Matter' are therefore intimately related.

In quantum theory, however, the relationship is more subtle. Quantum effects
such as vacuum polarisation in QED induce interactions which effectively violate
the strong equivalence principle and cause the quantum vacuum in the presence of
gravity to act as a dispersive medium for the propagation of light. At least for low
frequencies, the phenomenon of gravitational birefringence occurs and moreover, for
some metrics and polarisations, photons may acquire *superluminal* phase velocities.
This forces a reassessment of the identification of *light cones* with *null cones* and
raises the question of how causality and quantum theory can be reconciled in general
relativity.

Research into this effect began with a key paper of Drummond and Hathrell[1] in
which they calculated the one-loop vacuum polarisation contribution to the QED
effective action in a background gravitational field. They found the following modification to the free Maxwell action:

$$\Gamma = \int dx \sqrt{-g} \left[-\tfrac{1}{4} F_{\mu\nu} F^{\mu\nu} \right.$$
$$\left. + \tfrac{1}{m^2} \left(a R F_{\mu\nu} F^{\mu\nu} + b R_{\mu\nu} F^{\mu\lambda} F^{\nu}{}_{\lambda} + c R_{\mu\nu\lambda\rho} F^{\mu\nu} F^{\lambda\rho} + d D_{\mu} F^{\mu\lambda} D_{\nu} F^{\nu}{}_{\lambda} \right) \right] \tag{1}$$

where a,b,c,d are constants of $O(\alpha)$ and m is the electron mass. The important feature is the direct coupling of the electromagnetic field to the curvature. This is an effective violation of the *strong* equivalence principle (SEP), which is the dynamical ansatz that the 'laws of physics' should be the same in the local inertial frames at each point in spacetime. (The *weak* equivalence principle, viz. the existence of such local inertial frames, is in contrast a fundamental assumption underlying the structure of general relativity.) More precisely, the SEP requires that electromagnetism is minimally coupled to gravity, i.e. through the connections only, independent of the curvature. The effective action Eq.(1) shows that while this principle may be consistently imposed at the classical level, it is necessarily violated in quantum electrodynamics. It is the quantum violation of the SEP that allows the physical light cones to differ from the geometric null cones.

The new SEP-violating interactions in Eq.(1) affect the propagation of light and modify the physical light cones. Using the techniques of geometric optics, Drummond and Hathrell showed that the new light cones are given by

$$k^2 \; - \; \frac{2b}{m^2} R_{\mu\lambda} k^\mu k^\lambda \; + \; \frac{8c}{m^2} R_{\mu\nu\lambda\rho} k^\mu k^\lambda a^\nu a^\rho \; = \; 0 \tag{2}$$

where k_μ is the wave vector and a_μ the polarisation. Using the Einstein equations, this can be expressed as the sum of a 'matter' and a purely 'gravitational' contribution as follows:

$$k^2 \; - \; \frac{8\pi}{m^2} (2b + 4c) T_{\mu\lambda} k^\mu k^\lambda \; + \; \frac{8c}{m^2} C_{\mu\nu\lambda\rho} k^\mu k^\lambda a^\nu a^\rho \; = \; 0 \tag{3}$$

The first term involves the projection of the energy-momentum tensor which appears in the weak energy condition; this contribution is universal, appearing in the modified light cone condition for light propagating subluminally in a variety of backgrounds such as classical electromagnetic fields or finite temperature [2,3,4,5]. The second term depends on the Weyl tensor and is uniquely gravitational; since it depends explicitly on the polarisation, the modified light cones exhibit *gravitational birefringence*.

It is also instructive to express Eqs.(2),(3) in the Newman-Penrose formalism. Introducing a null tetrad with basis vectors ℓ_μ, n_μ, m_μ and $\bar{m}_\mu$ together with the corresponding components of the Ricci and Weyl tensors $\Phi_{00} = -\frac{1}{2} R_{\mu\nu} \ell^\mu \ell^\nu$, $\Psi_0 = - C_{\mu\nu\lambda\rho} \ell^\mu m^\nu \ell^\lambda m^\rho$ etc. (for details, see ref.[4]), and choosing ℓ_μ to coincide with the direction of propagation, i.e. $k_\mu = \omega \ell_\mu$, we find

$$k^2 \; - \; \frac{(4b + 8c)\omega^2}{m^2} \Phi_{00} \; \pm \; \frac{4c\omega^2}{m^2} (\Psi_0 + \Psi_0^*) \; = \; 0 \tag{4}$$

This representation makes it clear that the contribution from the Weyl tensor changes sign for the two physical transverse polarisations a_μ. It follows immediately that for Ricci-flat spacetimes, both timelike and spacelike values of k^2 are possible. In other words, Eqs.(2),(3),(4) necessarily imply the existence of *superluminal* propagation. Physical photons no longer follow the geometrical null cones, but instead propagate on the effective, polarisation-dependent, light cones defined above.

Many examples of the Drummond-Hathrell effect in a variety of gravitational wave, black hole and cosmological spacetimes have been studied [1,2,6,7]. The (Ricci-flat) black hole cases are particularly interesting, and it is found that for photons propagating orbitally, the light cones are modified such that superluminal propagation occurs. For radial geodesics in Schwarzschild spacetime, however, and the corresponding principal null geodesics in Reissner-Nordström and Kerr, the light cones are unchanged. The reason is simply that if we choose the standard Newman-Penrose tetrad in which ℓ^μ is tangent to the principal null geodesic, the only non-vanishing component of the Weyl tensor is Ψ_2 since these black hole spacetimes are all Petrov type D, whereas the modification to the light cone condition involves only Ψ_0. Superluminal propagation is also predicted in the (Weyl-flat) FRW cosmological spacetimes, with the correction to the speed of light increasing as $1/t^2$ towards the initial singularity. Another interesting case involves the Bondi-Sachs metric describing gravitational radiation from an isolated source, where the magnitude of the superluminal effect is related to the *peeling theorem* for the Weyl tensor.

The existence of superluminal propagation in QED in curved spacetime of course raises immediate questions as to the realisation of causality. The purpose of this article is to review our research programme on photon propagation in gravitational fields with particular emphasis on the issue of causality and the consistency of quantum field theory with classical gravitation. We begin, in section 2, by considering carefully the implications of the Drummond-Hathrell effective action as it stands, reviewing the *bimetric* interpretation of the light cones, the realisation of *stable causality* in general relativity, possible time machine constructions and the consequences for event horizons.

However, the action Eq.(1) is only the lowest-order term in an expansion of the full effective action in powers of derivatives. That is, results derived from it are valid only in a low-frequency approximation $\lambda \gg \lambda_c$, where λ_c is the electron Compton wavelength. The inclusion of terms of higher orders in derivatives shows that the propagation is in fact *dispersive*[8]. In section 3, we discuss the precise definition of the 'speed of light' and present a proof that the wavefront velocity, which is the relevant speed of light for causality, can be identified as the high-frequency limit of the phase velocity, $v_{\rm ph}(\infty)$. This means that a resolution of the issues raised in section 2 concerning causality depends on an explicit calculation of the light cones for *high-frequency* propagation. In section 4, we present a recently derived extension of the effective action valid to third order in curvatures and field strengths and to all orders in derivatives[9]. The resulting light cone is derived and some potential special features are described. Finally, however, we argue on the basis of a detailed comparison with photon propagation in background magnetic fields that a further, non-perturbative contribution to the effective action may ultimately control the high-frequency limit and we close with an assessment of the prospects for a final resolution of the question of dispersion and causality for QED in curved spacetime.

2 Causality and Superluminal Propagation

In this section, we discuss the implications for causality of photon propagation based on the Drummond-Hathrell action Eq.(1) and its associated light cones, setting aside for the moment the issue of dispersion.

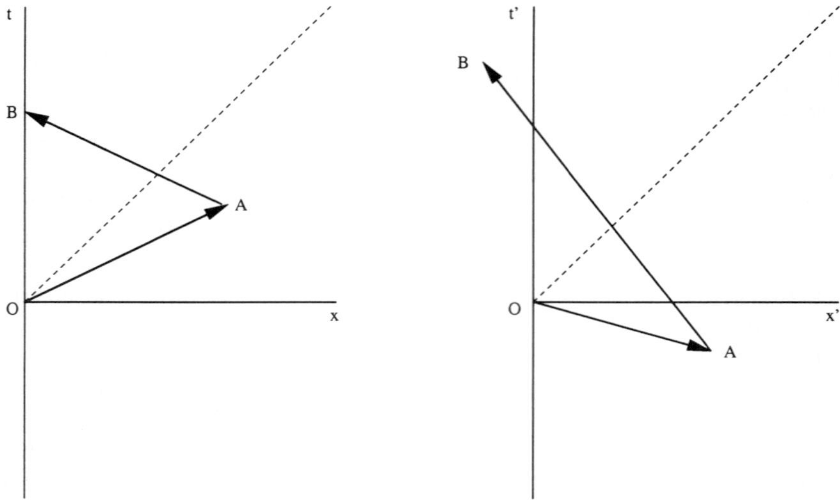

Figure 1. A superluminal $(v > 1)$ signal OA which is forwards in time in frame $\mathcal{S}$ is backwards in time in a frame $\mathcal{S}'$ moving relative to $\mathcal{S}$ with speed $u > \frac{1}{v}$. However, the return path with the same speed in $\mathcal{S}$ arrives at B in the future light cone of O, independent of the frame.

2.1 Superluminal Propagation in Special and General Relativity

It is generally understood that superluminal propagation in special relativity leads to unacceptable violations of causality. Indeed the absence of tachyons is traditionally employed as a constraint on fundamental theories. We therefore begin by reviewing some basic features of superluminal propagation in order to sharpen these ideas in preparation for our subsequent discussion of causality in general relativity.

The first important observation is that given a superluminal signal we can always find a reference frame in which it is travelling backwards in time. This is illustrated in Fig. 1. Suppose we send a signal from O to A at speed $v > 1$ (in $c = 1$ units) in frame $\mathcal{S}$ with coordinates (t, x). In a frame $\mathcal{S}'$ moving with respect to $\mathcal{S}$ with velocity $u > \frac{1}{v}$, the signal travels backwards in t' time, as follows immediately from the Lorentz transformation.[a]

The important point for our considerations is that this *by itself* does not necessarily imply a violation of causality. For this, we require that the signal can be returned from A to a point in the past light cone of O. However, if we return the signal from A to B with the same speed in frame $\mathcal{S}$, then of course it arrives at B in the future cone of O. The situation is physically equivalent in the Lorentz boosted frame $\mathcal{S}'$ – the return signal travels forward in t' time and arrives at B in the future cone of O. This, unlike the assignment of spacetime coordinates, is a frame-independent statement.

The problem with causality arises from the scenario illustrated in Fig. 2. Clearly,

---

[a]From the Lorentz transformations, we have $t'_A = \gamma(u)t_A(1 - uv)$ and $x'_A = \gamma(u)x_A(1 - \frac{u}{v})$ For the situation realised in Fig. 1, we require both $x'_A > 0$ and $t'_A < 0$, that is $\frac{1}{v} < u < v$, which admits a solution only if $v > 1$.

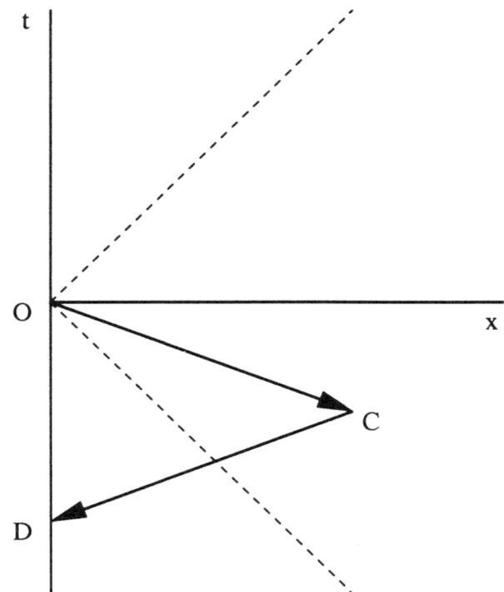

Figure 2. A superluminal ($v > 1$) signal OC which is backwards in time in frame $\mathcal{S}$ is returned at the same speed to point D in the past light cone of O, creating a closed time loop.

if a backwards-in-time signal OC is possible in frame $\mathcal{S}$, then a return signal sent with the same speed will arrive at D in the past light cone of O creating a closed time loop OCDO. The crucial point is that local Lorentz invariance of the laws of motion implies that if a superluminal signal such as OA is possible, then so is one of type OC, since it is just given by an appropriate Lorentz boost (as in Fig. 1). The existence of *global* inertial frames then guarantees the existence of the return signal CD (in contrast to the situation in Fig. 1 viewed in the $\mathcal{S}'$ frame).

The moral is that *both* conditions must be met in order to guarantee the occurrence of unacceptable closed time loops – the existence of a superluminal signal *and* global Lorentz invariance. Of course, since global Lorentz invariance (the existence of global inertial frames) is the essential part of the structure of special relativity, we recover the conventional wisdom that in this theory, superluminal propagation is indeed in conflict with causality.[b]

The reason for presenting this elementary discussion is to emphasise that the situation is crucially different in general relativity. The *weak* equivalence principle, which we understand as the statement that a *local* inertial frame exists at each point in spacetime, implies that general relativity is formulated on a Riemannian manifold. However, local Lorentz invariance alone is not sufficient to establish the link between superluminal propagation and causality violation. This is usually established by adding a second, dynamical, assumption. The *strong* equivalence principle (SEP) states that the laws of physics should be identical in the local frames at different points in spacetime, and that they should reduce to their special relativistic forms at the origin of each local frame. It is the SEP which takes over the role of the existence of global inertial frames in special relativity in establishing

---

[b]Notice that this is not in contradiction with the occurrence of superluminal propagation in flat spacetime with Casimir plates[10,11,12], since the modification of the spacetime geometry by the plates removes *global* Lorentz invariance.

50

the incompatibility of superluminal propagation and causality.

However, unlike the weak equivalence principle, which underpins the essential structure of general relativity, the SEP is merely a simplifying assumption about the dynamics of matter coupled to gravitational fields. Mathematically, it requires that matter or electromagnetism is *minimally coupled* to gravity, i.e. with interactions depending only on the connections but not the local curvature. This ensures that at the origin of a local frame, where the connections may be Lorentz transformed locally to zero, the dynamical equations recover their special relativistic form. In particular, the SEP is violated by interactions which explicitly involve the curvature, such as those occurring in the Drummond-Hathrell action (1) and the consequent modified light cones.

The question of whether this specific realisation of superluminal propagation is in conflict with causality is discussed in section 2.4 using the concept of *stabele causality* described in ref.[13]. Notice though that by violating the SEP, we have evaded the *necessary* association of superluminal motion with causality violation that held in special relativity. Referring back to the figures, what is established is the existence of a signal of type OA, which as we saw, does not by itself imply problems with causality even though frames such as $\mathcal{S}'$ exist *locally* with respect to which motion is backwards in time. However, since the SEP is broken, even if a local frame exists in which the signal looks like OC, it does *not* follow that a return path CD is allowed. The signal propagation is fixed, determined locally by the spacetime curvature.

2.2 Geometric Optics

The most direct way to deduce the form of the light cones for QED in curved spacetime is to use geometric optics. This starts from the ansatz

$$(A_\mu + i\epsilon B_\mu + \ldots)\exp\left(i\frac{\vartheta}{\epsilon}\right) \tag{5}$$

in which the electromagnetic field is written as a slowly-varying amplitude and a rapidly-varying phase. The parameter ϵ is introduced as a device to keep track of the relative order of magnitude of terms, and the Bianchi and Maxwell equations are solved order-by-order in ϵ.

The wave vector is identified as the gradient of the phase, $k_\mu = \partial_\mu \vartheta$. We also write $A_\mu = A a_\mu$, where A represents the amplitude itself while a_μ specifies the polarisation, which satisfies $k_\mu a^\mu = 0$.

Solving the usual Maxwell equation $D_\mu F^{\mu\nu} = 0$, we find at $O(1/\epsilon)$,

$$k^2 = 0 \tag{6}$$

while at $O(1)$,

$$k^\mu D_\mu a^\nu = 0 \tag{7}$$

and

$$k^\mu D_\mu(\ln A) = -\frac{1}{2}D_\mu k^\mu \tag{8}$$

Eq.(6) shows immediately that k^μ is a null vector. From its definition as a gradient, we also see

$$k^\mu D_\mu k^\nu = k^\mu D^\nu k_\mu = \frac{1}{2} D^\nu k^2 = 0 \tag{9}$$

Light rays, or equivalently photon trajectories, are the integral curves of k^μ, i.e. the curves $x^\mu(s)$ where $dx^\mu/ds = k^\mu$. These curves therefore satisfy

$$0 = k^\mu D_\mu k^\nu = \frac{d^2 x^\nu}{ds^2} + \Gamma^\nu_{\mu\lambda} \frac{dx^\mu}{ds} \frac{dx^\lambda}{ds} \tag{10}$$

This is the geodesic equation. We conclude that for the usual Maxwell theory in general relativity, light follows null geodesics. Eqs.(9),(7) show that both the wave vector and the polarisation are parallel transported along these null geodesic rays, while Eq.(8), whose r.h.s. is just (minus) the optical scalar θ, shows how the amplitude changes as the beam of rays focuses or diverges.

2.3 Bimetricity

The same method is applied to the modified Maxwell equation derived from the effective action Eq.(1):

$$D_\mu F^{\mu\nu} - \frac{1}{m^2}\left[2bR_{\mu\lambda}D^\mu F^{\lambda\nu} + 4cR_\mu{}^\nu{}_{\lambda\rho}D^\mu F^{\lambda\rho} \right] = 0 \tag{11}$$

from which the new light cone condition

$$k^2 - \frac{2b}{m^2}R_{\mu\lambda}k^\mu k^\lambda + \frac{8c}{m^2}R_{\mu\nu\lambda\rho}k^\mu k^\lambda a^\nu a^\rho = 0 \tag{12}$$

follows immediately. Since this new light cone relation is still homogeneous and quadratic in k^μ, we can write it as

$$\mathcal{G}^{\mu\nu} k_\mu k_\nu = 0 \tag{13}$$

defining $\mathcal{G}^{\mu\nu}$ as the appropriate function of the curvature and polarisation.

Now notice that in the discussion of the free Maxwell theory, we did not need to distinguish between the photon momentum p^μ, i.e. the tangent vector to the light rays, and the wave vector k_μ since they were simply related by raising the index using the spacetime metric, $p^\mu = g^{\mu\nu}k_\nu$. In the modified theory, however, there is an important distinction. The wave vector, defined as the gradient of the phase, is a covariant vector or 1-form, whereas the photon momentum/tangent vector to the rays is a true contravariant vector. The relation is non-trivial. In fact, given k_μ, we should define the corresponding 'momentum' as

$$p^\mu = \mathcal{G}^{\mu\nu} k_\nu \tag{14}$$

and the light rays as curves $x^\mu(s)$ where $\frac{dx^\mu}{ds} = p^\mu$. This definition of momentum satisfies

$$G_{\mu\nu}p^\mu p^\nu = \mathcal{G}^{\mu\nu} k_\mu k_\nu = 0 \tag{15}$$

where $G \equiv \mathcal{G}^{-1}$ defines a new *effective metric* which determines the light cones mapped out by the geometric optics light rays. (Indices are always raised or lowered

52

using the true metric $g_{\mu\nu}$.) The *ray velocity* $v_{\rm ray}$ corresponding to the momentum p^μ, which is the velocity with which the equal-phase surfaces advance, is given by (defining components in an orthonormal frame)

$$v_{\rm ray} = \frac{|\underline{p}|}{p^0} = \frac{d|\underline{x}|}{dt} \qquad (16)$$

along the ray. This is in general different from the *phase velocity*

$$v_{\rm ph} = \frac{k^0}{|\underline{k}|} \qquad (17)$$

This shows that photon propagation for QED in curved spacetime can be characterised as a *bimetric* theory[c] – the physical light cones are determined by the effective metric $G_{\mu\nu}$ while the geometric null cones are fixed by the spacetime metric $g_{\mu\nu}$.

2.4 Stable Causality

The bimetric formulation is the most natural language in which to discuss whether the superluminal velocities predicted by the Drummond-Hathrell action are compatible with our usual idea of causality. We have already seen that with SEP violation in general relativity, the arguments that in special relativity led to the incompatibility of superluminal propagation and causality are no longer valid. Superluminal motion *may* be possible – the question is to find a criterion to decide whether it *is*[d].

One special case where causality is realised in a particularly simple way is in globally hyperbolic spacetimes, where the manifold admits a foliation into a set of spacelike Cauchy surfaces with fibres given by timelike geodesics. It is not hard to imagine that the same structure could be preserved using the effective metric $G_{\mu\nu}$ to define 'spacelike' or 'timelike', especially if $G_{\mu\nu}$ is only perturbatively different from the actual spacetime metric $g_{\mu\nu}$. But this would be a global question and the preservation of global hyperbolicity is not *a priori* guaranteed.

The clearest criterion for causality in general involves the concept of *stable causality* discussed, for example, in the monograph of Hawking and Ellis[13]. Proposition 6.4.9 states the required definition and theorem:
• A spacetime manifold $(\mathcal{M}, g_{\mu\nu})$ is *stably causal* if the metric $g_{\mu\nu}$ has an open neighbourhood such that $\mathcal{M}$ has no closed timelike or null curves with respect to any metric belonging to that neighbourhood.
• Stable causality holds everywhere on $\mathcal{M}$ if and only if there is a globally defined function f whose gradient $D_\mu f$ is everywhere non-zero and timelike with respect to $g_{\mu\nu}$.

---

[c]Bimetric theories of gravity have an extensive literature. See ref.[14] for an elegant recent construction and references therein for earlier work.

[d]The equivalent discussion of causality for superluminal propagation in Minkowski spacetime with Casimir plates is given in ref.[12]. This also provides a nice example of the distinction between ray and phase velocities discussed above.

According to this theorem, the absence of causality violation in the form of closed timelike or lightlike curves is assured if we can find a globally defined function f whose gradient is timelike *with respect to the effective metric $G_{\mu\nu}$ for light propagation*. f then acts as a global time coordinate.

To see how this criterion can be applied to a particular example, one for which stable causality *is* preserved by the new light cone metric, consider the cosmological Friedmann-Robertson-Walker spacetime. Since the FRW metric is Weyl flat, the modified light cone condition Eq.(3) reads simply

$$k^2 = \zeta\, T_{\mu\nu}k^\mu k^\nu \tag{18}$$

where $\zeta = \frac{8\pi}{m^2}(2b + 4c)$ and the energy-momentum tensor is

$$T_{\mu\nu} = (\rho + P)n_\mu n_\nu - Pg_{\mu\nu} \tag{19}$$

with n^μ specifying the time direction in a comoving orthonormal frame. ρ is the energy density and P is the pressure, which in a radiation-dominated era are related by $\rho - 3P = 0$. The phase velocity is independent of polarisation and is found to be superluminal[e]:

$$v_{\rm ph} = \frac{k^0}{|\underline{k}|} = 1 + \frac{1}{2}\zeta(\rho + P) \tag{21}$$

At first sight, this looks surprising given that $k^2 > 0$, its sign fixed by the weak energy condition $T_{\mu\nu}k^\mu k^\nu \geq 0$. However, if instead we consider the momentum along the rays, $p^\mu = \mathcal{G}^{\mu\nu}k_\nu$, we find

$$p^2 = g_{\mu\nu}p^\mu p^\nu = -\zeta(\rho + P)(p^0)^2 \tag{22}$$

and

$$v_{\rm ray} = \frac{|\underline{p}|}{p^0} = 1 + \frac{1}{2}\zeta(\rho + P) \tag{23}$$

The effective metric $G = \mathcal{G}^{-1}$ is (in the orthonormal frame)

$$G = \begin{pmatrix} 1 + \zeta\rho & 0 & 0 & 0 \\ 0 & -(1 - \zeta P) & 0 & 0 \\ 0 & 0 & -(1 - \zeta P) & 0 \\ 0 & 0 & 0 & -(1 - \zeta P) \end{pmatrix} \tag{24}$$

In this case, therefore, we find equal and superluminal velocities $v_{\rm ph} = v_{\rm ray}$ and $p^2 < 0$ is manifestly spacelike as required.

Is stable causality preserved? In this case the answer is yes, since we may still use the cosmological time coordinate t as the globally defined function f. We need

---

[e]In the radiation dominated era, where $\rho(t) = \frac{3}{32\pi}t^{-2}$, we have

$$v_{\rm ph} = 1 + \frac{1}{16\pi}\zeta\, t^{-2} \tag{20}$$

which, as already observed in ref.[1], increases towards the early universe. Although this expression is only reliable in the perturbative regime where the correction term is small, it is intriguing that QED predicts a rise in the speed of light in the early universe. It is interesting to speculate whether this superluminal effect persists for high curvatures near the initial singularity and whether it could play a role in resolving the horizon problem in cosmology only check that $D_\mu t$ defines a timelike vector with respect to the effective metric $G_{\mu\nu}$. This is true provided $G_{00} > 0$, which is certainly satisfied by Eq.(24). So at least in this case, superluminal propagation is compatible with causality.

2.5 Time Machines?

Although we have seen that causality is not necessarily violated by superluminal propagation, it is important to look for counter-examples where the Drummond-Hathrell effect may create a time machine. One imaginitive suggestion was put forward by Dolgov and Novikov (DN)[15], involving two gravitating sources in relative motion. This scenario therefore has some echoes of the Gott cosmic string time machine[16]; both are reviewed in ref.[17].

The DN proposal is to consider first a gravitating source with a superluminal photon following a trajectory which we may take to be radial. Along this trajectory, the metric interval is effectively two-dimensional and DN consider the form

$$ds^2 = A^2(r)dt^2 - B^2(r)dr^2 \tag{25}$$

(An explicit realisation is given by radial superluminal signals in the Bondi-Sachs spacetime, described in ref.[7].) The photon velocity in the (t, r) coordinates is taken to be $v = 1 + \delta v$, so the effective light cones lie perturbatively close to the geometric ones. The trajectory is forward in time with respect to t.

DN now make a coordinate transformation corresponding to a frame in relative motion to the gravitating source, rewriting the metric interval along the trajectory as

$$ds^2 = A^2(t', r')\big(dt'^2 - dr'^2\big) \tag{26}$$

The transformation is[f]

$$\begin{aligned} t' &= \gamma(u)\big(t - ur - uf(r)\big) \\ r' &= \gamma(u)\big(r - ut + f(r)\big) \end{aligned} \tag{27}$$

with

$$f(r) = \int dr \left(\frac{B}{A} - 1\right) \tag{28}$$

Now, a superluminal signal with velocity

$$v = 1 + \delta v = \frac{B}{A}\frac{dr}{dt} \tag{29}$$

emitted at (t_1, r_1) and received at (t_2, r_2) travels forward in t time (for small, positive δv) with interval

$$t_2 - t_1 = \int_{r_1}^{r_2} dr\, (1 - \delta v)\frac{B}{A} \tag{30}$$

---

[f] This transformation comprises two steps. First, since any 2-dim metric is conformally flat, we can bring the metric Eq.(25) into standard form $ds^2 = \Omega^2\big(d\tilde{t}^2 - d\tilde{r}^2\big)$. Then, a boost with velocity u is made on the flat coordinates $(\tilde{t}, \tilde{r})$ to give the DN coordinates (t', r').

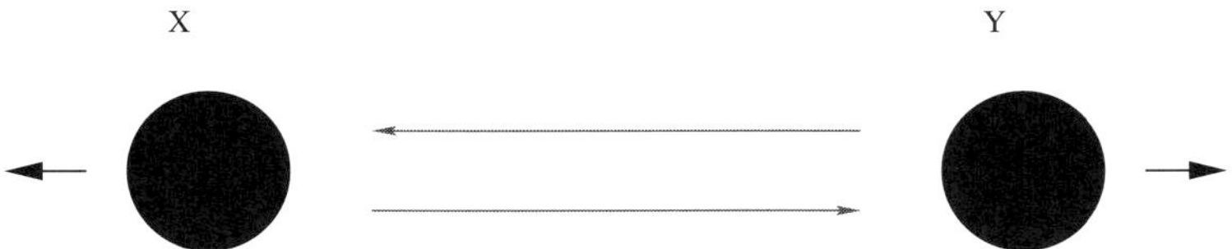

Figure 3. The Dolgov-Novikov time machine proposal. A superluminal signal from X, described as backwards-in-time in a relevant frame, is sent towards a second gravitating source Y moving relative to X and returned symmetrically.

As DN show, however, this motion is backwards in t' time for sufficiently large u, since the equivalent interval is

$$t'_2 - t'_1 = \gamma(u) \int_{r_1}^{r_2} dr \, (1 - u - \delta v) \frac{B}{A} \tag{31}$$

The required frame velocity is $u > 1 - \delta v$, i.e. since δv is small, $u > \frac{1}{v}$.

The situation so far is therefore identical in principle to the discussion of superluminal propagation illustrated in Fig. 1. In DN coordinates the outward superluminal signal is certainly propagating backwards in time, but a reverse path with the same perturbatively superluminal velocity would necessarily go sufficiently forwards in time to arrive back within the future light cone of the emitter.

At this point, however, DN propose to introduce a second gravitating source moving relative to the first, as illustrated in Fig. 3. They now claim that a superluminal photon emitted with velocity $v(r)$ in the region of X will travel backwards in time (according to the physically relevant coordinate t') to a receiver in the region of Y. A signal is then returned symmetrically to be received at its original position in the vicinity of X, arriving, according to DN, in its past. This would then be analogous to the situation illustrated in Fig. 2.

However, as we emphasised in section 2.1, we are *not* free to realise the scenario of Fig. 2 in the gravitational case, because the SEP-violating superluminal propagation proposed by Drummond and Hathrell is pre-determined, fixed by the local curvature. The t' frame may describe back-in-time motion for the outward leg, but it does not follow that the return path is similarly back-in-time *in the same frame*. The appropriate special relativistic analogue is the scenario of Fig. 1, not Fig. 2. This critique of the DN time machine proposal has already been made by Konstantinov[18] and further discussion of the related effect in flat spacetime with Casimir plates is given in ref.[12]. The relative motion of the two sources, which at first sight seems to allow the backwards-in-time coordinate t' to be relevant and to be used symmetrically, does not in fact alleviate the problem.

The true situation seems rather more to resemble Fig. 4. With the gravitating sources X and Y sufficiently distant that spacetime is separated into regions where it is permissible to neglect one or the other, a signal sent from the vicinity of X towards Y and back would follow the paths shown. But it is clear that this is no more than stitching together an outward plus inward leg near source X with an inward plus outward leg near Y. Since both of these are future-directed motions, in the sense of Fig. 1, their combination cannot produce a causality-violating trajectory. If, on the

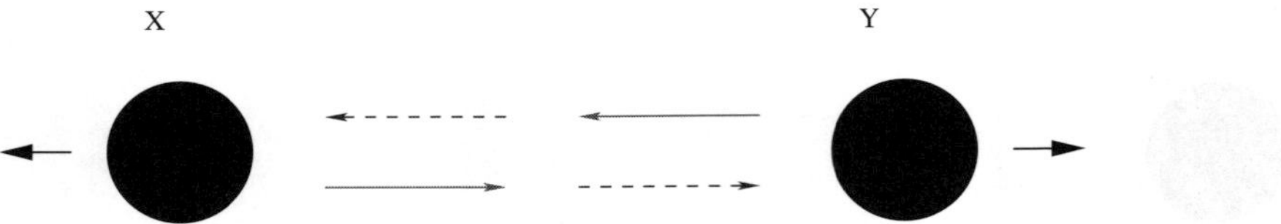

Figure 4. A decomposition of the paths in Fig. 3 for well-separated sources

other hand, we consider X and Y to be sufficiently close that this picture breaks down, we lose our ability to analyse the Drummond-Hathrell effect, since we would need the full collision metric for the gravitating sources which is not known for physically realisable examples.

We therefore conclude that the Dolgov-Novikov time machine does not work. The essential idea of trying to realise the causality-violating special relativistic scenario of Fig. 2 by using two gravitational sources in relative motion does not in the end succeed, precisely because the physical Drummond-Hathrell light cones are fixed by the local curvature. Once more it appears that in general relativity with SEP-violating interactions, superluminal photon propagation and causality can be compatible.

2.6 The Event Horizon

We have seen that when quantum effects are taken into account, the physical light cones need not coincide with the geometrical null cones. This immediately raises the question of black hole event horizons – do the physical horizons for light propagation also differ from the geometrical horizons, and are they polarisation dependent? If so, this would have profound repercussions for phenomena such as Hawking radiation.

The answer is best seen using the Newman-Penrose form of the light cone, viz.

$$k^2 - \frac{(4b+8c)\omega^2}{m^2}\Phi_{00} \pm \frac{4c\omega^2}{m^2}(\Psi_0 + \Psi_0^*) = 0 \tag{32}$$

If we define the tetrad with ℓ^μ as an outward-directed null vector orthogonal to the horizon 2-surface, then a fundamental theorem on horizons states that both Φ_{00} and Ψ_0 are zero precisely at the horizon. The detailed proof, which is given in ref.[19], involves following the convergence and shear of the generators of the horizon. In physical terms, however, it is easily understood as the requirement that the flow of both matter (given by the Ricci term) and gravitational radiation (given by the Weyl term) are zero across the horizon.

It follows that for outward-directed photons with $k_\mu = \omega\ell_\mu$, the quantum corrections vanish at the horizon and the light cone coincides with the null cone. The geometrical event horizon is indeed the true horizon for physical photon propagation[4,20]. Again, no conflict arises between superluminal propagation and essential causal properties of spacetime.

3 Causality, Characteristics and the 'Speeds of Light'

So far, our analysis of photon propagation has been based entirely on the leading-order, Drummond-Hathrell effective action Eq.(1). However, as we show in section 4, the full effective action contains terms to all orders in a derivative expansion and these must be taken into account to go beyond the low-frequency approximation. Photon propagation in QED in curved spacetime is therefore dispersive and we must understand how to identify the 'speed of light' which is relevant for causality.

3.1 'Speeds of Light'

An illuminating discussion of wave propagation in a simple dispersive medium is given in the classic work by Brillouin[21]. This considers propagation of a sharp-fronted pulse of waves in a medium with a single absorption band, with refractive index $n(\omega)$:

$$n^2(\omega) = 1 - \frac{a^2}{\omega^2 - \omega_0^2 + 2i\omega\rho} \tag{33}$$

where a, ρ are constants and ω_0 is the characteristic frequency of the medium. Five[g] distinct velocities are identified: the *phase velocity* $v_{\rm ph} = \frac{\omega}{|\underline{k}|} = \Re\frac{1}{n(\omega)}$, *group velocity* $v_{\rm gp} = \frac{d\omega}{d|\underline{k}|}$, *signal velocity* $v_{\rm sig}$, *energy-transfer velocity* $v_{\rm en}$ and *wavefront velocity* $v_{\rm wf}$, with precise definitions related to the behaviour of contours and saddle points in the relevant Fourier integrals in the complex ω-plane. Their frequency dependence is illustrated in Fig. 5. As the pulse propagates, the first disturbances to arrive are very small amplitude waves, 'frontrunners', which define the wavefront velocity $v_{\rm wf}$. These are followed continuously by waves with amplitudes comparable to the initial pulse; the arrival of this part of the complete waveform is identified in ref.[21] as the signal velocity $v_{\rm sig}$. As can be seen from Fig. 5, it essentially coincides with the more familiar group velocity for frequencies far from ω_0, but gives a much more intuitively reasonable sense of the propagation of a signal than the group velocity, whose behaviour in the vicinity of an absorption band is relatively eccentric.[h] As the figure makes clear, the phase velocity itself also does not represent a 'speed of light' relevant for considerations of signal propagation or causality.

The appropriate velocity to define light cones and causality is in fact the wavefront velocity $v_{\rm wf}$. (Notice that in Fig. 5, $v_{\rm wf}$ is a constant, equal to c, independent of the frequency or details of the absorption band.) This is determined by the boundary between the regions of zero and non-zero disturbance (more generally, a discontinuity in the first or higher derivative of the field) as the pulse propagates. Mathematically, this definition of wavefront is identified with the characteristics of the partial differential equation governing the wave propagation[24]. Our problem is therefore to determine the velocity associated with the characteristics of the wave

---

[g]In fact, if we take into account the distinction discussed in section 2 between the *phase velocity* $v_{\rm ph}$ and the *ray velocity* $v_{\rm ray}$, and include the fundamental speed of light constant c from the Lorentz transformations, we arrive at *seven* distinct definitions of 'speed of light'.

[h]Notice that it is the group velocity which is measured in quantum optics experiments which find light speeds of essentially zero[22] or many times c [23]. A particularly clear description in terms of the effective refractive index is given in ref.[22].

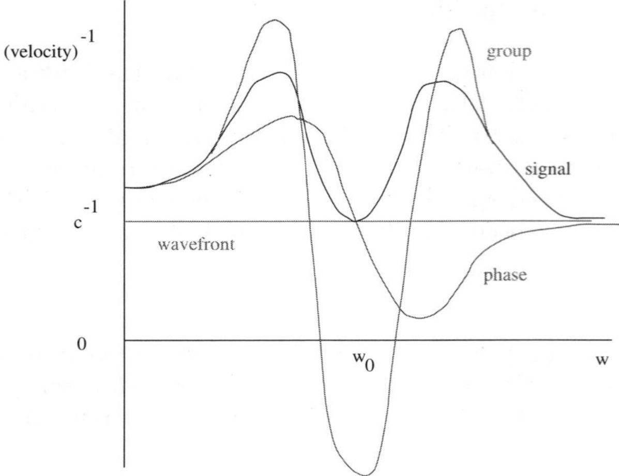

Figure 5. Sketch of the behaviour of the phase, group and signal velocities with frequency in the model described by the refractive index Eq.(33). The energy-transfer velocity (not shown) is always less than c and becomes small near ω_0. The wavefront speed is identically equal to c.

operator derived from the modified Maxwell equations of motion appropriate to the new effective action.

Notice that a very complete and rigorous discussion of the wave equation in curved spacetime has already been given in the monograph by Friedlander[25], in which it is proved (Theorem 3.2.1) that the characteristics are simply the null hypersurfaces of the spacetime manifold, in other words that the wavefront always propagates with the fundamental speed c. However, this discussion assumes the standard form of the (gauge-fixed) Maxwell wave equation (cf. ref.[25], eq.(3.2.1)) and does *not* cover the modified wave equation derived from the action Eq.1, precisely because of the extra curvature couplings which lead to the effective metric $G_{\mu\nu}$ and superluminal propagation.

3.2 Characteristics, Wavefronts and the Phase Velocity $v_{\rm ph}(\infty)$

Instead, the key result which allows a derivation of the wavefront velocity is derived by Leontovich[26]. In this paper[i], an elegant proof is presented for a very general set of PDEs that the wavefront velocity associated with the characteristics is identical to the $\omega \to \infty$ limit of the phase velocity, i.e.

$$v_{\rm wf} = \lim_{\omega \to \infty} \frac{\omega}{|\underline{k}|} = \lim_{\omega \to \infty} v_{\rm ph}(\omega) \tag{34}$$

The proof is rather formal, but is of sufficient generality to apply to our discussion of photon propagation using the modified effective action of section 4. We reproduce the essential details below.

---

[i] I am very grateful to A. Dolgov, V. Khoze and I. Khriplovich for their help in obtaining and interpreting ref.[26].

The first step is to recognise that any second order PDE can be written as a system of first order PDEs by considering the first derivatives of the field as independent variables. Thus, if for simplicity we consider a general second order wave equation for a field $u(t, x)$ in one space dimension, the system of PDEs we need to solve is

$$a_{ij} \frac{\partial \phi_j}{\partial t} + b_{ij} \frac{\partial \phi_j}{\partial x} + c_{ij} \phi_j = 0 \tag{35}$$

where $\phi_i = \{u, \frac{\partial u}{\partial t}, \frac{\partial u}{\partial x}\}$.

Making the 'geometric optics' ansatz

$$\phi_i = \varphi_i \exp i(wt - kx) \tag{36}$$

where the frequency-dependent phase velocity is $v_{\mathrm{ph}}(k) = \omega(k)/k$, and substituting into Eq.(35) we find

$$\left(i\omega a_{ij} - ikb_{ij} + c_{ij} \right) \varphi_j = 0 \tag{37}$$

The condition for a solution,

$$\det\left[a_{ij} v_{\mathrm{ph}}(k) - b_{ij} - \frac{i}{k} c_{ij} \right] = 0 \tag{38}$$

then determines the phase velocity.

On the other hand, we need to find the characteristics of Eq.(35), i.e. curves $\mathcal{C}$ on which Cauchy's theorem breaks down and the evolution is not uniquely determined by the initial data on $\mathcal{C}$. The derivatives of the field may be discontinuous across the characteristics and these curves are associated with the wavefronts for the propagation of a sharp-fronted pulse. The corresponding light rays are the 'bicharacteristics'. (See, for example, ref.[24] chapters $5.1, 6.1$ for further discussion.)

We therefore consider a characteristic curve $\mathcal{C}$ in the (t, x) plane separating regions where $\phi_i = 0$ (ahead of the wavefront) from $\phi_i \neq 0$ (behind the wavefront). At a fixed point (t_0, x_0) on $\mathcal{C}$, the absolute derivative of ϕ_i along the curve, parametrised as $x(t)$, is just

$$\frac{d\phi_i}{dt} = \left.\frac{\partial \phi_i}{\partial t}\right|_0 + \left.\frac{\partial \phi_i}{\partial x}\right|_0 \frac{dx}{dt} \tag{39}$$

where $dx/dt = v_{\mathrm{wf}}$ gives the wavefront velocity. Using this to eliminate $\frac{\partial \phi_i}{\partial t}$ from the PDE Eq.(35) at (t_0, x_0), we find

$$\left(-a_{ij} \frac{dx}{dt} + b_{ij} \right) \left.\frac{\partial \phi_j}{\partial x}\right|_0 + a_{ij} \frac{d\phi_j^{(0)}}{dt} + c_{ij} \phi_j^{(0)} = 0 \tag{40}$$

Now since $\mathcal{C}$ is a wavefront, on one side of which ϕ_i vanishes identically, the second two terms above must be zero. The condition for the remaining equation to have a solution is simply

$$\det\left[a_{ij} v_{\mathrm{wf}} - b_{ij} \right] = 0 \tag{41}$$

which determines the wavefront velocity v_{wf}. The proof is now evident. Comparing Eqs.(38) and (41), we clearly identify

$$v_{\mathrm{wf}} = v_{\mathrm{ph}}(k \to \infty) \tag{42}$$

The wavefront velocity in a gravitational background is therefore not given *a priori* by c. Taking vacuum polarisation into account, there is no simple non-dispersive medium corresponding to the vacuum of classical Maxwell theory in which the phase velocity represents a true speed of propagation; for QED in curved spacetime, even the vacuum is dispersive. In order to discuss causality, we therefore have to extend the original Drummond-Hathrell results for $v_{\rm ph}(\omega \sim 0)$ to the high frequency limit $v_{\rm ph}(\omega \to \infty)$, as already emphasised in their original work.

A further subtle question arises if we write the standard dispersion relation for the refractive index $n(\omega)$ in the limit $\omega \to \infty$:

$$n(\infty) = n(0) - \frac{2}{\pi} \int_0^\infty \frac{d\omega}{\omega} \Im n(\omega) \tag{43}$$

For a conventional dispersive medium, $\Im n(\omega) > 0$, which implies that $n(\infty) < n(0)$, or equivalently $v_{\rm ph}(\infty) > v_{\rm ph}(0)$. Evidently this is satisfied by Fig. 5. The key question though is whether the usual assumption of positivity of $\Im n(\omega)$ holds in the present situation of the QED vacuum in a gravitational field. If so, then (as already noted in ref.[1]) the superluminal Drummond-Hathrell results for $v_{\rm ph}(0)$ would actually be *lower bounds* on the all-important wavefront velocity $v_{\rm ph}(\infty)$. However, it is not clear that positivity of $\Im n(\omega)$ holds in the gravitational context. Indeed it has been explicitly criticised by Dolgov and Khriplovich in refs.[27,28], who point out that since gravity is an inhomogeneous medium in which beam focusing as well as diverging can happen, a growth in amplitude corresponding to $\Im n(\omega) < 0$ is possible. The possibility of $v_{\rm ph}(\infty) < v_{\rm ph}(0)$, and in particular $v_{\rm ph}(\infty) = c$, therefore does not seem to be convincingly ruled out by the dispersion relation Eq.(43).

4 Dispersion

After these general considerations, we now return to QED in curved spacetime and use the full effective action to study dispersion and investigate the high-frequency limit.

4.1 *Effective Action for Photon-Gravity Interactions*

The local effective action in QED to $O(RFF)$ keeping terms of all orders in derivatives is derived in ref.[9]. The result is:

$$\begin{aligned}
\Gamma = \int dx \sqrt{-g} \Big[&-\tfrac{1}{4} F_{\mu\nu} F^{\mu\nu} + \tfrac{1}{m^2} \Big(D_\mu F^{\mu\lambda} \, \overrightarrow{G_0} \, D_\nu F^\nu{}_\lambda \\
&+ \overrightarrow{G_1} \, RF_{\mu\nu} F^{\mu\nu} + \overrightarrow{G_2} \, R_{\mu\nu} F^{\mu\lambda} F^\nu{}_\lambda + \overrightarrow{G_3} \, R_{\mu\nu\lambda\rho} F^{\mu\nu} F^{\lambda\rho} \Big) \\
&+ \tfrac{1}{m^4} \Big(\overrightarrow{G_4} \, RD_\mu F^{\mu\lambda} D_\nu F^\nu{}_\lambda + \overrightarrow{G_9} \, R_{\mu\nu\lambda\rho} D_\sigma F^{\sigma\rho} D^\lambda F^{\mu\nu} \\
&\quad + \overrightarrow{G_5} \, R_{\mu\nu} D_\lambda F^{\lambda\mu} D_\rho F^{\rho\nu} + \overrightarrow{G_6} \, R_{\mu\nu} D^\mu F^{\lambda\rho} D^\nu F_{\lambda\rho} \\
&\quad + \overrightarrow{G_7} \, R_{\mu\nu} D^\mu D^\nu F^{\lambda\rho} F_{\lambda\rho} + \overrightarrow{G_8} \, R_{\mu\nu} D^\mu D^\lambda F_{\lambda\rho} F^{\rho\nu} \Big) \Big]
\end{aligned} \tag{44}$$

This was found by adapting a background field action valid to third order in generalised curvatures due to Barvinsky, Gusev, Zhytnikov and Vilkovisky[29] (see also ref.[30]) and involves re-expressing their more general result in manifestly local form by an appropriate choice of basis operators.

In this formula, the $\overrightarrow{G_n}$ $(n \geq 1)$ are form factor functions of three operators:

$$\overrightarrow{G_n} \equiv G_n\left(\frac{D^2_{(1)}}{m^2}, \frac{D^2_{(2)}}{m^2}, \frac{D^2_{(3)}}{m^2}\right) \tag{45}$$

where the first entry $(D^2_{(1)})$ acts on the first following term (the curvature), etc. $\overrightarrow{G_0}$ is similarly defined as a single variable function. These form factors are found using heat kernel methods and are given by 'proper time' integrals of known algebraic functions. Their explicit expressions can be found in ref.[9]. Evidently, Eq.(44) reduces to the Drummond-Hathrell action if we neglect all the higher order derivative terms.

4.2   Dispersion and the Light Cone

The next step is to derive the equation of motion analogous to Eq.(11) from this generalised effective action and to apply geometric optics to find the corresponding light cone. This requires a very careful analysis of the relative orders of magnitudes of the various contributions to the equation of motion arising when the factors of D^2 in the form factors act on the terms of $O(RF)$. These subtleties are explained in detail in ref.[8]. The final result for the new effective light cone has the form

$$k^2 - \frac{1}{m^2}F\left(\frac{k.D}{m^2}\right)R_{\mu\lambda}k^\mu k^\lambda + \frac{1}{m^2}G\left(\frac{k.D}{m^2}\right)R_{\mu\nu\lambda\rho}k^\mu k^\lambda a^\nu a^\rho = 0 \tag{46}$$

where F and G are known functions with well-understood asymptotic properties[8]. Clearly, for agreement with Eq.(12), we have $F(0) = 2b$, $G(0) = 8c$.

The novel feature of this new light cone condition is that F and G are functions of the *operator* $k.D$ acting on the Ricci and Riemann tensors.[j] So although the asymptotic behaviour of F and G as functions is known, this information is not really useful unless the relevant curvatures are eigenvalues of the operator. On the positive side, however, $k.D$ does have a clear geometrical interpretation – it simply describes the variation along a null geodesic with tangent vector k^μ.

The utility of this light cone condition therefore seems to hinge on what we know about the variations along null geodesics of the Ricci and Riemann (or Weyl) tensors. It may therefore be useful to re-express Eq.(46) in Newman-Penrose form:

$$k^2 - \frac{\omega^2}{m^2}\tilde{F}\left(\frac{\omega\ell.D}{m^2}\right)\Phi_{00} \pm \frac{\omega^2}{2m^2}G\left(\frac{\omega\ell.D}{m^2}\right)(\Psi_0 + \Psi_0^*) = 0 \tag{47}$$

where $\tilde{F} = 2F + G$.

Unfortunately, we have been unable to find any results in the relativity literature for $\ell.D\Phi_{00}$ and $\ell.D\Psi_0$ which are valid in a general spacetime. In particular, this

---

[j]Note that because these corrections are already of $O(\alpha)$, we can freely use the usual Maxwell relations $k.Dk^\nu = 0$ and $k.Da^\nu = 0$ in these terms; we need only consider the effect of the operator $k.D$ acting on $R_{\mu\nu}$ and $R_{\mu\nu\lambda\rho}$.

is *not* one of the combinations that are constrained by the Bianchi identities in Newman-Penrose form (as displayed for example in ref.[31], chapter 1, Eq.(321)). To try to build some intuition, we have therefore looked at particular cases. The most interesting is the example of photon propagation in the Bondi-Sachs metric[32,33] which we recently studied in detail[7].

The Bondi-Sachs metric describes the gravitational radiation from an isolated source. The metric is

$$ds^2 = -W\,du^2 - 2e^{2\beta}du\,dr + r^2 h_{ij}(dx^i - U^i du)(dx^j - U^j du) \tag{48}$$

where

$$h_{ij}dx^i dx^j = \frac{1}{2}(e^{2\gamma}+e^{2\delta})d\theta^2 + 2\sinh(\gamma-\delta)\sin\theta d\theta d\phi + \frac{1}{2}(e^{-2\gamma}+e^{-2\delta})\sin^2\theta d\phi^2 \tag{49}$$

The metric is valid in the vicinity of future null infinity $\mathcal{I}^+$. The family of hypersurfaces $u = const$ are null, i.e. $g^{\mu\nu}\partial_\mu u \partial_\nu u = 0$. Their normal vector ℓ_μ satisfies

$$\ell_\mu = \partial_\mu u \qquad \Rightarrow \qquad \ell^2 = 0, \qquad \ell^\mu D_\mu \ell^\nu = 0 \tag{50}$$

The curves with tangent vector ℓ^μ are therefore null geodesics; the coordinate r is a radial parameter along these rays and is identified as the luminosity distance. The six independent functions $W, \beta, \gamma, \delta, U^i$ characterising the metric have expansions in $\frac{1}{r}$ in the asymptotic region near $\mathcal{I}^+$, the coefficients of which describe the various features of the gravitational radiation.

In the low frequency limit, the light cone is given by the simple formula Eq.(4) with $\Phi_{00} = 0$. The velocity shift is quite different for the case of outgoing and incoming photons[7]. For outgoing photons, $k^\mu = \ell^\mu$, and the light cone is

$$k^2 \;=\; \pm\, \frac{4c\omega^2}{m^2}\left(\Psi_0 + \Psi_0^*\right) \;\sim\; O\!\left(\frac{1}{r^5}\right) \tag{51}$$

while for incoming photons, $k^\mu = n^\mu$,

$$k^2 \;=\; \pm\, \frac{4c\omega^2}{m^2}\left(\Psi_4 + \Psi_4^*\right) \;\sim\; O\!\left(\frac{1}{r}\right) \tag{52}$$

Now, it is a special feature of the Bondi-Sachs spacetime that the absolute derivatives of each of the Weyl scalars $\Psi_0, \ldots, \Psi_4$ along the ray direction ℓ^μ vanishes, i.e. $\Psi_0, \ldots, \Psi_4$ are parallel transported along the rays[33,34]. In this case, therefore, we have:

$$\ell \cdot D\,\Psi_0 \;=\; 0 \qquad\qquad\qquad \ell \cdot D\,\Psi_4 \;=\; 0 \tag{53}$$

but there is no equivalent simple result for either $n \cdot D\,\Psi_4$ or $n \cdot D\,\Psi_0$.

Although it is just a special case, Eq.(53) nevertheless leads to a remarkable conclusion. The full light cone condition Eq.(47) applied to outgoing photons in the Bondi-Sachs spacetime now reduces to

$$k^2 \;\pm\; \frac{\omega^2}{2m^2}\,G(0)\left(\Psi_0 + \Psi_0^*\right) \;=\; 0 \tag{54}$$

since $\ell \cdot D\Psi_0 = 0$. In other words, the low-frequency Drummond-Hathrell prediction of a superluminal phase velocity $v_{\rm ph}(0)$ is *exact* for all frequencies. There is no dispersion, and the wavefront velocity $v_{\rm ph}(\infty)$ is indeed superluminal.

This is potentially a very important result. Based on the improved effective action Eq.(44), we have shown there is at least one example in which the wavefront truly propagates with superluminal velocity. Quantum effects have indeed shifted the light cone into the geometrically spacelike region.

4.3 Non-Perturbative Effective Action and High-Frequency Propagation

Unfortunately, there is one final twist to the story which could invalidate the above conclusion. If instead of a gravitational field, we consider photon propagation in a constant background magnetic field, we find the following birefringent modification to the light cones:[35,36,8]

$$k^2 \;-\; \frac{\alpha}{2\pi}\left(\frac{eB\omega}{m^2}\right)^2 \int_{-1}^{1} du \int_{0}^{\infty} ds\; s\; N_{\parallel,\perp}(u,z)\; e^{-is\left(1+s^2\Omega^2 P(u,z)\right)} \;=\; 0 \quad (55)$$

where $z = \frac{eB}{m^2}s$ and $\Omega = \frac{eB}{m^2}\frac{\omega}{m}$. Exact expressions are known for the functions N and P as well as their weak-field expansions.

Now, in the weak-field, low-frequency regime, $\Omega \ll 1$ and we can simply expand the exponential to leading order in $\Omega^2 P$. The low-frequency limit of the phase velocity for the two polarisations is the well-known result[2,37]

$$v_{\mathrm{ph}}(0) \;=\; 1 - \frac{\alpha}{4\pi}\left(\frac{eB}{m^2}\right)^2 \left[\frac{14}{45}_{\parallel}, \frac{8}{45}_{\perp}\right] \quad (56)$$

and is in each case subluminal. At high frequencies, however, Eq.(55) is dominated by the rapidly-varying phase factor $\exp(-is^3\Omega^2 P)$ and the correction to the phase velocity tends to zero (from a superluminal value) with a non-analytic $\Omega^{-4/3}$ behaviour:

$$v_{\mathrm{ph}}(\omega \to \infty) \;=\; 1 + \frac{\alpha}{4\pi}\left(\frac{eB}{m^2}\right)^2 \left[c_{\parallel}, c_{\perp}\right] \Omega^{-\frac{4}{3}} \quad (57)$$

where $c_{\parallel}, c_{\perp}$ are known positive constants[8]. The phase velocity therefore has the standard form for a dispersive medium illustrated in Fig. 5. In particular, the wavefront velocity $v_{\mathrm{wf}} = v_{\mathrm{ph}}(\infty) = c$.

The important lesson for the gravitational case is this. If we had simply used an effective action for QED in a background electromagnetic field keeping terms up to $O(F^4)$ in the field strength and to all orders in derivatives, generalising the Euler-Heisenberg Lagrangian, we would have accurately found the leading low-frequency dependence of $v_{\mathrm{ph}}(\omega)$ but would have completely missed the non-analytic high-frequency behaviour. This arises from terms in the full effective action which are *non-perturbative* in the background field and give rise to the phase factor in Eq.(55).

If the gravitational case is similar, this would imply that the modified light cone can be written heuristically as

$$k^2 \;+\; \frac{\alpha}{\pi}\int_{0}^{\infty} ds\; \mathcal{N}(s,R)\; e^{-is\left(1+s^2\Omega^2\mathcal{P}(s,R)\right)} \;=\; 0 \quad (58)$$

where both $\mathcal{N}$ and $\mathcal{P}$ can be expanded in powers of curvature, and derivatives of curvature, presumably associated with factors of ω as in the last section. The frequency dependent factor Ω would be $\Omega \sim \frac{R}{m^2}\frac{\omega}{m} \sim O(\frac{\lambda_c^3}{\lambda L^2})$, where '$R$' denotes some generic curvature component and L is the typical curvature scale. If this is true, then an expansion of the effective action to $O(RFF)$, even including higher derivatives, would not be sufficient to reproduce the full, non-perturbative contribution $\exp(-is^3\Omega^2\mathcal{P})$. The Drummond-Hathrell action would correspond to the leading order term in the expansion of Eq.(58) in powers of $\frac{R}{m^2}$ neglecting derivatives, while our improved effective action of section 4.1 sums up all orders in derivatives while retaining the restriction to leading order in curvature.

The omission of the non-perturbative contribution would be justified only in the limit of small Ω, i.e. for $\frac{\lambda_c^3}{\lambda L^2} \ll 1$. Neglecting this therefore prevents us from accessing the genuinely high frequency limit $\lambda \to 0$ needed to find the asymptotic limit $v_{\mathrm{ph}}(\infty)$ of the phase velocity. Moreover, assuming Eq.(58) is indeed on the right lines, it also seems inevitable that for high frequencies (large Ω) the rapid phase variation in the exponent will drive the entire heat kernel integral to zero, ensuring the wavefront velocity $v_{\mathrm{wf}} = c$.

4.4 Outlook

At present, it is not clear how to make further progress. The quantum field theoretic calculation required to find such non-perturbative contributions to the effective action and confirm an $\exp(-is^3\Omega^2\mathcal{P})$ structure in Eq.(58) appears difficult, although some technical progress in this area has been made recently in ref.[38] and work in progress[39]. One of the main difficulties is that since a superluminal effect requires some anisotropy in the curvature, it is not sufficient just to consider constant curvature spacetimes. (Recall that the Ricci scalar term in the effective action Eq.(1) does not contribute to the modified light cone Eq.(2).) A possible approach to this problem, which would help to control the plethora of indices associated with the curvatures, might be to reformulate the heat kernel calculations directly in the Newman-Penrose basis. On the other hand, perhaps a less ambitious goal would be to try to determine just the asymptotic form of the non-perturbative contribution in the $\Omega \to \infty$ limit.

A final resolution of the dispersion problem for QED in curved spacetime has therefore still to be found. If the perturbative expansion of the effective action to $O(RFF)$ is sufficient, then as we have seen there exist at least some examples where the wavefront velocity is really superluminal. In this case, all the issues concerning causality discussed in section 2 would apply to QED. Perhaps more likely, however, is the scenario described above, where the high-frequency dispersion is driven by non-perturbative contributions to the effective action such that the wavefront velocity remains precisely c. It would then be interesting to see exactly how the phase velocity behaves as a function of ω (c.f. Fig. 5) and whether the explanation advanced in section 3.2 for the non-validity of the standard refractive index dispersion relation Eq.(43) is correct.

Finally, of course, even if it does turn out that $v_{\mathrm{wf}} = c$ for QED itself, the discussion of causality in this paper may still be relevant to photon propagation in more speculative theories, including bimetric theories of gravity[14], string-inspired non-linear electrodynamics[20], Lorentz and CPT violating effective Lagrangians[40] and non-commutative gauge theories.

Acknowledgments

This work is supported in part by PPARC grant PP/G/O/2000/00448.

References

1. I.T. Drummond and S. Hathrell, Phys. Rev. D22 (1980) 343.
2. R.D. Daniels and G.M. Shore, Nucl. Phys. B425 (1994) 634.
3. J. I. Latorre, P. Pascual and R. Tarrach, Nucl. Phys. B437 (1995) 60.
4. G.M. Shore, Nucl. Phys. B460 (1996) 379.
5. W. Dittrich and H. Gies, Phys. Lett. B431 (1998) 420-429; Phys. Rev. D58 (1998) 025004.
6. R.D. Daniels and G.M. Shore, Phys. Lett. B367 (1996) 75.
7. G.M. Shore, Nucl. Phys. B605 (2001) 455.
8. G.M. Shore, Nucl. Phys. B633 (2002) 271.
9. G.M. Shore, Nucl. Phys. B646 (2002) 281.
10. K. Scharnhorst, Phys. Lett. B236 (1990) 354.
11. G. Barton, Phys. Lett. B237 (1990) 559.
12. S. Liberati, S. Sonego and M. Visser, Annals Phys. 298 (2002) 167.
13. S.W. Hawking and G.F.R. Ellis, *The Large Scale Structure of Spacetime*, Cambridge University Press, 1973.
14. I.T. Drummond, Phys. Rev. D63 (2001) 043503.
15. A.D. Dolgov and I.D. Novikov, Phys. Lett. B442 (1998) 82.
16. J.R. Gott, Astrophys. J. 288 (1985) 422.
17. G.M. Shore, gr-qc/0210048, *to be published in* Int. J. Mod. Phys. A
18. M.Yu. Konstantinov, gr-qc/9810019; Russ. Phys. J. 45 (2002) 23.
19. S.W. Hawking, '*The Event Horizon*', 1972 Les Houches lectures, ed. B. De Witt, Gordon and Breach, 1972.
20. G.W. Gibbons and C.A.R. Herdeiro, Phys. Rev. D63 (2001) 064006.
21. L. Brillouin, *Wave Propagation and Group Velocity*, Academic Press (London) 1960.
22. L.V. Hau, Scientific American 7 (2001) 66.
23. L.J. Wang, A. Kuzmich and A. Dogoriu, Nature 406 (2000) 277.
24. R. Courant and D. Hilbert, *Methods of Mathematical Physics, Vol II*, Interscience, New York, 1962.
25. F.G. Friedlander, *The Wave Equation on a Curved Spacetime*, Cambridge University Press, 1975.
26. M.A. Leontovich, *in* L.I. Mandelshtam, *Lectures in Optics, Relativity and Quantum Mechanics*, Nauka, Moscow 1972 *(in Russian)*.
27. A.D. Dolgov and I.B. Khriplovich, Sov. Phys. JETP 58(4) (1983) 671.
28. I.B. Khriplovich, Phys. Lett. B346 (1995) 251.
29. A.O. Barvinsky, Yu.V. Gusev, G.A. Vilkovisky and V.V. Zhytnikov, Print-93-0274 (Manitoba), 1993.
30. A.O. Barvinsky, Yu.V. Gusev, G.A. Vilkovisky and V.V. Zhytnikov, J. Math. Phys. 35 (1994) 3525; J. Math. Phys. 35 (1994) 3543; Nucl. Phys. B439 (1995) 561.

31. S. Chandresekhar, *The Mathematical Theory of Black Holes*, Clarendon, Oxford, 1985.
32. H. Bondi, M.G.J. van der Burg and A.W.K. Metzner, Proc. Roy. Soc. A269 (1962) 21
33. R.K. Sachs, Proc. Roy. Soc. A270 (1962) 103.
34. R.A. d'Inverno, *Introducing Einstein's Relativity*, Clarendon, Oxford, 1992.
35. W. Tsai and T. Erber, Phys. Rev. D10 (1974) 492.
36. W. Tsai and T. Erber, Phys. Rev. D12 (1975) 1132.
37. S. Adler, Ann. Phys. (N.Y.) 67 (1971) 599.
38. A.O. Barvinsky and V.F. Mukhanov, Phys. Rev. D66 (2002) 065007.
39. Yu.V. Gusev, private communication.
40. V.A. Kostelecky, Phys. Rev. D58 (1998) 116002.

QUANTUM FLUCTUATIONS AND SIGNAL VELOCITY IN SUPERLUMINAL LIGHT PULSE PROPAGATION

L. J. WANG

NEC Laboratories America,Princeton, NJ 08540, USA
Email: Lwan@research.nj.nec.com

We review the recent work on producing a transparent, linear, anomalously dispersive medium. Experimentally, a light pulse propagating through the atomic vapor
cell has its peak reaching the exit side before entering it, resulting in a negative
transit time, and negative group velocity. We further review a recently proposed
operational definition of the velocity of information transport based on a close
analysis of quantum fluctuations.

1 Introduction

In the process of light propagating through a transparent medium, it is well known
that a number of velocities are involved. First, there is the "wave (phase) velocity,"
c/n, determined by the refractive index n of the medium. The envelope of the
pulse, however, travels at the group velocity $v_g = c/n_g$, determined by the group
index: $n_g = n + \nu dn/d\nu$. In a dispersive medium where the wave velocity depends
on the frequency, i.e., $dn/d\nu \neq 0$, the group velocity can be very different from the
phase velocity. The topic has been extensively discussed in the past [1,2]. Furthermore, Sommerfeld and Brillouin considered to define a "signal velocity," and have
concluded that it is impossible to do so for a noiseless "classical" analytical signal.
Hence, they concluded that the proper definition of a signal is the propagation of
an abrupt discontinuity; the resulting "front velocity" should be considered the
velocity of signal transfer. And finally, there is the problem of defining the velocity
of energy transfer [1-4]. Of course, in vacuum, all five velocities are equal to c. In
this paper, we will summarize our recent study of the superluminal group velocity
in a transparent anomalously dispersive medium. We also discuss the definition of
the signal velocity for a smoothly-varying optical pulse. The discussion of energy
velocity is beyond the scope of this present paper.

A pulse of light propagates at the group velocity in a dispersive medium. In
recent years, much attention has been paid to the case where an artificial steep
normal dispersion is realized in a medium showing electromagnetically induced
transparancy (EIT). Consequently, very slow or vanishingly small group velocity can
be achieved. On the other hand, if a medium can be obtained to be both transparent
and anomalously dispersive, i.e., $n + dn/d\nu < 1$, the group velocity $v_g = c/n_g$ will
exceed c ("superluminal" group velocity), or even become negative. Of course, it
is well known that inside an absorption line, the dispersion is anomalous, resulting
in a superluminal group velocity [1,2,5-7]. However, when absorption is large, the
medium becomes opaque. In other words, the pulse is no longer "propagating"
through the medium. Instead, the pulse is heavily absorbed and only a very small
portion is transmitted.

Generally, for all passive, transparent matter (e.g. glass, water) such that it is
absorptive at all frequencies in the electromagnetic spectrum, the medium's optical

68

dispersion is normal. Landau and Lifshitz [2] showed that under the condition

$$\Im\left[\chi(\nu)\right] \geq 0, \quad \text{for any } \nu \tag{1}$$

and in the special case for media with a magnetic permeability $\mu(\nu) = 1$, two inequalities hold simultaneously:

$$n_g(\nu) = \frac{d\left[n(\nu)\nu\right]}{d\nu} > n(\nu),$$

$$\text{and} \qquad n_g(\nu) = \frac{d\left[n(\nu)\nu\right]}{d\nu} > \frac{1}{n(\nu)}, \tag{2}$$

for a transparent region where $\Im\left[\chi(\nu)\right] \approx 0$. Here n_g is the group index. Obviously, we have $n_g > 1$. Consequently, the group velocity will be less than c for all transparent and passive media.

However, for media with gain, the general assumption in Eq.(1) no longer holds. In a series of papers [8,9,10,11,12,13], Chiao and coworkers showed theoretically that anomalous dispersion can occur inside a transparent material. It was predicted that, by using a gain doublet [12], it is possible to obtain a transparent anomalous dispersion region where the group velocity of a light pulse exceeds c.

Here we use gain-assisted linear anomalous dispersion to demonstrate superluminal light pulse propagation with a negative group velocity through a transparent atomic medium [14,15,16]. We place two Raman gain peaks closely to obtain an essentially lossfree anomalous dispersion region that results in a superluminal group velocity. The group velocity of a pulse in this region exceeds c and can even become negative, while the shape of the pulse is essentially preserved. We have measured [14,15] a negative group velocity index of $n_g = -315(\pm 5)$. Experimentally, a light pulse propagating through the atomic vapor cell exits from it earlier than propagating through the same distance in vacuum by a time difference that is 315 times the vacuum light propagation time L/c.

The experimental situation invites the question of what the velocity of a light signal is. As is well known, special relativity requires that any object, and in the case of light, a signal, can never travel faster than c, the speed of light in a vacuum. Traditionally, the apparent contradiction of a superluminal group velocity in an absorbing medium with the above statement was resolved by Sommerfeld and Brillouin in the early 1910's. A long time has passed since and our knowledge and understanding in physics has evolved. We find that it is now perfectly fitting to address this question in the modern context of optical communication.

As noted many years ago by Sommerfeld and Brillouin [1], group velocity is not the velocity of signal transmission. For a smoothly varying pulse that is described by an analytic function, the signal velocity cannot be defined, they argued. Because an analytic signal is entirely determined by its very small leading edge, there is no new information being carried by the peak. Furthermore, this leading edge of the pulse can in principle extend infinitely far back in time, making it impossible to assign a point marking the onset of a signal. They noted that the "front velocity," the velocity at which an infinitely sharp step-function-like disturbance of the light intensity propagates, should be used as the velocity of information transmission [1].

These ingenious arguments, however, are not immediately applicable in practice. First, it is impossible even in principle to realize the infinite bandwidth associated with a step-function "front." But more subtle questions arise when one considers a smoothly-varying pulse, where a tiny leading edge of a smooth pulse determines the entire pulse. In practice, one cannot extend the "arrival time" to any time before the detection of the *first* photon. Furthermore, if the tiniest leading edge of a smooth "superluminal" pulse determines the entire pulse, we must account for the effect that quantum fluctuations at the leading edge might have on the detection of the pulse [17,18]. From a practical point of view, the argument of Brillouin and Sommerfeld that a signal velocity cannot be defined for a smooth pulse is completely incompatible with reality. For the simple reason that all useful signals, from that of a signal from a lighthouse to the modern-day optical network communication, all signals are carried by bandwidth limited pulses. The purely academically minded definition of an infinitely sharp "front" should be viewed as no more than a mathematical trick to resolve an apparent conflict with special relativity. In practice when a signal is involved, all that matters is the signal-to-noise ratio, the capacity of the channel, and the signal integrity measured by the "bit-error-rate."

Recently we have suggested [19] an operational definition of the signal velocity and applied it to the observed superluminal propagation of a light pulse in a gain medium [14,15]. Previous considerations of quantum noise in this context focused on the motion of the peak of a wave packet, and on the observability of the superluminal velocity of the peak at the one- or few-photon level [17,18]. Here we showed that quantum noise associated with the amplifying medium can act in effect to retard the observed signal. In order to achieve a given signal-to-noise ratio (SNR) at the output of an amplifying medium, a larger signal is required, resulting in a retardation of the signal. This retardation is found in numerical simulation to be larger than the propagation time reduction due to anomalous dispersion, leading to a signal velocity $\leq c$. The operational definition given and the conclusions reached here are independent of the intensity of the input pulse.

The paper is adopted from the proceeding of a lecture the author gave during a conference on electromagnetic waves held in Beijing, China, in May, 2002, and is organized as the following. We first summarize the realization of a transparent, linear, anomalously dispersive medium. Pulse propagation and the "rephasing" process in such a medium is then reviewed. The experimental details and observational results are briefly reviewed and the details can be found in references [14,15]. Finally, we review the operational approach to defining a signal velocity for a smooth pulse propagating through such an anomalously dispersive medium [19].

2   Transparent Anomalous Dispersion

Let us start by considering a classical Lorentz oscillator model of the refractive index. The electric displacement is given by:

$$D = \varepsilon_0 E + P = \varepsilon_0 (1 + \chi)E = \varepsilon_0 E(1 + N\alpha), \tag{3}$$

where N is the atomic density and α is the atomic polarizability. The polarization density $P = -Nex = \varepsilon_0 N\alpha E$ can be obtained using a simple Lorentz model.

In order to obtain the dipole polarization $p = -ex$ for a bound charge with an intrinsic angular frequency ω_0 and an angular damping rate $\Gamma = 2\gamma$, we start from the equation of motion of the electron:

$$\ddot{x} + \Gamma\dot{x} + \omega_0^2 x = -\frac{eE}{m} = -\frac{eE_0}{m}e^{-i\omega t}. \tag{4}$$

Hence, one obtains that,

$$x = \frac{eE}{m}\frac{1}{\omega^2 - \omega_0^2 + i\omega\Gamma} \approx \frac{eE}{2m\omega_0}\frac{1}{\omega - \omega_0 + i\Gamma/2}, \tag{5}$$

where the approximation is good as long as $\omega_0 \gg \Gamma$. We further obtain for the polarizability,

$$\alpha = -\frac{e^2}{2\varepsilon_0 m\omega_0}\frac{1}{\omega - \omega_0 + i\gamma}. \tag{6}$$

The dielectric susceptibility of the medium thus can be written:

$$\chi(\nu) = -\frac{Ne^2}{2\varepsilon_0 m\omega_0} \times \frac{1}{\omega - \omega_0 + i\gamma} = -f \times \frac{M}{\omega - \omega_0 + i\gamma}, \tag{7}$$

where $M = \omega_P^2/\omega_0$ with ω_P being the effective plasma frequency and f being the oscillator strength. When two absorption lines of frequencies ω_1 and ω_2 are placed nearby with equal oscillator strengths $f_1 = f_2 = 1$, the dielectric susceptibility can be written:

$$\chi(\nu) = -\frac{M}{\omega - \omega_1 + i\gamma} - \frac{M}{\omega - \omega_2 + i\gamma}. \tag{8}$$

For a narrow frequency region in the middle between the two absorption lines, a steep normal dispersion region occurs resulting in an ultra-slow group velocity [20].

Conversely, for gain lines, a negative oscillator strength $f = -1$ is assigned [9]. Hence between two closely placed gain lines, the effective dielectric constant can be obtained:

$$\epsilon(\omega) = 1 + \chi(\omega) = 1 + \frac{M}{\omega - \omega_1 + i\gamma} + \frac{M}{\omega - \omega_2 + i\gamma}. \tag{9}$$

For dilute gaseous medium, we obtain from Eq.(9) for the refractive index $n(\omega) = n'(\omega) + i\,n''(\omega) = 1 + \chi(\omega)/2$ and the real and imaginary parts of the refractive index are plotted in Fig.1(a). It is evident from Fig.1(a) that a steep anomalous dispersion region appears without the heavy absorption present. In fact, a residual gain persists. Furthermore, with the correct choice of experimental parameters, the steep drop of refractive index as a function of frequency can be made a mostly linear one in this region. Thus a light pulse with a frequency bandwidth within this narrow linear anomalous dispersion region will experience almost no change in pulse shape upon propagating through such a medium.

While the details of the experimental realization and parameters can be found in references [14,15], here we review the basics of the experiments. Illustrated in Fig.1(b), in a gaseous medium of atoms each of which has three levels: an excited state $|0\rangle$ and two ground states $|1\rangle$ and $|2\rangle$, we first prepare all atoms to be in a ground state $|1\rangle$ via optical pumping. For simplicity, let us first ignore the Doppler shift and assume that the atoms are at rest. We apply two strong continuous-wave (CW) Raman pump light beams E_1 and E_2 that propagate through the atomic

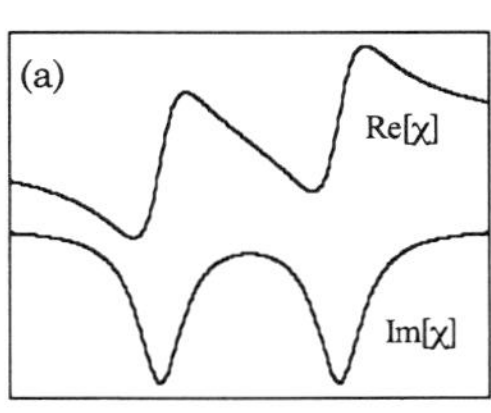
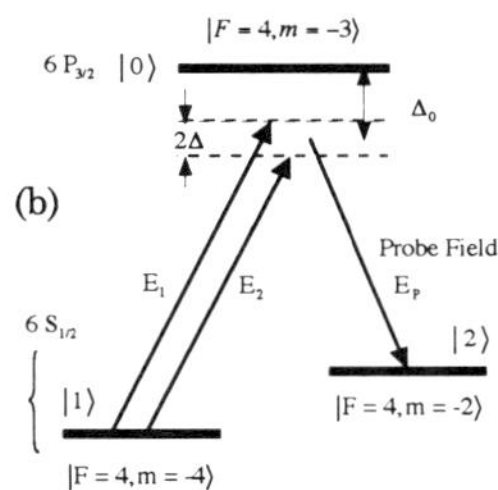

Figure 1. Gain-assisted anomalous dispersion. (a) frequency-dependent gain coefficient and refractive index, (b) schematic atomic level diagram.

medium. The frequencies of E_1 and E_2, ν_1 and ν_2, are different by a small amount 2Δ and both fields are detuned from the atomic transition frequency ν_{01} ($|1\rangle$ to $|0\rangle$) by a large average amount Δ_0. Since the Rabi frequencies associated with the fields E_1 and E_2 are small compared with the common detuning Δ_0, the atoms mostly remain in state $|1\rangle$. When a probe light beam E_P is introduced, a Raman transition can occur causing an atom to absorb a Raman pump photon from the fields E_1 or E_2 and emit a photon into the field E_P while in the mean time making a transition from $|1\rangle$ to $|2\rangle$. Obviously, there are two frequencies where the gain in the probe field is maximized. The maximum gain occurs when the probe field is resonant with the Raman transitions caused by either of the two pump fields E_1 and E_2. We hence obtain a medium whose optical susceptibility for the probe field is described by Eq.(9). Here we have

$$M_{1,2} = \frac{|\mu_{02}|^2}{2\hbar\varepsilon_0} \frac{|\Omega_{1,2}|^2}{|\Delta_0|^2} N, \tag{10}$$

with μ_{02}, $\Omega_{1,2}$, and N being the dipole moment of the $|0\rangle$ to $|2\rangle$ atomic transition, the Rabi frequencies of the Raman pump fields E_1 and E_2, and the effective atomic density difference of states $|1\rangle$ to $|2\rangle$, respectively. The quantum mechanical treatment of atomic polarization that yields Eq.(10) and discussion related to the Doppler broadening, as well as linewidth of the gain lines can be found in reference 14,15.

3 Pulse Propagation in an Anomalously Dispersive Medium, Pulse "rephasing" versus "reshaping"

Now we consider the propagation of a light pulse of an arbitrary shape but of long duration [a] through a transparent anomalous dispersing medium of a length L as illustrated in Fig.2. For a scalar light pulse that can be decomposed into its positive and negative frequency parts:

$$E(z,t) = E^{(+)}(z,t) + E^{(-)}(z,t), \tag{11}$$

---

[a] It is important to have a limited bandwidth within the anomalously dispersive region between the two gain lines.

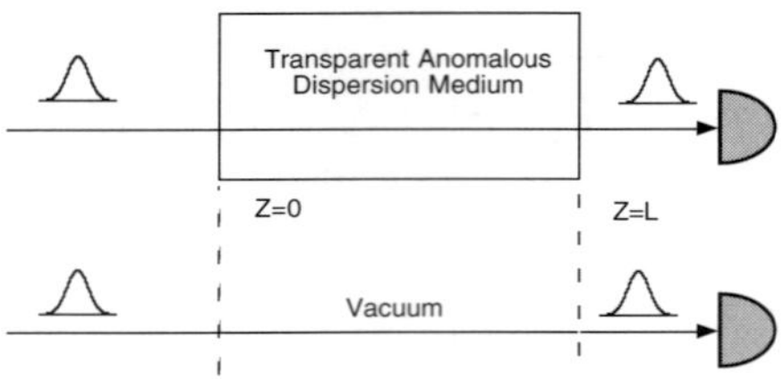

Figure 2. Pulse propagation through a medium of length L at a group velocity $v_g = c/(n + \nu\, dn/d\nu)$. and through vacuum for the same length.

we have for its Fourier decomposition:

$$E^{(+)}(z,t) = \frac{1}{\sqrt{2\pi}} e^{-i\,[\omega_0 t - k(\omega_0)z]}$$
$$\times \int d\Omega\, E^{(+)}(\Omega)\, e^{-i\,\{\Omega t - [k(\omega_0 + \Omega) - k(\omega_0)]z\}}, \qquad (12)$$

where ω_0 is the carrier frequency of the light pulse. Inside the transparent anomalous dispersion medium, if over the narrow bandwidth of the incident light pulse $E(\omega - \omega_0)$, the gain is small, the propagation is largely governed by the wave vector $k(\omega)$. We can expand the wave vector into a Taylor series:

$$k(\omega) = k(\omega_0) + \frac{1}{v_g}(\omega - \omega_0) + \frac{1}{2}\left.\frac{d^2 k}{d\omega^2}\right|_{\omega_0} \cdot (\omega - \omega_0)^2 + \cdots. \qquad (13)$$

When the higher order terms in Eq.(13) are negligible, i.e., the dispersion is essentially linear, from Eqs.(12) and (13) we obtain:

$$E^{(+)}(L,t) = g \cdot e^{-i\,(\omega_0 t - k_0 \cdot L)}\, E^{(+)}(0, t - L/v_g), \qquad (14)$$

where $g \approx 1$ is a gain factor. Hence, the intensity of the light pulse as a function of time measured with a detector, $I(L,t)$, is related to the incident pulse's time-dependent intensity by:

$$I(L,t) = I(0, t - L/v_g). \qquad (15)$$

Ordinarily, in a normal dispersion medium, the group velocity $v_g < c$. Hence, the output intensity of a pulse propagating through the medium is retarded by the propagation time L/v_g, resulting in a delay longer than the vacuum transit time L/c. In a transparent anomalous dispersion medium, the group velocity $v_g = c/[n + \nu\, dn/d\nu]$ can exceed c provided the anomalous dispersion is sufficiently strong such that: $n + \nu\, dn/d\nu < 1$. In this case, the group velocity becomes superluminal: $v_g > c$, resulting in a "*superluminal transit time:*" $L/v_g < L/c$.

Furthermore, when the transparent anomalous dispersion becomes stronger to yield $n + \nu\, dn/d\nu = 0$, the group velocity $v_g = c/[n + \nu\, dn/d\nu]$ approaches infinity, resulting in a "*zero transit time*", such that Eq.(15) gives $I(L,t) = I(0,t)$. In this case, the output pulse and the input pulse vary the same way in time and there is no time delay experienced by the pulse propagating through the medium.

Finally, when the transparent anomalous dispersion becomes very steep, the dispersive term $\nu\, dn/d\nu$, which is negative, becomes very large in its magnitude such that $|\nu\, dn/d\nu| \gg 1$, resulting in a negative group velocity $v_g = c/[n + \nu\, dn/d\nu] < 0$. In this case, Eq.(11) gives $I(L,t) = I(0, t + |L/v_g|)$, where the quantity $|L/v_g| = |n_g|\, L/c$ is positive, and can become very large compared to the vacuum transit time L/c. This means that the intensity at the output of the medium of length L, $I(L,t)$, will vary in time *earlier* than that of the input pulse $I(0,t)$. Thus in this case, a *"negative transit time"* can be observed. The time difference between the output pulse and the input pulse in the form of a pulse advance, is $|n_g|$ fold of the vacuum transit time L/c. Practically, since the shape of the pulse is not changed, this results in a rather counterintuitive phenomenon where a certain part of the light pulse has already exited the medium before the corresponding part of the incident light pulse even enters, by a time difference that is $|n_g|$ times the vacuum transit time L/c.

This rather counterintuitive effect is a result of the wave nature of light.

To bring about a physical insight into this phenomenon, we have analyzed the behavior of pulse propagation by examining the phase change of the various frequency components of the pulse [16]. In fact, Lord Kelvin first pointed out [1] that the peak of a light pulse is merely the point in space where at a give time, all of it various frequency components are "in-phase." Inside a medium at a time t, the phase of a frequency component becomes $\phi(\omega) = \omega t - k(\omega)z$. Hence, the space-time point where phases of all the frequency components "line-up" is the point where the phase is independent of the frequency:

$$\ell = c \cdot \frac{\partial \phi}{\partial \omega} = c \cdot [t - \frac{z}{U}] = 0, \tag{16}$$

where $U = c/n_g$ is the group velocity.

In the special case where $U < 0$, one can find that a "rephasing" peak of the pulse exits the medium before the incoming pulse arrives at the input port. A detailed description and an animation of the pulse behavior can be found in reference [16].

Some authors [21], however, insisted that our experimental results must be attributed to a "reshaping" or "differential-gain" effect. In their lines of reasoning, the medium is so adaptive that it can selectively "'amplify' the front of a pulse and 'absorb' its tail." Thus came the argument of the "differential gain," where the medium can respond to the derivatives of the pulse intensity variation [21]. Here we simply point out an experimental fact that this type of theory cannot explain. Namely, in the experiments reported in references [14,15] (and discussed below), the pulses used had a typical duration of about 4 μsec, and the typical atomic dwell-time inside the beam is merely 1 μsec. Hence, the 'front' and the 'tail' of the pulse see the same ground-state inverted atoms and the same steady-state CW-Raman pump beams. Consequently, it can never be that the 'front' is amplified while the 'tail' is absorbed. If the argument of reference [21] is correct, both the front and the tail will be amplified.

4   EXPERIMENTS

From an experimental point of view, one must satisfy the following requirements. First, a gain doublet must be obtained for which the anomalous dispersion between the gain lines can become linear to avoid any pulse distortion. In previous work, excited state population inversion was considered to obtain gain[11]. However, spontaneous emission and the short excited state life time would cause such gain doublets to be very difficult to sustain. It is important to have a steady state gain with a lifetime longer than the pulse duration τ to avoid transient effects and the associated complications. Second, the medium must be transparent since opaque anomalous dispersion has been long known and has resulted in controversies in terms of interpretations. While ideally the dispersion shown in Fig.1 is transparent, residual absorption and the associated loss are often present and cannot be simply discarded in experimental situations. Third, in order to show superluminal light pulse propagation in a linear regime, one must employ a very weak light pulse for which the photon number is far less than the atomic number in order to avoid Raman gain saturation. Fourth, in order to achieve a reasonable accuracy in the measurement, a system should be designed to demonstrate a negative group velocity. In this case, the pulse advancement under conventional experimental situation will be substantially large compared with commonly obtained accuracy (about 1ns). A number of other experimental conditions also have to be considered such as atomic density, polarization decay time, etc. and they will be discussed as well.

4.1   *Experimental Setup*

The experiment is performed using an atomic Cesium (Cs) vapor cell at 30°C and the main setup is shown in Fig.3. The cesium atoms are confined in a 6-cm-long Pyrex glass cell coated with paraffin for the purpose of maintaining atomic ground-state spin polarization. The atomic cell is placed inside a magnetic shield made of a thin layer of high-μ metal material inside which the Earth magnetic field is reduced to sub-milli-Gauss level. A Helmholtz coil (not shown in Fig.3) produces a uniform magnetic field inside the magnetic shield parallel to the propagation direction of all optical beams. This uniform field is approximately 1 Gauss serving the purpose of defining a quantization axis for optical pumping. Inside the magnetic shield, the air temperature is controlled using a heater servo system in order to control the temperature of the Cesium cell. During data acquisition, this control system is turned off to avoid any stray magnetic field. Having good thermal insulation, the temperature of the atomic cell remains the same during the data acquisition time.

In region-I of Fig.3, two optical pumping laser beams prepare almost all Cesium atoms into the ground-state hyperfine magnetic sublevel $6S_{1/2}$, $|F = 4, m = -4\rangle$ that serves as the state $|1\rangle$ in Fig.1(b). Laser-1 is a narrow linewidth diode laser locked to the 852-nm D_2 transition of Cs using a Lamb-dip technique and empties the $6S_{1/2}$, $F = 3$ hyperfine ground states. Laser-2 is a broadband tunable Ti:sapphire laser tuned to the 894-nm D_1 transition of Cesium. The linewidth of laser-2 covers transitions from both the $6S_{1/2}$, $F = 4$ and $F = 3$ hyperfine ground states to the $6P_{1/2}$ excited state. Both laser beams are initially linearly polarized and are turned into left-hand polarization ($\sigma-$) using a quarter-wave plate

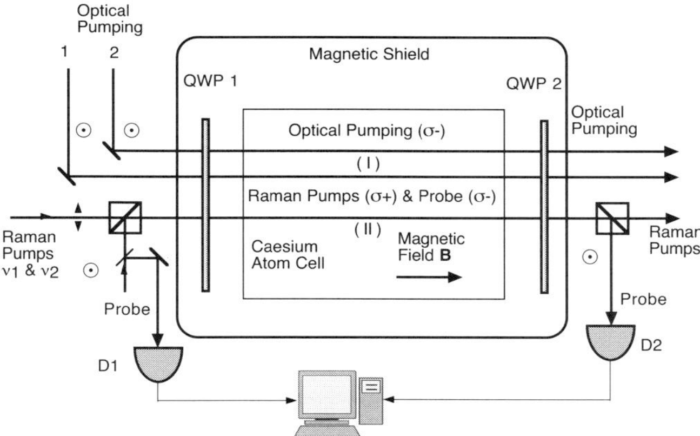

Figure 3. Schematic experimental set up. Two optical pumping beams tuned to the Cesium (Cs) atomic D1 and D2 transitions prepare the atoms in its ground state hyperfine sublevel $|F = 4, m = -4\rangle$. Two Raman pump beams and a Raman probe beam derived from a common narrow linewidth diode laser propagate colinearly parallel to a small magnetic field B through the atomic cell. Two $\lambda/4$-plates (QWP1 and 2) are used to prepare the three light beams into the corresponding circular polarization states and then separate them for analysis.

placed before the atomic cell. Inside the vapor cell, Cesium atoms collide with the paraffin-coated glass walls, the atoms change their velocities inside the Doppler profile. However, their ground state spin polarizations are not changed during collisions. Hence, all atoms inside the entire Doppler broadening profile are optically pumped into the ground state $|F = 4, m = -4\rangle$ quickly. The mean dephasing time of the ground state spin polarization of Cesium atoms in a paraffin coated cell is of order a fraction of a second.

In region-II, three light beams derived from the same narrow linewidth diode laser propagate collinearly through the cell. Two strong continuous-wave (CW) Raman pump beams are right-hand polarized ($\sigma+$) and a weak Raman probe beam is left-hand polarized ($\sigma-$). Using three acousto-optical modulators (AOM's), the frequency difference of the two Raman pump beams can be tuned continuously over a few MHz while the probe beam can also be tuned in frequency and can be operated in both CW or pulsed mode. The typical carrier frequency of the AOM's is 80 MHz and the linewidth is about 20 kHz. A residual optical beam that is shifted in frequency by 80 MHz generated from the same AOM which modulates the probe beam is also available for the refractive index measurement.

4.2 Experimental Methods and Results

First, we operate the Raman probe beam in a tunable CW mode to measure the gain and refractive index of the atomic system as a function of the probe frequency detuning. Fig.4 shows the measured gain coefficient and the refractive index. In order to obtain the gain coefficient, we first measure the intensity of the transmitted probe beam as a function of probe frequency. We then extract the gain coefficient.

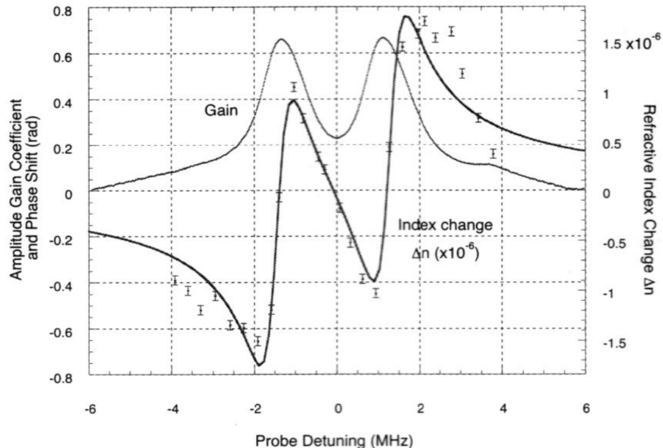

Figure 4. Measured refractive index and gain coefficient. The superposed curve over the index data is obtained using Eq. (30) with parameters $\nu_0, \Delta\nu$, and γ obtained experimentally. The length of the cell is 6 cm.

The refractive index is measured using a radio-frequency (rf) interferometric technique. The superimposed curve is obtained from Eq.(4) using parameters obtained from the gain measurement. From Fig.4, we readily see that a negative change of $\Delta n = -1.8 \times 10^{-6}$ in the refractive index occurs over a narrow probe frequency range of $\Delta\nu = 1.9$ MHz between the two gain lines. Using the expression of the group-velocity index, we obtain the result $n_g = -330(\pm 30)$ in that frequency region. The 10% error reflects the accuracy of the phase measurement.

Next, a pulsed Raman probe beam is employed to observe the superluminal propagation. A near Gaussian probe pulse with a 2.4 μsec FWHM is generated by applying a biased sinusoidal electronic signal to the probe beam A/O modulator. The repetition rate is 50 kHz. A portion of the pulsed probe beam is divided at a beam-splitter before the atomic cell and aligned onto photodiode D1 as a reference. Because the total number of atoms in the probe volume limits the maximum net energy gain of the probe pulse, we use a very weak probe beam ($\approx 1\mu$W) in order to avoid saturation and hence to optimize the anomalous dispersion. A high sensitivity avalanche photodiode, reverse-biased below breakdown, serves as detector D2 to measure the weak probe pulse that propagates through the atomic cell. The photoelectric current produced by detector D2 is converted to a voltage signal using a 500-Ω load resistor and recorded by a digitizing oscilloscope using a synchronized output signal from the pulse generator as the trigger. Pulses from detector D1 are also recorded.

In order to measure the pulse propagation time, we first tune the diode laser that produces the Raman pump and probe beams far off-resonance from the 852 nm Cesium D2 lines (by 2.5 GHz) to measure the time-dependent probe-pulse

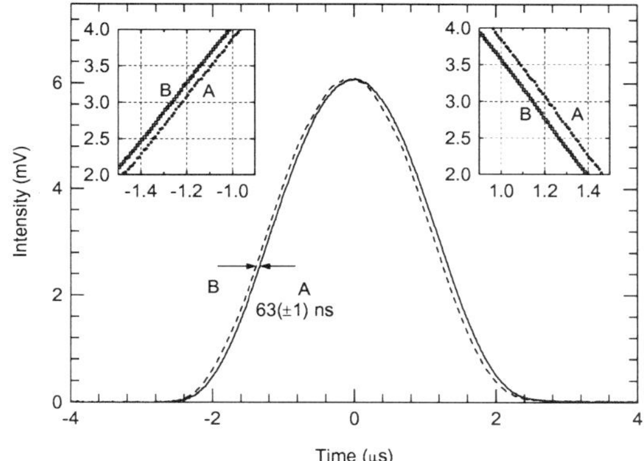

Figure 5. Measured pulse advancement for a light pulse traversing through the Cesium vapor. (A), light pulse far off-resonance from the Cesium D2 transitions propagating at speed c through 6 cm of vacuum. (B), same light pulse propagating through the same Cs-cell near resonance with a negative group velocity $-c/315$. Insets show the front and trail parts of the pulses. A and B are both the average of 1000 pulses. Off-resonance pulse (A) is normalized to the magnitude of (B).

intensity. When the laser is placed far off-resonance, the atoms have no effect and the probe pulse propagates at the speed c inside the cell. We then tune the diode laser back to within the Doppler absorption profile and lock it on its side. Using the same synchronized pulse generator output signal as the trigger, we record the time-dependent probe pulse intensity measured by detector D2. We verify that no systematic drift is present by tuning the laser off-resonance again by the same amount and record the probe pulse signal; the two off-resonance pulses are identical to within less than 1 ns. Probe pulses both on and off-resonance are shown in Fig.5 (average of approximately 1000 pulses).

Curve A in Fig.5 shows the light pulses far off-resonance from the Cesium D2 transitions propagating at speed c through 6 cm of vacuum. Curve B shows the same light pulse propagating through the same Cs-cell near resonance at a negative group velocity $-c/315$. Probe pulses on resonance show a 40% transmittance and this is due to the broadband absorption of those atoms reverse pumped away from the state $|F = 4, m = -4\rangle$. It is evident that there is almost no change in the pulse shape. The front edges and the trailing edges of the pulses are shown in the insets; both edges are shifted forward by the same amount.

Using a least square fitting procedure, we obtain a pulse advancement shift of 63 (± 1) nsec. Compared with the 0.2 nsec propagation time for light to traverse the 6-cm length of the atomic cell in vacuum, the 63 nsec advancement gives an effective group index of $n_g = 315(\pm 5)$. This is in good agreement with that inferred from the refractive index measurement. The pulses measured with detector D1 are also recorded in the sequence of the off-, on-, off-resonance pulse propagation measurements and are found to be identical to within 1.5 ns.

We note here that the measured superluminal pulse propagation inside the transparent anomalous dispersion medium is a linear effect. We further estimate the photon number per pulse and the interacting atomic number to show that there is no saturation effect present. In the experiment, the measured voltage signal peak strength is $V_p = \alpha\xi\,RG\,\dot{N}_{ph}\hbar\omega_0$, where $\alpha \approx 0.5 A/W$ is the photo responsivity of the avalanche photo detector and $\xi = 0.2$ is an effective efficiency of the detection imaging system. R=500 Ω is the load resistance. $G \approx 80$ is the avalanche gain; $\hbar\omega_0 = 1.5eV$ is the photon energy. Hence, we obtain the peak photon rate $\dot{N}_{ph} \approx 5 \times 10^{12}/sec$. Using the $2.4\,\mu sec$ FWHM as the pulse duration, each probe pulse contains approximately 1.2×10^7 photons. On the other hand, inside the volume of the probe light beam $\pi w_0^2\,L$, there are on average $N\,\pi w_0^2\,L$ atoms at any given moment, where $N \approx 10^{11} cm^{-3}$ is the atomic density. The beam radius is approximately $90\,\mu m$ and the Cesium cell is of a length 6 cm, the atoms inside the beam volume are approximately 1.4×10^8 at any given moment. However, since atoms are coming in and out of this volume within an average time of $2\,w_0/V_R \approx 1\,\mu sec$, during the $2.4\,\mu sec$ pulse duration, there are approximately 3.4×10^8 atoms inside the volume of the light pulse, much larger than the photon number per pulse. Hence, gain saturation effects are insignificant and the observed superluminal pulse propagation is a linear effect.

The anomalous dispersion and the resulting advance in pulse propagating through the medium depends on a few parameters. Particularly, the pulse advance $|\Delta T|$ is linear to the atomic density N and the Raman pumping Rabi frequency $|\Omega_{1,2}|^2$ which in turn is linear to the Raman pump power. These parameter dependence, as well as the dependence on detuning, can be tested and detailed experimental results and theoretical analyses are given in reference [15].

5   Signal velocity and quantum fluctuation

In order to properly analyze the signal velocity of light, let us start by considering the detection of a signal carried by a light pulse shown in Fig.2. We assign a time window T centered about a pre-arranged time t_0 at the detector and monitor the photo-current produced by the detector. We assume that there is a background level of irradiation that causes a constant photo-current i_0 when no light pulse was sent. We further assume that an increased photo current $i_1(t)$ is registered as a result of a light pulse being received. If the detector's integrated photo current $\int dt\,i_1(t)$ rises above the background level during that time by a certain number of the level of noise fluctuation, we can confidently determine that a signal has been received. The time when this preset level of confidence is reached in the detection is to be defined as the time the signal has arrived, assuming perfect detectors as allowed by physical laws.

Hence, the realistic observable that should be considered for the definition of the arrival of the signal carried by a light pulse is the time-dependent integrated photon number in the pulse

$$\hat{S}(L,t) = \eta \int_{t_0-T/2}^{t} dt_1 \hat{E}^{(-)}(L,t_1)\hat{E}^{(+)}(L,t_1) \,, \tag{17}$$

where $\hat{E}^{(+)}(L,t_1)$ and $\hat{E}^{(-)}(L,t_1)$ are respectively the positive- and negative-frequency parts of the reduced electric field operator at the exit point $(z = L)$

of the medium. $t_0 = T_c + L/c$ where T_c is the time corresponding to the pulse peak. $T/2$ is half the time window assigned to the pulse, typically a few times the pulse width. η is a constant containing the quantum efficiency and will be taken as unity for the rest of the analysis. The expectation value $\langle \hat{S}(L,t) \rangle$ is proportional to the number of photons that have arrived at the detector at time t. If $\langle \hat{S}_1(L,t) \rangle$ and $\langle \hat{S}_0(L,t) \rangle$ are respectively the expectation values of $\hat{S}(L,t)$ with and without an input pulse, then the photocurrent difference for an ideal detector is $\langle \hat{S}_1(L,t) \rangle - \langle \hat{S}_0(L,t) \rangle$. Since the second-order variance of the integrated photon number, $\langle \Delta^2 \hat{S}(L,t) \rangle$, characterizes the noise power due to quantum fluctuations, we define an optical signal-to-noise ratio in accord with standard signal detection practice[22]

$$SNR(L,t) = \frac{(\langle \hat{S}_1(L,t) \rangle - \langle \hat{S}_0(L,t) \rangle)^2}{\langle \Delta^2 \hat{S}(L,t) \rangle} . \tag{18}$$

As discussed above, we define the arrival time t_s of a signal as the time at which $SNR(L,t)$ reaches a prescribed threshold level determined by the allowed error rate.

The positive-frequency part of the reduced electric field operator can be written as [23]

$$\hat{E}^{(+)}(z,t) = \frac{e^{-i\omega_o(t-z/c)}}{\sqrt{2\pi}} \int_{-\infty}^{\infty} d\omega\, \hat{a}(\omega) e^{-i\omega(t-z/v_g)}, \tag{19}$$

where ω_o is the carrier frequency of the pulse, and $[\hat{a}(\omega), \hat{a}^\dagger(\omega')] = \delta(\omega - \omega')$. Eq.(19) assumes plane-wave propagation in the z direction and that the group-velocity approximation is valid.

In the experiment discussed above the anomalously dispersive medium is a phase-insensitive linear amplifier for which [24]

$$\hat{a}_{\text{out}}(\omega) = g(\omega)\hat{a}_{\text{in}}(\omega) + \sqrt{|g(\omega)|^2 - 1}\,\hat{b}^\dagger(\omega) , \tag{20}$$

where $\hat{a}_{\text{in}}$ and $\hat{a}_{\text{out}}$ refer respectively to the input $(z = 0)$ and output $(z = L)$ ports of the amplifier and the operator $\hat{b}(\omega)$ is a bosonic operator $([\hat{b}(\omega), \hat{b}^\dagger(\omega')] = \delta(\omega-\omega'))$ that commutes with all operators $\hat{a}_{\text{in}}(\omega)$ and $\hat{a}_{\text{in}}^\dagger(\omega)$ and whose appearance in Eq.(20) is required among other things to preserve the commutation relations for the field operators $\hat{a}_{\text{out}}$ and $\hat{a}_{\text{out}}^\dagger$. $|g(\omega)|^2$ is the power gain factor given by Eq.(8).

We first consider the case of propagation over the distance L in a vacuum where $g(\omega) = 1$. We assume that the initial state $|\psi\rangle$ of the field is a coherent state such that $\hat{a}(\omega)|\psi\rangle = \alpha(\omega)|\psi\rangle$ for all ω, where $\alpha(\omega)$ is a c-number. For such a state we may write $\hat{E}^{(+)}(0,t)|\psi\rangle = \alpha(t)|\psi\rangle$, where $\alpha(t) \equiv \pi^{-1/4}(N_p/\tau)^{1/2} \exp(-(t - T_c)^2/2\tau^2)$, N_p is the average number of photons in the initial pulse of duration τ. We obtain after a straightforward calculation that

$$SNR_{\text{vac}}(L,t) = \langle \hat{S}_1(L,t) \rangle_{\text{vac}} = SNR_{\text{vac}}(0, t - L/c). \tag{21}$$

Clearly, the point $SNR_{vac}(L,t) = \text{const.}$ propagates at the velocity c without excess noise.

Next we treat the case of pulse propagation over the distance L in the anomalously dispersive medium, using Eq.(20) with $g(\omega) \neq 1$ and the same initially coherent field. We obtain in this case

$$\langle \hat{S}_1(L,t) \rangle - \langle \hat{S}_0(L,t) \rangle = |g(0)|^2 \langle \hat{S}_1(0, t - L/v_g) \rangle_{\text{vac}} \tag{22}$$

where $\langle \hat{S}_0(L,t) \rangle = (1/2\pi) \int_{t_0-T/2}^{t} dt_1 \int d\omega [|g(\omega)|^2 - 1]$ is the photon number in the absence of any pulse input to the medium. The fact that $\langle \hat{S}_0(L,t) \rangle > 0$ is due to amplified spontaneous emission (ASE) [22]; in the experiment of interest the ASE is due to a spontaneous Raman process.

For a probe pulse with sufficiently small bandwidth, the gain factor becomes

$$|g(0)|^2 = e^{4\pi M\gamma/(\Delta\nu^2 + \gamma^2) \cdot L/\lambda} , \tag{23}$$

and the effective signal $\langle \hat{S}_1(L,t) \rangle - \langle \hat{S}_0(L,t) \rangle$ is proportional to the input signal $\langle \hat{S}_1(0, t - L/v_g) \rangle_{\text{vac}}$ with time delay L/v_g determined by the group velocity v_g. In the anomalously dispersive medium $v_g = c/(n + \nu dn/d\nu)$ and can be $> c$ or even negative, resulting in a time delay

$$\frac{L}{v_g} = \left[1 - \nu_o M \cdot \frac{\Delta\nu^2 - \gamma^2}{(\Delta\nu^2 + \gamma^2)^2} \right] \frac{L}{c} , \tag{24}$$

which is shorter than the time delay the pulse would experience upon propagation through the same length in vacuum, or can become negative. In other words, the effective signal intensity defined here can be reached sooner than in the case of propagation in vacuum.

In order to determine with confidence when a signal is received, however, one must evaluate the SNR. Again using the commutation relations for the field operators, we obtain for the fluctuating noise background

$$\begin{aligned}
\langle \Delta^2 \hat{S}(L,t) \rangle &\equiv \langle \hat{S}^2(L,t) \rangle - \langle \hat{S}(L,t) \rangle^2 \\
&= |g(0)|^2 \langle \hat{S}_1(0, t - L/v_g) \rangle_{\text{vac}} + \langle \hat{S}_0(L,t) \rangle \\
&\quad + 2|g(0)|^2 \text{Re}[\int_{t_0-T/2}^{t} dt_1 \int_{t_0-T/2}^{t} dt_2 \\
&\quad \times \alpha^*(t_1 - L/v_g)\alpha(t_2 - L/v_g)F(t_1 - t_2)] \\
&\quad + \int_{t_0-T/2}^{t} dt_1 \int_{t_0-T/2}^{t} dt_2 |F(t_1 - t_2)|^2 .
\end{aligned} \tag{25}$$

Here

$$F(t) = \frac{1}{2\pi} \int_{-\infty}^{\infty} d\omega [|g(\omega)|^2 - 1]e^{-i\omega t} \tag{26}$$

is a correlation function for the amplified spontaneous emission noise. The four terms in Eq. (25) can be attributed to amplified shot noise, spontaneous emission noise, beat noise, and ASE self-beat noise, respectively [25]. Figure 6 shows the evolution of these noise terms within the time window T. Clearly, amplified shot noise dominates when the input pulse is strong.

Using Eqs. (22) and (25), we compute $SNR_{(\text{med})}(L,t)$ for the propagation through the anomalously dispersive medium. In Fig.7 we plot the results of such

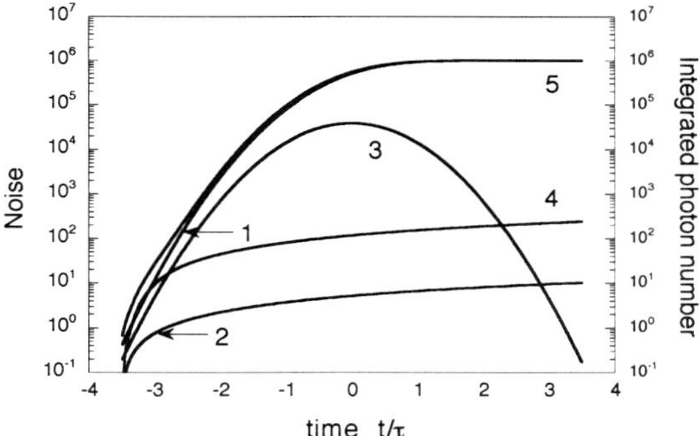

Figure 6. Evolution of quantum noise terms. Curves 1 to 5 indicate noises associated with terms 1 to 4 in Eq.(10), and the total noise, respectively. Parameters used in the figure are adopted from the experiments reported in References [15] and [16]. There are 10^6 photons per pulse. Noise retards the detection of the signal by reducing the SNR. (Figure adopted from reference [19]).

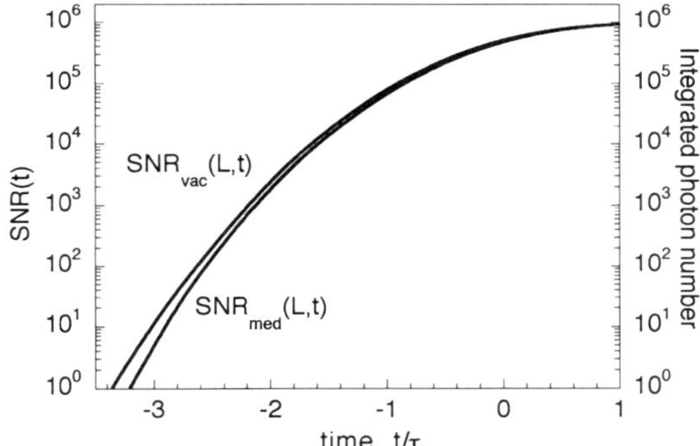

Figure 7. Signal-to-noise ratios for light pulses propagating through the gain-assisted anomalous dispersion medium $SNR_{\mathrm{med}}(L,t)$, and through the same distance in a vacuum $SNR_{\mathrm{vac}}(L,t)$. Parameters used are appropriate to the experimental situation of reference [15,16,19]. (Figure adopted from reference [19]).

computations for $SNR_{\mathrm{(med)}}(L,t)$ as a function of time on the output signal. For reference we also show SNR for the identical pulse propagating over the same length in vacuum. It is evident that the pulse propagating in vacuum always maintains a higher SNR. In other words, for the experiments of interest here [14,15], the signal arrival time defined here is delayed, even though the pulse itself on average is advanced compared with propagation over the same distance in vacuum.

To further examine the signal velocity, we require that at a time t' the SNR of a pulse propagating through the medium be equal to that of the same pulse propagating through a vacuum at a time t:

$$SNR_{(\text{med})}(L, t') = SNR_{(\text{vac})}(L, t). \tag{27}$$

Hence, we obtain a time difference $\delta t = t' - t$ that marks the retardation due to quantum noise. $\Delta t = t' - t + L/c$ gives the propagation time of the light signal, and $L/\Delta t$ gives the signal velocity.

6 summary

In conclusion, we note that the observed superluminal and negative group velocities are a result of the wave nature of light. The measured negative and superluminal group velocities of a light pulse propagating through a transparent anomalous dispersion medium are due to the physical effect of "rephasing" [14,15,16]. Specifically, inside a medium with refractive index n, the effective wavelength of a light ray is modified: $\lambda' = \lambda/n$, where λ is the vacuum wavelength. It is easy to derive:

$$\frac{d(\lambda/n)}{d\lambda} = \frac{n_g}{n^2}. \tag{28}$$

Under the condition $n_g < 0$, we have $d(\lambda/n)/d\lambda < 0$. Hence a longer wavelength (redder) component of the incident pulse becomes a shorter wavelength ray inside the medium, and vice versa. This results in an unusual situation where the phases of the different frequency components of a pulse become aligned at the exit surface of the medium earlier than even in the case of the same pulse propagating through the same distance in a vacuum.

Finally, we note that a superluminal group velocity is not at odds with causality or special relativity [1-4,8-19]. Simply put, causality only requires that the signal velocity, instead of the group velocity, be limited by c; the signal velocity is different from the group velocity, as first noted again by Sommerfeld and Brillouin [1]. We further reviewed an operational definition of the signal velocity [19]. In the experimental cases where the medium is transparent, we must consider the excess noise due to the gain lines. We found that, in these cases the excess spontaneous emission noise from the gain resonances retards the onset of the signal by retarding the time at which a prescribed signal-to-noise ratio (SNR) is reached [19]. Hence, it was concluded that the quantum noise associated with an amplifier (which is related to the "no-cloning" theorem and the linear nature of quantum mechanics) is associated with the basic requirement of causality that states that no "signal" can be transmitted faster than c.

In this review, we have only summarized the physical problems related with various velocities of light. It is also worth pointing out that there are immediate applications associated with dispersion. These interesting applications will be discussed elsewhere.

Acknowledgments

The author wishes to thank A. Dogariu, A. Kuzmich, P. W. Milonni, and R. Y. Chiao for helpful discussions.

Present address: Max-Planck Research Group, Institute of Optics, Information, & *Photonics, Günther-Scharowsky Str.1, 91058 Erlangen, Germany.* *Email: Lwan@optik.uni-erlangen.de*

References

1. L. Brillouin, *Wave Propagation and Group Velocity* (Academic, New York, 1960).
2. L. D. Landau and E. Lifshitz, *Electrodynamics of Continuous Media,* Chapter 9, 10, (Pergamon, Oxford, 1960).
3. G. Diener, Superluminal group velocities and information transfer. Phys. Lett. A **223**, 327-331 (1996).
4. G. Diener, Energy transport in dispersive media and superluminal group velocities. Phys. Lett. A **235**, 118-124 (1997).
5. C. G. B. Garrett and D. E. McCumber, Propagation of a gaussian light pulse through an anomalous dispersion medium. Phys. Rev. A **1**, 305-313 (1970).
6. S. Chu and S. Wong, Linear pulse propagation in an absorbing medium. Phys. Rev. Lett. **48**, 738-741 (1982); A. Katz and R. R. Alfano, Phys. Rev. Lett. **49**, 1292 (1982), and S. Chu and S. Wong, *ibid.*, 1293.
7. B. Segard and B. Macke, Observation of negative velocity pulse propagation. Phys. Lett. **109A**, 213-216 (1985).
8. R. Y. Chiao, in *Amazing Light: A Volume Dedicated to Charles Hard Townes on His 80th Birthday,* edited by R. Y. Chiao (Springer-Verlag, New York, 1996), p. 91.
9. R. Y. Chiao, Superluminal (but causal) propagation of wavepackets in transparent media with inverted atomic population. Phys. Rev. A **48**, R34-37 (1993).
10. M. W. Mitchell and R. Y. Chiao, Causality and negative group delays in a simple bandpass amplifier. Am. J. Phys. **66**, 14-19 (1998).
11. E. L. Bolda, R. Y. Chiao, and J. C. Garrison, Two theorems for the group velocity in dispersive media. Phys. Rev. A **48**, 3890-3894 (1993).
12. A. M. Steinberg and R. Y. Chiao, Dispersionless, highly superluminal propagation in amedium with a gain doublet. Phys. Rev. A **49**, 2071-2075 (1994).
13. R. Y. Chiao, et. al. in *Proceeding of the 8th Conference on Coherence and Quantum Optics,* edited by N. P. Bigelow, J. H. Eberly, and E. Wolf (Planum, New York, *to be published*).
14. L. J. Wang, A. Kuzmich, and A. Dogariu, Gain-assisted superluminal pulse propagation. Nature (London) **406**, 277-279 (2000).
15. A. Dogariu, A. Kuzmich, and L. J. Wang, Transparent anomalous dispersion and superluminal light pulse propagation at a negative group velocity. Phys. Rev. A, **63** 053806 (2001).
16. A. Dogariu, A. Kuzmich, H. Cao, and L. J. Wang, Superluminal light pulse propagation and rephasing in a transparent anomalous dispersion medium. Opt. Express **8**, 344-350 (2001).
17. Y. Aharonov, B. Reznik, and A. Stern, Quantum limitations on superluminal propagation. Phys. Rev. Lett. **81**, 2190-2193 (1998).

18. B. Segev, P. W. Milonni, J. F. Babb, and R. Y. Chiao, Quantum noise and superluminal propagation. Phys. Rev. **A62**, 022114/1-15 (2000).

19. A. Kuzmich, A. Dogariu, L. J. Wang, P. W. Milonni, and R. Y. Chiao, Signal velocity, causality, and quantum noise in superluminal light propagation. Phys. Rev. Lett., **86** 3925-3929 (2001).

20. S. E. Harris, Electromagnetically induced transparency. Phys. Today **50 (7)**, 36 (1997), and references therein.

21. P. Sprangle, J. R. Penano, B. Hafizi, Apparent superluminal propagation of a laser pulse in a dispersive medium. Phys. Rev. E. **64**, 026504/1-5 (2001),

22. E. Desurvire, *Erbium-Doped Fiber Amplifiers: Principles and Applications* (Wiley, New York, 1994), Chapter 2. B. E. A. Saleh and M. C. Teich, *Fundamentals of Photonics* (Wiley, New York, 1991), Chapter 16.

23. L. Mandel and E. Wolf, *Optical Coherence and Quantum Optics* (Cambridge, 1995).

24. C. M. Caves, Quantum limits on noise in linear amplifiers. Phys. Rev. **D26**, 1817 (1982). See also H. A. Haus and J. A. Mullen, Phys. Rev. **128**, 2407 (1962).

25. Y. Yamamoto, Noise and error rate performance of semiconductor laser amplifiers in PCM-IM optical transmission systems. IEEE J. Quantum Electron. **16**, 1073-1081 (1980).

TIME AND MATTER IN THE INTERACTION BETWEEN GRAVITY AND QUANTUM FLUIDS: ARE THERE MACROSCOPIC QUANTUM TRANSDUCERS BETWEEN GRAVITATIONAL AND ELECTROMAGNETIC WAVES?

RAYMOND Y. CHIAO

University of California, Department of Physics, Berkeley, CA 94720-7300, USA
E-mail: chiao@physics.berkeley.edu

WALTER J. FITELSON

University of California, Space Sciences Laboratory, Berkeley, CA 94720-7450, USA
E-mail: walt@ssl.berkeley.edu

Measurements of the tunneling time are briefly reviewed. Next, time and matter in general relativity and quantum mechanics is examined. In particular, the question arises: How does gravitational radiation interact with a coherent quantum many-body system (a "quantum fluid")? A minimal coupling rule for the coupling of the electron spin to curved spacetime in general relativity implies the possibility of a coupling between electromagnetic (EM) and gravitational (GR) radiation mediated by a quantum Hall fluid. This suggests that quantum transducers between these two kinds of radiation fields might exist. We report here on a first attempt at a Hertz-type experiment, in which a high-T_c superconductor (YBCO) was the material used as a quantum transducer to convert EM into GR microwaves, and a second piece of YBCO in a separate apparatus was used to back-convert GR into EM microwaves. An upper limit on the conversion efficiency of YBCO was measured to be 1.6×10^{-5}.

1 Introduction

In this conference in Venice on "Time and Matter," one of us (RYC), was invited to speak on the tunneling time problem: How quickly does a particle traverse a barrier in the quantum process of tunneling? A. M. Steinberg, P. G. Kwiat, and RYC have used a photon-pair emission light source (spontaneous parametric down-conversion) for measuring the single-photon tunneling time, using the "click" of a Geiger counter as the registration of when one photon, which had succeeded in tunneling through the barrier, reached the detector, relative to a second, vacuum-traversing photon, which was born at the same time as the first photon (hence its "twin"). The arrival time of the tunneled photon was measured with respect to that of its twin, which had traversed a distance equal to the tunnel barrier thickness, but in the vacuum, by means of the difference in the two "click" times of two Geiger counters. These two Geiger counters were used in the coincidence detection of the two photons, with one counter placed behind the tunnel barrier, and the other counter placed behind the vacuum, in conjunction with a Hong-Ou-Mandel interferometer.

By means of this two-photon interferometer, we achieved the sub-picosecond time resolution necessary for measuring the tunneling time of a photon relative to the vacuum-traversal time of its twin. The result was that the Wigner theory of tunneling time was confirmed to be the one that applied to our experiment. The surprising result was that when a photon succeeded in tunneling (which is rare), it arrived earlier than its twin which had traversed the vacuum, as indicated by the fact that the "clicks" of the Geiger counter registering the arrival of the tunneling photons occurred earlier on the average than the Geiger counter "clicks" registering the arrival of the vacuum-traversing twin photons, as if the tunneling photons had traversed the tunnel barrier superluminally. The effective group velocity of the tunneling single-photon wavepacket was measured to be 1.7 ± 0.2 times the vacuum speed of light.

Since our tunneling-time work has already been adequately reviewed,[1] here we shall concentrate instead on a different question concerning "Time and Matter," namely, the role of time in the interaction of gravity, in particular, of gravitational radiation, with matter in the form of quantum fluids, i.e., many-body systems which exhibit off-diagonal long-range order (ODLRO), such as superconductors, superfluids, atomic Bose-Einstein condensates (BECs), and quantum Hall fluids.[2] As we shall see, under the proper circumstances, there results the interesting possibility that such quantum fluids could in principle mediate the conversion of EM into GR waves, and vice versa.

2   Quantum fluids as antennas for gravitational radiation

Can quantum fluids circumvent the problem of the tiny rigidity of classical matter, such as that of the normal metals used in Weber bars, in their feeble responses to gravitational radiation? One consequence of the tiny rigidity of classical matter is the fact that the speed of sound in a Weber bar is typically five orders of magnitude less than the speed of light. In order to transfer energy coherently from a gravitational wave by classical means, for example, by acoustical modes inside the bar to some local detector, e.g., a piezoelectric crystal glued to the middle of the bar, the length scale L of the Weber bar is limited to a distance scale on the order of the speed of sound times the period of the gravitational wave, i.e., an acoustical wavelength λ_{sound}, which is typically five orders of magnitude smaller than the gravitational radiation wavelength λ to be detected. This makes the Weber bar, which is thereby limited in its length to $L \simeq \lambda_{sound}$, much too short an antenna to couple efficiently to free space.

However, rigid quantum objects, such as a two-dimensional electron gas in a strong magnetic field which exhibits the quantum Hall effect, in what Laughlin has called an "incompressible quantum fluid",[3] are not limited by these classical considerations, but can have macroscopic quantum phase coherence on a length scale L on the same order as (or even much greater than) the gravitational radiation wavelength λ. The origin of this rigidity is that the phase of the wavefunction must remain rigidly single-valued everywhere inside the quantum fluid, whenever the many-body system is perturbed by gravity waves whose time variations are slow compared to the time scale of the gap time $\hbar/E_{gap}$, where E_{gap} is the energy gap separating the ground state from all excited states. Then the wavefunction will remain adiabatically, and hence rigidly, in its ground state during these time variations. Since the radiation efficiency of a quadrupole antenna scales as the length of the antenna L to the fourth power when $L << \lambda$, such quantum antennas should be much more efficient in coupling to free space than classical ones like the Weber bar by a factor of $(\lambda/\lambda_{sound})^4$.

Weinberg gives a measure of the radiative coupling efficiency η_{rad} of a Weber bar of mass M, length L, and velocity of sound v_{sound}, in terms of a branching ratio for the emission of gravitational radiation by the Weber bar, relative to the emission of heat, i.e., the ratio of the *rate* of emission of gravitational radiation Γ_{grav} relative to the *rate* of the decay of the acoustical oscillations into heat Γ_{heat}, which is given by[4]

$$\eta_{rad} \equiv \frac{\Gamma_{grav}}{\Gamma_{heat}} = \frac{64 G M v_{sound}^4}{15 L^2 c^5 \Gamma_{heat}} \simeq 3 \times 10^{-34} \ , \tag{1}$$

where G is Newton's constant. The quartic power dependence of the efficiency η_{rad} on the velocity of sound v_{sound} arises from the quartic dependence of the coupling efficiency to free space of a quadrupole antenna upon its length L, when $L << \lambda$.

The long-range quantum phase coherence of a quantum fluid allows the typical size L of a quantum antenna to be comparable to the wavelength λ. Thus the phase rigidity of the quantum fluid allows us in principle to replace the velocity of sound v_{sound} by the speed of light c. Therefore, quantum fluids can be more efficient than Weber bars, based on the v_{sound}^4 factor alone, by twenty orders of magnitude, i.e.,

$$\left(\frac{c}{v_{sound}} \right)^4 \simeq 10^{20} \ . \tag{2}$$

Hence quantum fluids could be much more efficient receivers of this radiation than Weber bars for detecting astrophysical sources of gravitational radiation. This has previously been suggested to be the case for superfluids and superconductors.[5,6]

Another important property of quantum fluids lies in the fact that they can possess an extremely low dissipation coefficient Γ_{heat}, as can be inferred, for example, by the existence of persistent currents in superfluids that can last for indefinitely long periods of time. Thus the impedance matching of the quantum antenna to free space,[7] or equivalently, the branching ratio η_{rad} can be much larger than that calculated above for the classical Weber bar. Since it is difficult to calculate Γ_{heat}, we need to measure η_{rad} experimentally.

3 Minimal-coupling rule for a quantum Hall fluid

The electron, which possesses charge e, rest mass m, and spin $s = 1/2$, obeys the Dirac equation. The nonrelativistic, interacting, fermionic many-body system, such as that in the quantum Hall fluid, should obey the minimal-coupling rule which originates from the covariant-derivative coupling of the Dirac electron to curved spacetime, viz. (using the Einstein summation convention),[4,8]

$$p_\mu \to p_\mu - e A_\mu - \frac{1}{2} \Sigma_{AB} \omega_\mu^{AB} \tag{3}$$

where p_μ is the electron's four-momentum, A_μ is the electromagnetic four-potential, Σ_{AB} are the Dirac γ matrices in curved spacetime with tetrad (or vierbein) A, B indices, and ω_μ^{AB} are the components of the spin connection

$$\omega_\mu^{AB} = e^{A\nu} \nabla_\mu \, e^B{}_\nu \tag{4}$$

where $e^{A\nu}$ and $e^B{}_\nu$ are tetrad four-vectors, which are sets of four orthogonal unit vectors of spacetime, such as those corresponding to a local inertial frame.

The vector potential A_μ leads to a quantum interference effect, in which the gauge-invariant Aharonov-Bohm phase becomes observable. Similarly, the spin connection ω_μ^{AB}, in its Abelian holonomy, should also lead to a quantum interference effect, in which the gauge-invariant Berry phase[9] becomes observable. The following Berry phase picture of a spin coupled to curved spacetime leads to an intuitive way of understanding why there could exist a coupling between a classical GR wave and a classical EM wave mediated by the quantum Hall fluid.

Due to its gyroscopic nature, the spin vector of an electron undergoes *parallel transport* during the passage of a GR wave. The spin of the electron is constrained to lie inside the space-like submanifold of curved spacetime. This is due to the fact that we can always transform to a co-moving frame, such that the electron is at rest at the origin of this frame. In this frame, the spin of the electron must be purely a space-like vector with no time-like component. This imposes an important *constraint* on the motion of the electron's spin, such that whenever the space-like submanifold of spacetime is disturbed by the passage of a gravitational wave, the spin must remain at all times *perpendicular* to the local time axis. If the spin vector is constrained to follow a conical trajectory during the passage of the gravitational wave, the electron picks up a Berry phase proportional to the solid angle subtended by this conical trajectory after one period of the GR wave.

In a manner similar to the persistent currents induced by the Berry phase in systems with off-diagonal long-range order,[10] such a Berry phase induces an electrical current in the quantum Hall fluid, which is in a macroscopically coherent ground state.[11] This current generates an EM wave. Thus a GR wave can be converted into an EM wave. By reciprocity, the time-reversed process of the conversion from an EM wave to a GR wave must also be possible.

In the nonrelativistic limit, the four-component Dirac spinor is reduced to a two-component spinor. While the precise form of the nonrelativistic Hamiltonian is not known for the many-body system in a weakly curved spacetime consisting of electrons in a strong magnetic field, I conjecture that it will have the form

$$H = \frac{1}{2m} \left(p_i - eA_i - \frac{1}{2}\sigma_{ab}\Omega_i^{ab} \right)^2 + V \tag{5}$$

where i is a spatial index, a, b are spatial tetrad incides, σ_{ab} is a two-by-two matrix-valued tensor representing the spin, and $\sigma_{ab}\Omega_i^{ab}$ is the nonrelativistic form of $\Sigma_{AB}\omega_\mu^{AB}$. Here H and V are two-by-two matrix operators on the two-component spinor electron wavefunction in the nonrelativistic limit. The potential energy V includes the Coulomb interactions between the electrons in the quantum Hall fluid. This nonrelativistic Hamiltonian has the form

$$H = \frac{1}{2m} (\mathbf{p} - \mathbf{a} - \mathbf{b})^2 + V , \tag{6}$$

where the particle index, the spin, and the tetrad indices have all been suppressed. Upon expanding the square, it follows that for a quantum Hall fluid of uniform density, there exists a cross-coupling or interaction Hamiltonian term of the form

$$H_{int} \sim \mathbf{a} \cdot \mathbf{b} , \tag{7}$$

which couples the electromagnetic **a** field to the gravitational **b** field. In the case of time-varying fields, $\mathbf{a}(t)$ and $\mathbf{b}(t)$ represent EM and GR radiation, respectively.

In first-order perturbation theory, the quantum adiabatic theorem predicts that there will arise the cross-coupling energy between the two radiation fields mediated by this quantum fluid

$$\Delta E \sim \langle \Psi_0 | \mathbf{a} \cdot \mathbf{b} | \Psi_0 \rangle \tag{8}$$

where $|\Psi_0\rangle$ is the unperturbed ground state of the system. For the adiabatic theorem to hold, there must exist an energy gap E_{gap} (e.g., the quantum Hall energy gap) separating the ground state from all excited states, in conjunction with a time variation of the radiation fields which must be slow compared to the gap time $\hbar/E_{gap}$. This suggests that under these conditions, there might exist an interconversion process between these two kinds of classical radiation fields mediated by this quantum fluid, as indicated in Fig.1.

The question immediately arises: EM radiation is fundamentally a spin 1 (photon) field, but GR radiation is fundamentally a spin 2 (graviton) field. How is it possible to convert one kind of radiation into the other, and not violate the conservation of angular momentum? The answer: The EM wave converts to the GR wave *through a medium*. Here specifically, the medium of conversion consists of a strong DC magnetic field applied to a system of electrons. This system possesses an axis of symmetry pointing along the magnetic field direction, and therefore transforms like a spin 1 object. When coupled to a spin 1 (circularly polarized) EM radiation field, the total system can in principle produce a spin 2 (circularly polarized) GR radiation field, by the addition of angular momentum. However, it remains an open question as to how strong this interconversion process is between EM and GR radiation. Most importantly, the size of the conversion efficiency of this transduction process needs to be determined by experiment.

We can see more clearly the physical significance of the interaction Hamiltonian $H_{int} \sim \mathbf{a} \cdot \mathbf{b}$ once we convert it into second quantized form and express it in terms of the creation and annihilation operators for the positive frequency parts of the two kinds of radiation fields, as in the theory of quantum optics, so that in the rotating-wave approximation

$$H_{int} \sim a^\dagger b + b^\dagger a \ , \tag{9}$$

where the annihilation operator a and the creation operator $a^\dagger$ of the single classical mode of the plane-wave EM radiation field corresponding the **a** term, obey the commutation relation $[a, a^\dagger] = 1$, and where the annihilation operator b and the creation operator $b^\dagger$ of the single classical mode of the plane-wave GR radiation field corresponding to the **b** term, obey the commutation relation $[b, b^\dagger] = 1$. (This represents a crude, first attempt at quantizing the gravitational field, which applies only in the case of weak, linearized gravity.) The first term $a^\dagger b$ then corresponds to the process in which a graviton is annihilated and a photon is created inside the quantum fluid, and similarly the second term $b^\dagger a$ corresponds to the reciprocal process, in which a photon is annihilated and a graviton is created inside the quantum fluid.

One may ask whether there exists *any* difference in the response of quantum fluids to tidal fields in gravitational radiation, and the response of classical matter,

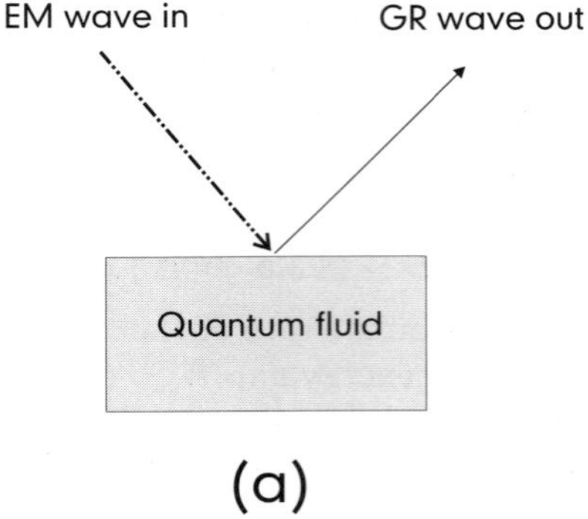

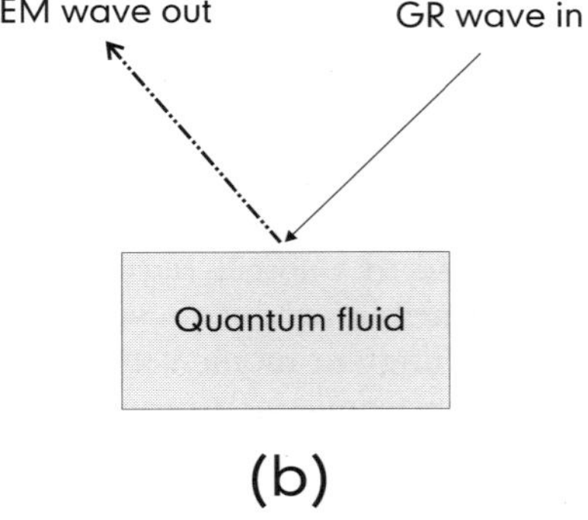

Figure 1. Quantum transducer between electromagnetic (EM) and gravitational (GR) radiation, consisting of a quantum fluid, such as the quantum Hall fluid, which possesses charge and spin. The minimal-coupling rule for an electron coupled to curved spacetime via its charge and spin, results in two processes. In (a) an EM plane wave is converted upon reflection from the quantum fluid into a GR plane wave; in (b), which is the reciprocal or time-reversed process, a GR plane wave is converted upon reflection from the quantum fluid into an EM plane wave.

such as the lattice of ions in a superconductor, for example, to such fields. The essential difference between quantum fluids and classical matter is the presence or absence of macroscopic quantum interference. In classical matter, such as in the lattice of ions of a superconductor, decoherence arising from the environment destroys any such quantum interference. Hence, the response of quantum fluids and of classical matter to these fields will therefore differ from each other.[2]

In the case of superconductors, Cooper pairs of electrons possess a macroscopic phase coherence, which can lead to an Aharonov-Bohm-type interference absent in the ionic lattice. Similarly, in the quantum Hall fluid, the electrons will also possess macroscopic phase coherence,[11] which can lead to Berry-phase-type interference absent in the lattice. Furthermore, there exist ferromagnetic superfluids with intrinsic spin,[12] in which an ionic lattice is completely absent, such in superfluid helium 3. In such ferromagnetic quantum fluids, there exists no ionic lattice to give rise to any classical response which could prevent a quantum response to tidal gravitational radiation fields. The Berry-phase-induced response of the ferromagnetic superfluid arises from the spin connection (see the above minimal-coupling rule, which can be generalized from an electron spin to a nuclear spin), and leads to a purely quantum response to this radiation. The Berry phase induces time-varying macroscopic quantum flows in this ferromagnetic ODLRO system,[10] which transports time-varying orientations of the nuclear magnetic moments, and thus generates EM waves. This ferromagnetic superfluid can therefore also in principle interconvert GR into EM radiation, and vice versa, in a manner similar to the case discussed above for the ferromagnetic quantum Hall fluid. Thus there may be more than one kind of quantum fluid which can serve as a transducer between EM and GR waves.

Like superfluids, the quantum Hall fluid is an example of a quantum fluid which differs from a classical fluid in its current-current correlation function in the presence of GR waves. In particular, GR waves can induce a transition of the quantum Hall fluid out of its ground state *only* by exciting a quantized, collective excitation, such as the vortex-like $\frac{1}{3}e$ quasi-particle, across the quantum Hall energy gap. This collective excitation would involve the correlated motions of a macroscopic number of electrons in this coherent quantum system. Hence the quantum Hall fluid, like the other quantum fluids, should be effectively incompressible and dissipationless, and is thus a good candidate for a quantum antenna and transducer.

There exist other situations in which a minimal-coupling rule similar to the one above, arises for *scalar* quantum fields in curved spacetime. DeWitt[13] suggested in 1966 such a coupling in the case of superconductors. Speliotopoulos noted in 1995[14] that a cross-coupling term of the form $H_{int} \sim \mathbf{a} \cdot \mathbf{b}$ arose in the long-wavelength approximation of a certain quantum Hamiltonian derived from the geodesic deviation equations of motion using the transverse-traceless gauge for GR waves.

Speliotopoulos and I have been working on the problem of the coupling of a scalar quantum field to curved spacetime in a general laboratory frame, which avoids the use of the long-wavelength approximation.[15] In general relativity, there exists in general no global time coordinate that can apply throughout a large system, since for nonstationary metrics, such as those associated with gravitational radiation, the local time axis varies from place to place in the system. It is therefore necessary to set up operationally a general laboratory frame by which an observer can measure the motion of slowly moving test particles in the presence of weak, time-varying gravitational radiation fields.

For either a classical or quantum test particle, the result is that its mass m should enter into the Hamiltonian through the replacement of $\mathbf{p} - e\mathbf{A}$ by $\mathbf{p} - e\mathbf{A} - m\mathbf{N}$, where $\mathbf{N}$ is the small, local tidal velocity field induced by gravitational radiation on a test particle located at X_a relative to the observer at the origin (i.e., the center of mass) of this frame, where, for the small deviations h_{ab} of the metric from that of flat spacetime,

$$N_a = \frac{1}{2} \int_0^{X_a} \frac{\partial h_{ab}}{\partial t} dX^b. \tag{10}$$

Due to the quadrupolar nature of gravitational tidal fields, the velocity field $\mathbf{N}$ for a plane wave grows linearly in magnitude with the distance of the test particle as seen by the observer located at the center of mass of the system. Therefore, in order to recover the standard result of classical GR that only *tidal* gravitational fields enter into the coupling of radiation and matter, one expects in general that a new characteristic length scale L corresponding to the typical size of the distance X_a separating the test particle from the observer, must enter into the determination of the coupling constant between radiation and matter. For example, L can be the typical size of the detection apparatus (e.g., the length of the arms of the Michelson interferometer used in LIGO), or of the transverse Gaussian wave packet size of the gravitational radiation, so that the coupling constant associated with the Feynman vertex for a graviton-particle interaction becomes proportional to the *extensive* quantity $\sqrt{G}L$, instead of an *intensive* quantity involving only $\sqrt{G}$. For the case of superconductors, treating Cooper pairs of electrons as bosons, we would expect the above arguments would carry over with the charge e replaced by $2e$ and the mass m replaced by $2m$.

4 An experiment using YBCO as transducers between GR and EM waves

4.1 Motivation and idea of the experiment

Motivated by the above theoretical considerations, we performed an experiment using a high T_c superconductor, yttrium barium copper oxide (YBCO), as one such possible quantum transducer, in a first attempt to observe the predicted quantum transduction process from EM to GR waves, and vice versa. We chose YBCO because it allowed us to use liquid nitrogen as the cryogenic fluid for cooling the sample down below $T_c = 90$ K to achieve macroscopic quantum coherence, which is much simpler to use than liquid helium. Although we did not observe a detectable conversion signal in this first experiment, we did establish an upper bound on the transducer conversion efficiency of YBCO, and the techniques we used in this experiment could prove to be useful in future experiments.

The idea of the experiment was as follows: Use a first YBCO sample to convert EM into the GR radiation by shining microwaves onto it, and use a second sample to back-convert the GR radiation generated in the far field by the first sample back into EM radiation of the original frequency. In this way, GR radiation could be generated by the first YBCO sample as the *source* of such radiation inside a first closed metallic container, and GR radiation could be detected by the second sample as the *receiver* of such radiation inside a second closed metallic container, in a Hertz-type experiment.

The electromagnetic coupling between the two halves of the apparatus containing the two YBCO samples, called the "Emitter" and the "Receiver," respectively, could be prevented by means of two Faraday cages, i.e., the two closed metallic cans which completely surrounded the two samples and their associated microwave equipment. See Fig.2. The Faraday cages consisted of two empty one-gallon paint cans with snugly fitting cover lids, whose inside walls, cover lids, and can bottoms,

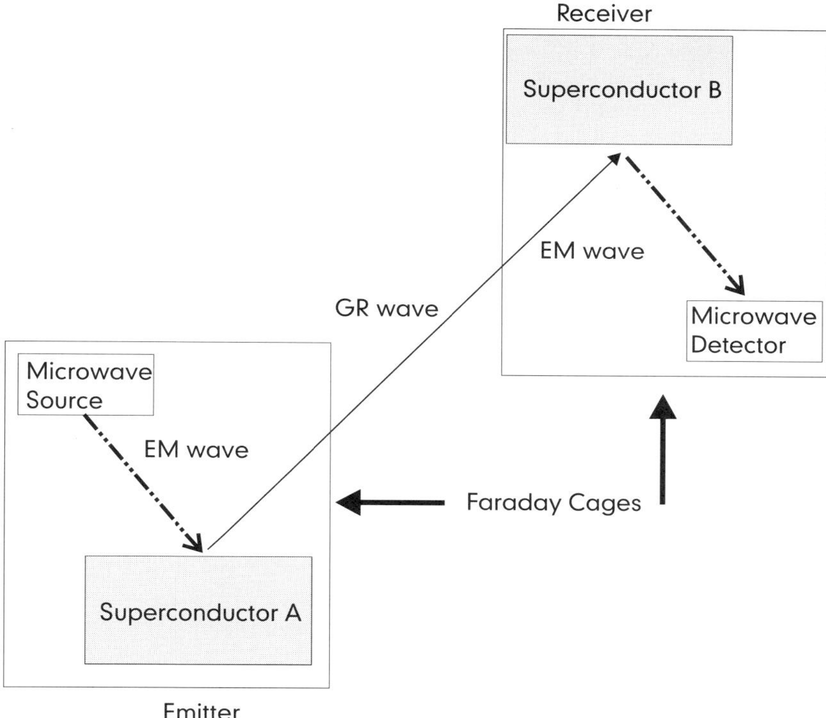

Figure 2. Simplified schematic of a Hertz-type experiment, in which gravitational radiation at 12 GHz could be emitted and received using two superconductors. The "Microwave Source" generated electromagnetic radiation at 12 GHz ("EM wave"), which impinged on Superconductor A, could be converted upon reflection into gravitational radiation ("GR wave"). The GR wave, but not the EM wave, could pass through the "Faraday Cages." In the far field of Superconductor A, Superconductor B could reconvert upon reflection the GR wave back into an EM wave at 12 GHz, which could then be detected by the "Microwave Detector."

were lined on their interiors with a microwave-absorbing foam-like material (Eccosorb AN70), so that any microwaves incident upon these walls were absorbed. Thus multiply-reflected EM microwave radiation within the cans could thereby be effectively eliminated.

The electromagnetic coupling between the two cans with their cover lids on, was measured to be extremely small (see below). Since the Faraday cages were made out of normal metals, and the Eccosorb materials were also not composed of any macroscopically coherent quantum matter, these shielding materials should have been essentially transparent to GR radiation. Therefore, we would expect that GR radiation should have been able to pass through from the source can to the receiver can without much attenuation.

A simplified schematic outlining the Hertz-type experiment is shown in Fig.2, in which gravitational radiation at 12 GHz could be emitted and received using two superconductors. The "Microwave Source" in this Figure generated electromagnetic radiation at 12 GHz ("EM wave"), which was directed onto Superconductor A (the first piece of YBCO) immersed in liquid nitrogen, and would be converted upon reflection into gravitational radiation ("GR wave").

The GR wave, but not the EM wave, could pass through the "Faraday Cages." In the far field of Superconductor A, Superconductor B (a second piece of YBCO), also immersed in liquid nitrogen, could reconvert upon reflection the GR wave back into an EM wave at 12 GHz, which could then be detected by the "Microwave Detector."

For a macroscopically coherent quantum state in YBCO to be produced, the frequency of the microwaves was chosen to be well below the superconducting gap frequency of YBCO. In order to satisfy this requirement, we chose for our experiment the convenient microwave frequency of 12 GHz (or a wavelength of 2.5 cm), which is three orders of magnitude less than gap frequency of YBCO.

Since the predicted conversion process is fundamentally quantum mechanical in nature, the signal would be predicted to disappear if either of the two samples were to be warmed up above the superconducting transition temperature. Hence the signal at the microwave detector should disappear once either superconductor was warmed up above its transition temperature, i.e., after the liquid nitrogen boiled away in either dewar containing the YBCO samples.

It should be emphasized that the predicted quantum transducer conversion process involves a linear relationship between the amplitudes of the two kinds of radiation fields (EM and GR), since we are considering the linear response of the first sample to the incident EM wave during its generation of the outgoing GR wave, and also the linear response of the second sample to the incident GR wave during its generation of the outgoing EM wave. Time-reversal symmetry, which has been observed to be obeyed by EM and GR interactions at low energies for classical fields, would lead us to expect that these two transducer conversion processes obey the principle of reciprocity, so that the reverse process should have an efficiency equal to that of the forward process. However, it should be noted that although time-reversal symmetry for EM interactions has been extensively experimentally tested, it has not been as well tested for GR interactions.

Thus, assuming that the two samples are identical, we expect that the overall power conversion efficiency of this Hertz-type experiment η_{Hertz} should be

$$\eta_{Hertz} = \eta_{EM \to GR} \cdot \eta_{GR \to EM} = \eta^2 \tag{11}$$

where $\eta_{EM \to GR}$ is the EM-to-GR power conversion efficiency by the first sample, and $\eta_{GR \to EM}$ is the GR-to-EM power conversion efficiency of the second sample. If the two samples are closely similar to each other, we expect that $\eta_{EM \to GR} = \eta_{GR \to EM} = \eta$, where η is the transducer power conversion efficiency of a given sample. Hence, the overall efficiency should be $\eta_{Hertz} = \eta^2$.

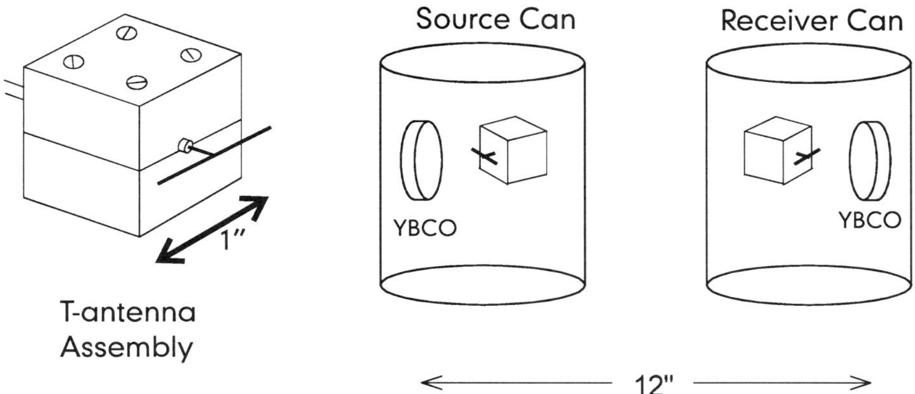

Figure 3. The T-antenna (expanded view on the left) used as antennas inside the "Source Can" and the "Receiver Can." The YBCO samples were oriented so that a GR microwave beam could be directed from one YBCO sample to the other along a straight line of sight.

5 Experimental details

5.1 The T antennas

In the case of the quantum Hall fluid considered earlier, the medium would have a strong magnetic field applied to it, so that the conservation of total angular momentum during the conversion process between the spin-1 EM field and the spin-2 GR field, could be satisfied by means of the angular momentum exchange between the fields and the anisotropic quantum Hall medium. Here, however, our isotropic, compressed-powder YBCO medium did not have a magnetic field applied to it in our initial experiments, so that it was necessary to satisfy the conservation of angular momentum in another way: One must first convert the EM field into an angular-momentum 2, quadrupolar, far-field radiation pattern.

This was accomplished by means of a T-shaped electromagnetic antenna, which generated in the far field an quadrupolar EM field pattern that matched that of the quadrupolar GR radiation field pattern. In order to generate a quadrupolar EM radiation field, it is necessary to use an antenna with structure possessing an even-parity symmetry. This was implemented by soldering onto the central conductor of a SMA coaxial cable a one-wavelength-long wire extending symmetrically on either side of the central conductor in opposite directions, in the form of a T-shaped antenna (see Fig.3).

A one-inch cube aluminum block assembly was placed at approximately a quarter of a wavelength behind the "T," so as to reflect the antenna radiation pattern into the forwards direction, and also to impedance-match the antenna to free space. The aluminum block assembly consisting of two machined aluminum half-blocks which could be clamped tightly together to fig snugly onto the outer conductor of the SMA coaxial cable, so as to make a good ohmic contact with it. The joint between the two aluminum half-blocks was oriented parallel to the bar of the "T."

Thus the block formed a good ground plane for the antenna. The resonance frequency of this T-antenna assembly was tuned to be 12 GHz, and its Q was measured to be about 10, using a network analyzer (Hewlett Packard model HP8720A).

Measurements of the radiative coupling between two such T antennas placed directly facing each other at a fixed distance, while varying their relative azimuthal angle, showed that extinction between the antennas occured at a relative azimuthal angle of 45° between the two "T"s, rather than at the usual 90° angle expected for dipolar antennas. Furthermore, we observed that at a mutual orientation of 90° between the two T antennas (i.e., when the two "T"s were crossed with respect to each other), a *maximum* in the coupling between the antennas, in contrast to the *minimum* expected in the coupling between two crossed linear dipole antennas. This indicates that our T antennas were indeed functioning as quadrupole antennas. Thus, they would generate a quadrupolar pattern of EM radiation fields in the far field, which should be homologous to that of GR radiation.

5.2 *The 12 GHz microwave source*

For generating the 12 GHz microwave beam of EM radiation, which we used for shining a beam of quadrupolar radiation on the first YBCO sample, we started with a 6 GHz "brick" oscillator (Frequency West model MS-54M-09), with an output power level of 13 dBm at 6 GHz. This 6 GHz signal was amplified, and then doubled in a second harmonic mixer (MITEQ model MX2V080160), in order to produce a 12 GHz microwave beam with a power level of 7 dBm. The 12 GHz microwaves was fed into the T antenna that shined a quadrupolar-pattern beam of EM radiation at 12 GHz onto the first YBCO sample immersed in a liquid nitrogen dewar inside the source can. The sample was oriented so as to generate upon reflection a 12 GHz GR radiation beam directed towards the second YBCO sample along a line of sight inside the receiver can (see Fig.3).

The receiver can contained the second YBCO sample inside a liquid nitrogen dewar, oriented so as to receive the beam of GR, and back-convert it into a beam of EM radiation, which was directed upon reflection towards a second T antenna. A low-noise preamp (Astrotel model PMJ-LNB KU, used for receiving 12 GHz microwave satellite communications), which had a noise figure of 0.6 dB, was used as the first-stage amplifier of the received signal. This noise temperature determined the overall sensitivity of the measurement. This front-end LNB (Low-Noise Block) assembly, besides having a low-noise preamp, also contained inside it an internal mixer that down-converted the amplified 12 GHz signal into a standard 1 GHz IF (Intermediate Frequency) band. We then fed this IF signal into a commercial satellite signal level meter (Channel Master model 1005IFD), which both served as the DC power supply for the LNB assembly by supplying a DC voltage back through the center conductor of a F-style IF coax cable into the LNB assembly, and also provided amplification of the IF signal. Its output was then fed into a spectrum analyzer (Hewlett-Packard model 8559A).

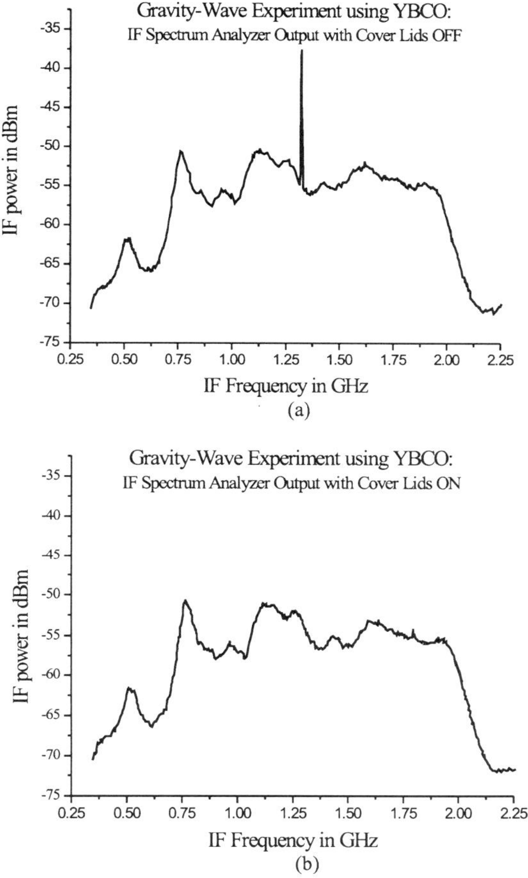

Figure 4. Data from the Hertz-type gravity-wave experiment using YBCO superconductors as transducers between EM and GR radiation. In (a), the cover lids were off both the source and the receiver cans, so that a small leakage signal (the central spike) could serve to test the system. In (b), both cover lids were on the cans, but no detectable signal of coupling between the cans could be seen above the noise. Both YBCO samples were immersed in liquid nitrogen for these data.

5.3 The liquid nitrogen dewars

In order for the YBCO samples (1 inch diameter, 1/4 inch thick pieces of high-density YBCO) to become superconducting, we cooled these samples to 77K by immersing them in liquid nitrogen. The dewars needed for holding this cryogenic fluid together with the YBCO samples consisted of a stack of styrofoam cups; the dead air space between the cups, which were glued together at their upper lips, served as good thermal insulation.

The samples were epoxied in a vertical orientation into a slot in a styrofoam piece which fit snugly into the bottom of the top cup of the stack, and the cups also fit snugly into a hole in the top layer of Eccosorb foam pieces placed at the bottom of the can. Also, since styrofoam was transparent to microwave radiation, these cup stacks also served as convenient *dielectric* dewars for holding the YBCO samples in liquid nitrogen. At the beginning of a run, we would pour into these cups liquid nitrogen, which would last about a hour before it boiled away. The temperatures of the samples were monitored by means of thermocouples attached to the back of the samples.

6 Data

We show in part (a) of Fig.4 data showing the IF spectrum analyzer output of the signal from the receiver can with the cover lids *off* both the source can and the receiver can, which allowed a small leakage signal to be coupled between the two cans (to test whether the entire system was working properly), and in part (b), data with covers lids *on* both cans. Both YBCO samples were immersed in liquid nitrogen for both (a) and (b). The data in (b) show that the Eccosorb-lined Faraday cages were very effective in screening out any electromagnetic pickup. However, there is no detectable signal above the noise that would indicate any detectable coupling due to the quantum transducer conversion between EM and GR waves. Before taking these data, we tested *in situ* that when they immersed in liquid nitrogen, the YBCO samples were indeed in a superconducting state by the observation of a repulsion away from the YBCO of a small permanent magnet hung by means a string near the samples.

The sensitivity of the source-receiver system was calibrated in a separate experiment, in which we replaced the two T antennas by a low-loss cable directly connecting the source to the receiver, in series with 70 dB of calibrated attenuation. We could then measure the size of this directly coupled 12 GHz electromagnetic signal on the spectrum analyzer with respect to the noise rise, which served as a convenient measure of the minimum detectable signal strength. In the resulting spectrum, which was similar to that shown in Fig.4(a), we observed a -77 dBm central peak at 12 GHz, which was 25 dB above the noise rise. This implies that we could have seen a signal of -102 dBm of transducer-coupled radiation with a signal-to-noise ratio of about unity. Assuming that the T antennas were perfectly efficient in coupling to the YBCO samples, from the data shown in Fig.4 we would infer that the observed efficiency η_{Hertz} was less than 95 dB, and therefore from Eq.(11), that the quantum transducer efficiency η was less than 48 dB, i.e., $\eta < 1.6 \times 10^{-5}$.

7 Conclusions

Why did we even bother performing this transducer experiment, when we knew that Faraday cages were essentially perfect shields, and therefore that there seemingly should have been no coupling at all between the two cans? The first answer: Even classically, one expects a *nonzero* coupling between the cans due to the fact that accelerated electrons produce a *nonvanishing* amount of GR radiation, since each electron possesses a mass m, as well as a charge e. Therefore, whenever an electron's charge undergoes acceleration, so will its mass. Relativistic causality therefore necessitates that changes in the gravitational field of an electron in the radiation zone due to its acceleration must be retarded by the speed of light, just like the electromagnetic field in the radiation zone. This implies that there must exist a transducer power conversion efficiency of at least $Gm^2 \cdot 4\pi\varepsilon_0/e^2 = 2.4 \times 10^{-43}$, based on a naive classical picture in which each individual electron possesses a deterministic, Newtonian trajectory. Thus even in principle, the Faraday cages could not have provided a perfect shielding between the two cans. However, if this classical picture had been correct, there would have been no hope of actually observing this conversion process, based on the sensitivity of existing experimental techniques such as those described above.

The second answer: Quantum mechanically, there may exist due to the macroscopic quantum phase coherence of the superconductor, collective, many-body enhancements of the above classical conversion efficiency. Most importantly, as discussed earlier,[15] there must exist enhancements due to the fact that the intensive coupling constant $\sqrt{G}$ of the Feynman graviton-matter vertex should be replaced by the extensive coupling constant $\sqrt{GL}$, in order to account correctly for the *tidal* nature of GR waves.

The third answer: The justification for this experiment ultimately is that the ground state of a superconductor, which possesses spontaneous symmetry breaking, and therefore off-diagonal long-range order, is very similar to that of the physical vacuum, which is believed also to possess spontanous symmetry breaking through the Higgs mechanism. In this sense, therefore, the vacuum is "superconducting." The question thus arises: How does such a broken-symmetry ground, or "vacuum," state interact with a dynamically changing spacetime, such as that associated with a GR wave? More generally: How do we embed quantum fields in dynamically curved spacetimes? We believe that this question has never been explored before *experimentally*.

How then do we account for the lack of any observable quantum transducer conversion in our experiment? There are several possible reasons, the most important ones probably having to do with the material properties of the YBCO medium. One such possible reason is the earlier observations of unexplained residual microwave and far-infrared losses (of the order of 10^{-5} ohms per square at 10 GHz) in YBCO and other high T_c superconductors, which are independent of temperature and have a frequency-squared dependence,[16] which may be due to the fact that YBCO is a D-wave superconductor.[17] In D-wave superconductors, there exists a four-fold symmetry of nodal lines along which the BCS gap vanishes,[18] where the microwave attenuation may become large. Thus D-wave superconductors are quite unlike the classic, low-temperature S-wave superconductors with respect to their microwave losses. Since one of conditions for a good coupling of a quantum antenna and transducer to the GR wave sector is extremely low dissipative losses, the choice of YBCO as the material medium for the Hertz-type experiment may not have been a good one.

8 Acknowledgments

I thank Dung-Hai Lee, Jon-Magne Leinaas, Robert Littlejohn, Joel Moore, Richard Packard, Paul Richards, Daniel Solli, Achilles Speliotopoulos, Neal Snyderman, and Sandy Weinreb for stimulating discussions. This work was supported in part by the ONR.

References

1. R. Y. Chiao and A. M. Steinberg, Prog. in Optics **37**, 347 (1997); R. Y. Chiao and A. M. Steinberg, Physica Scripta **T76**, 61 (1998).
2. R. Y. Chiao, gr-qc/0211078v4, to appear in *Science and Ultimate Reality: Quantum Theory, Cosmology and Complexity*, J. D. Barrows, P. C. W. Davies, and C. L. Harper, Jr., editors (Cambridge University Press, Cambridge, 2003).
3. R. B. Laughlin, Phys. Rev. Lett. **50**, 1395 (1983).
4. S. Weinberg, *Gravitation and Cosmology: Principles and Applications of the General Theory of Relativity* (John Wiley and Sons, New York, 1972).
5. J. Anandan, Phys. Rev. Lett. **47**, 463 (1981); R. Y. Chiao, Phys. Rev. B **25**, 1655 (1982); J. Anandan and R. Y. Chiao, Gen. Rel. and Grav. **14**, 515 (1982); J. Anandan, Phys. Rev. Lett. **52**, 401 (1984); J. Anandan, Phys. Lett. **110A**, 446 (1985).
6. H. Peng and D. G. Torr, Gen. Rel. and Grav. **22**, 53 (1990); H. Peng, D. G. Torr, E. K. Hu, and B. Peng, Phys. Rev. B **43**, 2700 (1991).
7. R. Y. Chiao, gr-qc/0208024.
8. N. D. Birrell and P. C. W. Davies, *Quantum Fields in Curved Space* (Cambridge University Press, Cambridge, 1982).
9. M. V. Berry, Proc. Roy. Soc. London Ser. A **392**, 45 (1984); R. Y. Chiao and Y. S. Wu, Phys. Rev. Lett. **57**, 933 (1986); A. Tomita and R. Y. Chiao, Phys. Rev. Lett. **57**, 937 (1986); R. Y. Chiao and T. F. Jordan, Phys. Lett. A **132**, 77 (1988).
10. A. Stern, Phys. Rev. Lett. **68**, 1022 (1992); Y. Lyanda-Geller and P. M. Goldbart, Phys. Rev. A **61**, 043609 (2000).
11. S. M. Girvin and A. H. MacDonald, Phys. Rev. Lett. **58**, 1252 (1987); S. C. Zhang, T. H. Hansson, and S. Kivelson, *ibid.* **62**, 82 (1989).
12. D. D. Osheroff, R. C. Richardson, and D. M. Lee, Phys. Rev. Lett. **28**, 885 (1972); D. D. Osheroff, W. J. Gully, R. C. Richardson, and D. M. Lee, *ibid.* **29**, 920 (1972).
13. B. S. DeWitt, Phys. Rev. Lett. **16**, 1092 (1966).
14. A. D. Speliotopoulos, Phys. Rev. D **51**, 1701 (1995).
15. A. D. Speliotopoulos and R. Y. Chiao, gr-qc/0302045.
16. D. Miller, "Submillimeter residual losses in high-T_c superconductors," (Ph.D. thesis, U. C. Berkeley, 1993).
17. M. Tinkham, *Introduction to Superconductivity*, 2nd edition (McGraw-Hill, New York, 1996).
18. S. H. Pan, E. W. Hudson, K. M. Lang, H. Eisaki, S. Uchida, and J. C. Davis, Nature **403**, 746 (2000).

SECTION III
COHERENCE & DECOHERENCE

Decoherence Unlimited: From Zeno to Classical Motion
  E. Joos

Coherence and the Clock
  L. Stodolsky

Decoherence, Chaos and the Second Law
  W. H. Zurek

Contributed Papers

What Could We Have Been Missing While Pauli's Theorem Was in Force?
  E. A. Galapon

Simultaneity and the Concept of 'Particle'
  C. E. Dolby

DECOHERENCE UNLIMITED: FROM ZENO TO CLASSICAL MOTION

ERICH JOOS

Rosenweg 2, D-22869 Schenefeld
E-mail: ej@erichjoos.de

Motion in quantum theory as a unitary evolution has very different properties compared to classical physics. If the phase relations defining unitary dynamics are destroyed or unavailable, motion becomes impossible (Zeno effect). The most important mechanism is decoherence, arising from coupling of a quantum system to its environment. Macroscopic systems are not frozen, although strong decoherence is important to understand classical objects within the quantum framework. These two conflicting consequences of strong decoherence are analyzed and compared.

1 Introduction

The problem of motion and time has always been one of the most disputed in science and philosophy. In this contribution I will discuss a modern version of Zeno's "proof" that motion may be just an illusion. There are two main threads merging here – both arising from the still vividly debated problem of the interpretation of quantum theory. The modern (quantum) version of Zeno's paradox arises in the context of quantum measurement theory. Here quite simple arguments seem to show that a quantum system cannot evolve any more if it is observed often enough [1]. On the other hand, a major current line of research is the emergence of classicality from quantum theory through decoherence [2]. Macroscopic objects cannot escape being continuously "watched" by their natural environment, leading to strong "decoherence". Here a possible contradiction shows up: The objects in our macroscopic world obviously *are* moving around and there seems to be no "Zeno effect". How this puzzle can be solved will be discussed in the following sections.

2 The Quantum Zeno Effect

Let us start with a general argument concerning the decay probability in quantum theory. Let a system be described by some "undecayed" state $|\Psi(0)\rangle = |u\rangle$ at some initial time $t = 0$. The probability $P(t)$ to find it again in this "undecayed" state at a later time t is

$$P(t) = \left| \langle u | e^{-iHt} | u \rangle \right|^2 \tag{1}$$

where H is the Hamiltonian of the system. For small times we can expand $P(t)$, yielding

$$P(t) = 1 - (\Delta H)^2 t^2 + O(t^4) \tag{2}$$

with

$$(\Delta H)^2 = \langle u | H^2 | u \rangle - \langle u | H | u \rangle^2 . \tag{3}$$

The important feature to notice here is the *quadratic* time dependence of the survival probability. This may be compared with usual exponential decay law

$$P(t) = \exp(-\Gamma t), \tag{4}$$

which leads to a *linear* time dependence for small times,

$$P(t) = 1 - \Gamma t + \dots . \tag{5}$$

This raises the question, how these two differing results can be made compatible. Both look fundamental, but they obviously contradict each other. This conflict can be made even stronger, when we consider the case of repeated measurements in a short time interval.

Suppose we repeat the measurement N times during the interval $[0, t]$. Then the non-deacy (survival) probability is

$$P_N(t) \approx \left[1 - (\Delta H)^2 \left(\frac{t}{N} \right)^2 \right]^N > P(t), \tag{6}$$

which for large N gives

$$P_N(t) = 1 - (\Delta H)^2 \frac{t^2}{N} + \dots \xrightarrow{N \to \infty} 1. \tag{7}$$

This is the Zeno effect: Sufficiently dense measurements should halt any motion!

There is no Zeno effect, if the system decays according to the exponential decay law, since in this case trivially

$$P_N(t) = \left(\exp \left(-\Gamma \frac{t}{N} \right) \right)^N = \exp(-\Gamma t) = P(t). \tag{8}$$

The conclusion is, that any system showing a quadratic short-time behavior is very sensitive to measurements, whereas an exponentially decaying system does not care about whether its decay status is measured or not, that is, it behaves classically in this respect.

If a system is governed by the Schrödinger equation, as used in Equ. (1), the transition probability for small times *must* start quadratically, hence the exponential decay law can only be an approximation for larger times. [a] What happens in the limit of "continuous" observation? The Zeno argument seems to show that there will be no motion at all!

To gain a better understanding of what is going on here, I will discuss in the following why motion is slowed down by measurements. In addition, the measurement process itself will be described be a unitary evolution following the Schrödinger equation as the fundamental law of motion for quantum states. It will turn out, that the Zeno effect can be understood as a dynamical process and the collapse of the wave function is not required.

---

[a] There is a certain irony in this situation, since – at least in popular accounts – exponential ("random") decay is used as a major argument that classical physics has to be replaced by a new (quantum) theory. But there is no strict exponential decay law in quantum theory.

3   Interference, Motion and Measurement in Quantum Theory

Why does measurement slow down motion in quantum theory, but not in classical physics? The reason can be traced back to the very nature of quantum evolution. Quantum dynamics is unitary and can be viewed as a rotation in Hilbert space, see Fig. 1. If a system is going to move from a state $|a\rangle$ to another state $|b\rangle$, it has to go

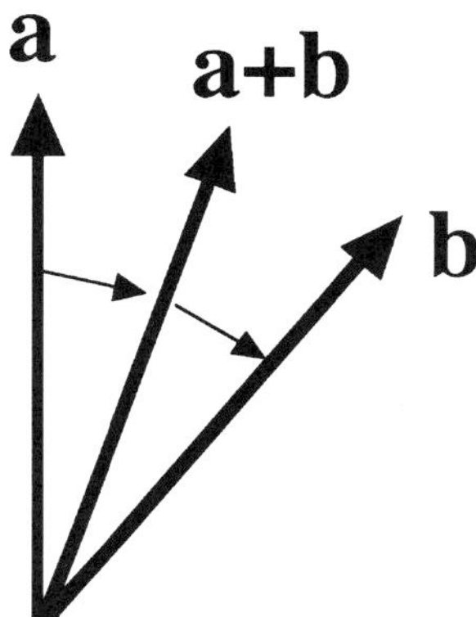

Figure 1. Evolution in quantum theory can be viewed as a rotation connecting an initial state $|a\rangle$ with a final state $|b\rangle$. For intermediate times a superposition $|a\rangle + |b\rangle$ is required for undisturbed motion.

through a superposition state $|a\rangle + |b\rangle$. An essential feature of such a superposition is the presence of interference (coherence). As is well known, such a superposition has properties which none of its components has – it is an entirely new state. [b] Unitary evolution from $|a\rangle$ to $|b\rangle$ requires all the phase relations contained in the intermediate state $|a\rangle + |b\rangle$. Phase relations are destroyed by measurements, so it is not surprising that motion becomes impossible in quantum theory if coherence is completely absent!

As an example consider the evolution of a two-state system from an initial state $|1\rangle$ as a two-step process connecting times $0, t$, and $2t$, as shown in Fig. 2. If a_{ij} are transition amplitudes (calculated from the Schrödinger equation) we have the chain

$$
\begin{aligned}
t = 0: \quad & |1\rangle \\
&\longrightarrow a_{11}|1\rangle + a_{12}|2\rangle \\
&\longrightarrow (a_{11}^2 + a_{12}a_{21})|1\rangle + (a_{12}a_{22} + a_{11}a_{12})|2\rangle .
\end{aligned}
\tag{9}
$$

The final probability for state $|2\rangle$ at time $2t$ is then

$$
\begin{aligned}
P_2 &= |a_{12}a_{22} + a_{11}a_{12}|^2 \\
&\approx V^2 (2t)^2
\end{aligned}
\tag{10}
$$

---

[b]This is the reason why stochastic models for quantum evolution are unsuccessful: A superposition cannot be replaced by an ensemble of its components.

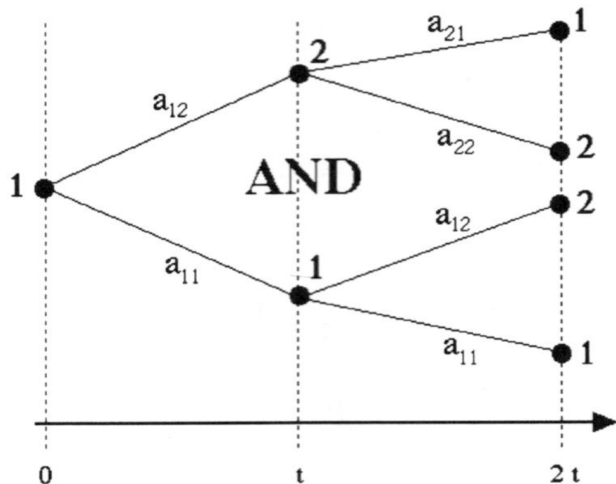

Figure 2. Evolution of a two-state system away from initial state $|1\rangle$. The amplitude (and therefore the probability) of state $|2\rangle$ at time $2t$ depends on the phases contained in the superposition of $|1\rangle$ and $|2\rangle$ at the intermediate time t, as in a double-slit experiment.

with

$$V = \langle 1| H |2\rangle. \tag{11}$$

Clearly the value for P_2 depends essentially on the presence of interference terms. In a sense unitary evolution is an ongoing double- (or multi-)slit experiment!

Now compare this evolution with the same process, but a measurement is made at the intermediate time t. This measurement may be described by a collapse producing an ensemble (that is, resulting in $|1\rangle$ *or* $|2\rangle$)), or dynamically by coupling to another degree of freedom. In the latter case an entangled state containing the

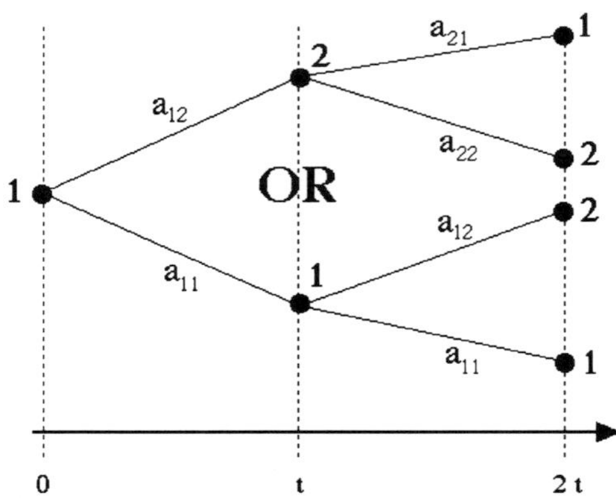

Figure 3. Evolution of a two-state system with measurement. The probability for state $|2\rangle$ at time $2t$ results solely from the transition probabilities to intermediate states at time t. The loss of phase relations leads to a decrease of the total transition probability.

system and the measuring device $|\Phi\rangle$ (or, more general the system's environment) ensues (more on this in the next section). The equations now look like

$$\begin{aligned}
t = 0 : \quad & |1\rangle |\Phi\rangle \\
& \longrightarrow (a_{11} |1\rangle + a_{12} |2\rangle) |\Phi\rangle \\
& \longrightarrow a_{11} |1\rangle |\Phi_1\rangle + a_{12} |2\rangle |\Phi_2\rangle \\
& \longrightarrow (a_{11}^2 |\Phi_1\rangle + a_{12}a_{21} |\Phi_2\rangle) |1\rangle + (a_{12}a_{22} |\Phi_1\rangle + a_{11}a_{12} |\Phi_2\rangle) |2\rangle
\end{aligned} \tag{12}$$

(the third line describes the new measurement step) and the transition probability is given by

$$\begin{aligned}
P_2 &= |a_{12}a_{22}|^2 + |a_{11}a_{12}|^2 \\
&\approx \frac{1}{2} V^2 (2t)^2.
\end{aligned} \tag{13}$$

Since the interference terms are missing, we lose half of the probability! Clearly then, if we divide the time interval not in two but into N steps, the transition probability is reduced by a factor $1/N$: the Zeno effect. This reduction is a sole consequence of entanglement, without any "disturbance" of the measured system, since the measurement is assumed ideal in this model. No coherence, no motion!

The Zeno effect can also be seen more formally from the von Neumann equation for the density matrix. If coherence is absent in a certain basis, the density matrix is diagonal, i.e.,

$$\rho_{nm} = \rho_{nn} \delta_{nm}. \tag{14}$$

But then no evolution is possible, since the von Neumann equation immediately yields

$$\frac{d}{dt} \rho_{nn} = \sum_k (H_{nk}\rho_{kn} - \rho_{nk}H_{kn}) \equiv 0. \tag{15}$$

4 Measurement as a Dynamical Process: Decoherence

To further analyze the Zeno effect I will consider a specific model for measurements of an N-state system. As a preparation, let me shortly review the dynamical description of a measurement process. In a dynamical description of measurement, the well-known loss of interference during measurement follows from a certain kind of interaction between a system and its environment.

Following von Neumann, consider an interaction between an N-state system and a "measurement device" in the form

$$|n\rangle |\Phi_0\rangle \longrightarrow \exp(-iHT) |n\rangle |\Phi_0\rangle = |n\rangle |\Phi_n\rangle \tag{16}$$

where $|n\rangle$ are the system states to be discriminated by the measurement device and $|\Phi_n\rangle$ are "pointer states" telling which state of the system has been found. H is an appropriate interaction leading after the completion of the measurement (at time T) to orthogonal states of the measuring device. Since in Equ. (16) the system state is not changed, this measurement is called "ideal" (recoil-free). A general initial state of the system will – via the superposition principle – lead to an entangled state,

$$\left(\sum_n c_n \, |n\rangle \right) |\Phi_0\rangle \longrightarrow \sum_n c_n \, |n\rangle \, |\Phi_n\rangle . \tag{17}$$

This correlated state is still pure and does therefore *not* represent an ensemble of measurement results (therefore such a model alone does not solve the measurement problem of quantum theory). The important point is that the phase relations between different n are *delocalized* into the larger system and are no longer available at the system alone. Therefore the system *appears* to be in one of the states $|n\rangle$, formally described by the diagonalization of its density matrix,

$$\begin{aligned} \rho \;=\;& \sum_{n,m} c_m^* c_n \, |n\rangle \, \langle m| \\ \longrightarrow\;& \sum_{n,m} c_m^* c_n \, \langle \Phi_m | \Phi_n \rangle \, |n\rangle \, \langle m| \\ =\;& \sum_n |c_n|^2 \, |n\rangle \, \langle n| , \end{aligned} \tag{18}$$

where the last line is valid, if the pointer (or environmental) states are orthogonal, $\langle \Phi_m | \Phi_n \rangle = 0$.

Any measurement-like interaction will therefore produce an apparent ensemble of system states. This process is now usually called "decoherence"[2]. Note that the origin of this effect is *not* a disturbance of the system. Quite to the contrary: the system states $|n\rangle$ remain unchanged, but they "disturb" (change) the environment!

5 Strong Decoherence of a Two-State System

As a first application of the von-Neumann measurement model let us look at an explicit scheme for a two-state system with Hamiltonian

$$\begin{aligned} H \;=\;& H_0 + H_{\mathrm{int}} \\ =\;& V(|1\rangle \, \langle 2| + |2\rangle \, \langle 1|) + E \, |2\rangle \, \langle 2| \\ & + \gamma \hat{p}(|1\rangle \, \langle 1| - |2\rangle \, \langle 2|). \end{aligned} \tag{19}$$

Here the momentum operator $\hat{p}$ leads to a shift of a pointer wavefunction $\Phi(x)$ "to the right" or "to the left", depending on the state of the measured system, γ represents a measure of the strength of this interaction. Because of the structure of the Hamiltonian, this interaction is recoil-free. This model can be solved exactly and shows the expected damped oscillations. In view of the Zeno effect we are mostly interested in the limit of strong coupling. Here the solutions (calculated in perturbation theory) show two interesting features, as displayed in Figs. 4 and 5 [3]. First, the transition probability from $|1\rangle$ to $|2\rangle$ depends in a complicated way on the coupling strength, but for large coupling always decreases with increasing interaction. This is the expected Zeno behavior.

If we look at the time dependence of the transition probability, we can see the quadratic behavior for very small times (as is required by the general theorem

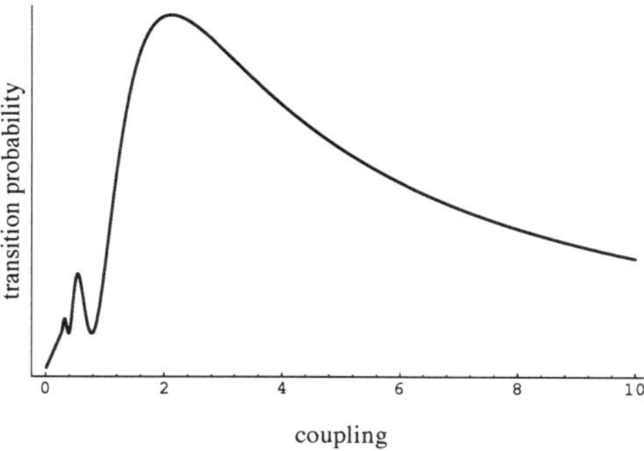

Figure 4. Transition probability as a function of the coupling strength in a two-state model. For strong coupling, transitions are always damped (Zeno effect).

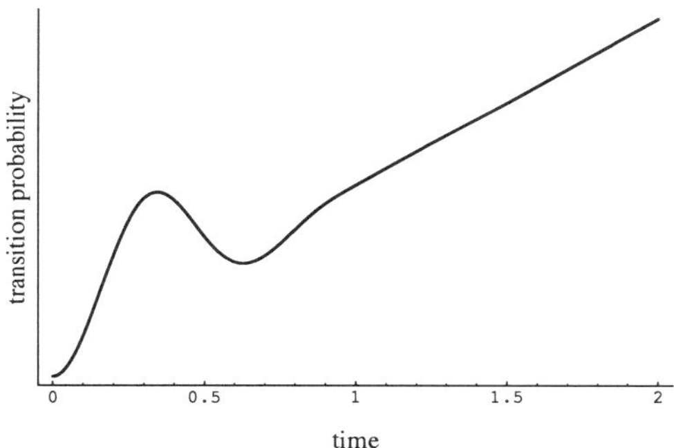

Figure 5. Transition probability as a function of time. If the measurement can be considered complete (here at $t \approx 1$), the transition probability grows linearly (constant transition rates)

Equ. (2)), but soon the transition probability grows linearly, as in an exponentially decaying system (the rate, however, still depends on the coupling strength).

A realization of the quantum Zeno effect has been achieved in an experiment [4] where to two-state system is represented in the form of an atomic transition and the measurement process is realized by coupling to a third state which emits fluorescence radiation, see. Fig. 6.

The Zeno effect also shows up in a curious way in a recent proposal of "interaction-free measurement".

Early ideas about "negative result" or "interaction-free" measurements [5] can be combined with the Zeno mechanism [6]. One of these schemes is exemplified in Fig. 7. If a horizontally polarized photon is sent through N polarization rotators (or repeatedly through the same one) each of which rotates the polarization by an angle $\Delta\Theta = \frac{\pi}{2N}$, the photon ends up with vertical polarization; i.e., the probability to find horizontal polarization is zero:

$$P_H = 0. \tag{20}$$

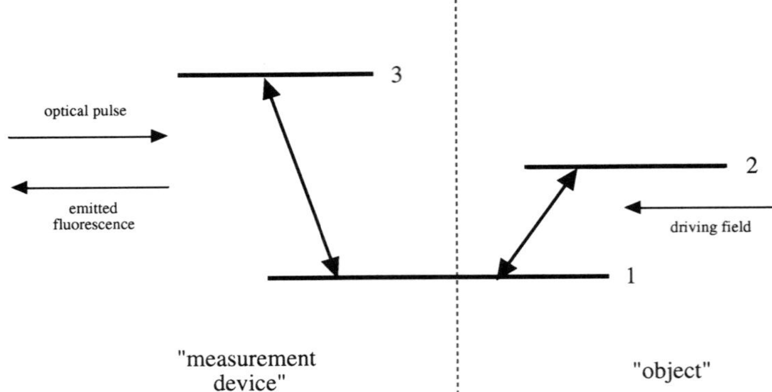

Figure 6. Zeno experiment in atomic physics. The two-state system under repeated observation is represented by a transition between states $|1\rangle$ and $|2\rangle$. Measurement is accomplished through an optical pulse leading to fluorescence from level $|3\rangle$, if the state $|1\rangle$ is present.

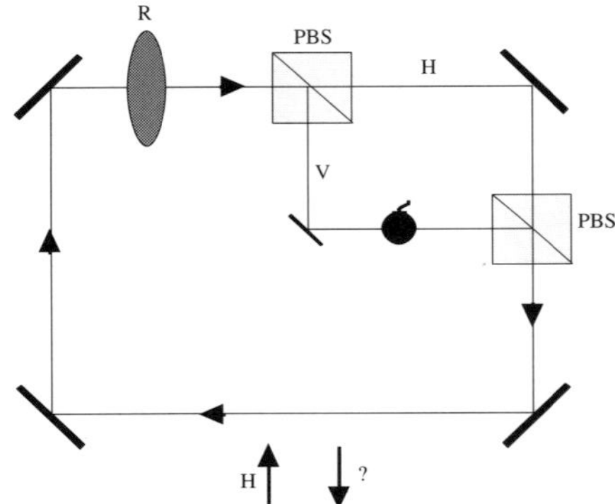

Figure 7. Scheme of "interaction-free interrogation" as a variant of the Zeno effect. Without the absorbing object (the bomb), the polarization of the injected photon (initially horizontal) is rotated by the rotator R by a small angle on every passage. The two polarizing beam splitters PBS have no effect, if properly adjusted, since horizontal and vertical components are recombined coherently. If an absorbing object is present, the vertical polarization is removed at every passage. Inspecting the photon after many cycles allows one to infer the presence of the object with high probability, while the photon is only very infrequently absorbed.

If this evolution is interrupted by a horizontal polarizer (absorber) the probability of transmission is (similar to Eqs. (6) and (7)) given by

$$P'_H = \cos^{2N} \Delta\Theta = \cos^{2N} \frac{\pi}{2N} \approx 1 - \frac{\pi^2}{4N} \longrightarrow 1. \tag{21}$$

To implement this idea, a photon is injected into the setup shown in Fig. 7 and goes N times through the rectangular path, as indicated. The initial polarization is rotated at R by an angle $\Delta\Theta = \frac{\pi}{2N}$ on each passage. In the absence of the absorbing object, the polarizing beam splitters, making up a Mach-Zehnder interferometer, are adjusted to have no effect. That is, the vertical component V is coherently recombined with the horizontal one (H) at the second beamsplitter to reproduce the rotated state of polarization. If, however, the "bomb" is present, the vertical component is absorbed at each step. After N cycles, the photon is now still horizontally polarized, thereby indicating the presence of the object with probability near one, or has been absorbed (with arbitrarily small probability). For details of experimental setups see [7].

One should be aware of the fact, that the term "interaction-free" is seriously misleading, since the Zeno mechanism is a consequence of *strong* interaction. Part of this conceptual confusion is related to the classical particle pictures often used in the interpretation of interference experiments, in particular "negative-result measurements".

6 Strong Decoherence of Many-State Systems

Why does the Zeno effect not show up in our macroscopic world? I will consider two examples of classical dynamics. The first is the motion of a massive object such as a dust particle or a planet. The second example will be a reconsideration of Pauli's rate equation, describing classical random processes, where interference apparently plays no role. In both cases it will turn out that (1) continuous measurement (i.e. decoherence) is an essential ingredient for deriving classical motion and (2) the Zeno effect plays no role.

6.1 Macroscopic objects

With hindsight it seems to be trivial observation that all macroscopic objects are strongly interacting with their natural environment. The consequences have been analyzed only rather late in the history of quantum theory [8,9]. One reason for this is certainly the prevailing Copenhagen orthodoxy. For generations students were told, that quantum theory should only be used for microscopic objects, while macroscopic bodies are necessarily described by classical physics.

The typical scenario is represented by scattering processes, where the state of the scattered objects, a photon or a molecule typically depends on the position of the macrosopic body. Quantitative estimates [9] show a strong effect, even in extreme situations, for example, a dust particle scattering only cosmic background radiation. For small distances, interference is damped according to

$$\rho(x, x', t) = \rho(x, x', 0)\exp[-\Lambda t(x - x')^2] \tag{22}$$

with a "localization rate" Λ given by

$$\Lambda = \frac{k^2 N v \sigma_{eff}}{V} \tag{23}$$

Here k is the wave vector of the scattered particle, Nv/V the incoming flux and σ_{eff} of the order of the total cross section. Some typical numbers are shown in the table.

Figure 8. Macroscopic objects can never be considered as isolated from their natural environment. Irreversible scattering processes lead to ever-increasing entanglement.

	Free electron	10^{-3} cm dust particle	Bowling ball
300 K air at 1 atm pressure	10^{31}	10^{37}	10^{45}
300 K air in lab vacuum	10^{18}	10^{23}	10^{31}
Sunlight (on earth)	10^{1}	10^{20}	10^{28}
300 K photons	10^{0}	10^{19}	10^{27}
Background radioactivity	10^{-4}	10^{15}	10^{23}
Quantum gravity	10^{-25}	10^{10}	10^{22}
GRW effect	10^{-7}	10^{9}	10^{21}
Cosmic background radiation	10^{-10}	10^{6}	10^{17}
Solar neutrinos	10^{-15}	10^{1}	10^{13}

Table 1. Localization rate Λ in $cm^{-2}s^{-1}$ for some scattering processes.

The equation of motion of, say, a dust particle, is then no longer the von Neumann-Schrödinger equation, but contains an additional scattering term,

$$i\frac{\partial \rho(x, x', t)}{\partial t} = \frac{1}{2m}\left(\frac{\partial^2}{\partial x'^2} - \frac{\partial^2}{\partial x^2}\right)\rho - i\Lambda(x - x')^2\rho. \qquad (24)$$

If one analyzes the solutions of this equation, one finds that, for example, the Ehrenfest theorems for mean position and momentum are still valid: The motion is *not* damped, although coherence between different positions is destroyed. There is no Zeno effect.

The above equation of motion is a special case of more general equations which are studied under the topic "Quantum Brownian Motion". These models include friction effects, a simple example is [10]

$$i\frac{\partial \rho(x, x', t)}{\partial t} = \left[\frac{1}{2m}\left(\frac{\partial^2}{\partial x'^2} - \frac{\partial^2}{\partial x^2}\right)\right.$$

$$\begin{aligned} -i\Lambda(x-x')^2 \\ + i\frac{\gamma}{2}(x-x')\left(\frac{\partial}{\partial x'} - \frac{\partial}{\partial x}\right)\Bigg] \rho(x,x',t) \end{aligned} \tag{25}$$

where

$$\Lambda = m\gamma k_B T. \tag{26}$$

In typical macroscopic situations, decoherence is much more important, however. The ratio of decoherence to relaxation rate can be estimated as

$$\frac{\text{decoherence rate}}{\text{relaxation rate}} \approx m k_B T (\Delta x)^2 = \left(\frac{\Delta x}{\lambda_{th}}\right)^2, \tag{27}$$

where λ_{th} is the thermal deBroglie wavelength of the macroscopic body. This ratio has the enormous value of about 10^{40} for a macroscopic situation (m=1 g, $\Delta x = 1$ cm) [11].

We can conclude from these models that

- Newton's reversible laws of motion can be derived (to a very good approximation) from strong *irreversible* decoherence.

- The appearance of classical objects has its origin in low-entropy condition in the early universe and the unique features of quantum nonlocality.

- Decoherence works *much* faster than friction in macroscopic situations.

- Although coherence is strongly suppressed, no Zeno effect (slowing down of motion) appears.

6.2 Rate equations

The exponential decay law $P(t) = \exp(-\lambda t)$ mentionend at the beginning is a special case of a general rate equation

$$\frac{d}{dt}P_\alpha = \sum_\beta A_{\alpha\beta} P_\beta = \sum_{\beta\neq\alpha} (A_{\alpha\beta}P_\beta - A_{\beta\alpha}P_\alpha). \tag{28}$$

Its quantum analogue, describing dynamics of "occupation probabilities" is usually called the "Pauli equation",

$$\frac{d}{dt}\rho_{\alpha\alpha} = \sum_\beta A_{\alpha\beta}\rho_{\beta\beta}. \tag{29}$$

An obvious feature of (29) is that interference terms do not play any dynamical role. On the other hand, this cannot be true exactly, since then the von Neumann equation would lead to Zeno freezing,

$$\frac{d}{dt}\rho_{\alpha\alpha} = \sum_\beta (H_{\alpha\beta}\rho_{\beta\alpha} - \rho_{\alpha\beta}H_{\beta\alpha}) \equiv 0. \tag{30}$$

To further analyze these matters let us assume that the properties α in the rate equation are macroscopic in the sense that they are *continuously observed by the environment.* The microscopic characterization is in the following assumed to be given entirely by the energy, further degeneracies are neglected for simplicity. The macroscopic feature α is measured by a "pointer" as in the two-state Zeno model above, see Fig. 9. The Hamiltonian then reads [3]

$$H = \sum_{\alpha E} E\,|\alpha E\rangle\,\langle\alpha E| + \sum_{\alpha E \neq \alpha' E'} V_{\alpha E,\alpha' E'}\,|\alpha E\rangle\,\langle\alpha' E'|$$

$$+ \sum_{\alpha E} \gamma(\alpha)\hat{p}\,|\alpha E\rangle\,\langle\alpha E| \tag{31}$$

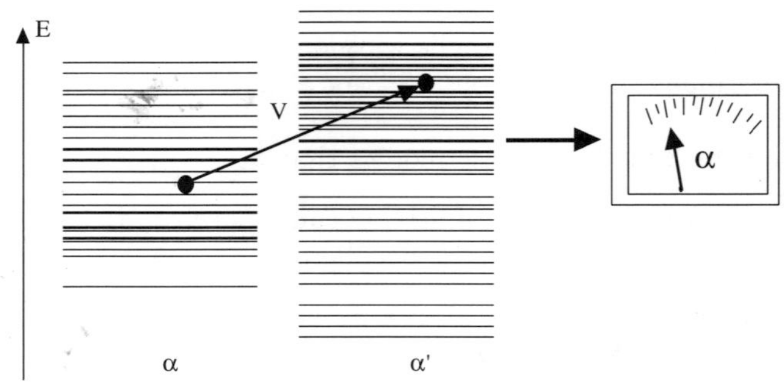

Figure 9. Transitions between groups of states are monitored by a pointer. The symbolic measurement device in the figure represents the interaction with the environment (which may or may not contain an experimental setup). Transition probabilities often follow Fermi's Golden rule (rates governed by transition matrix elements V and level densities at resonance energy), but may be influenced by the presence of the environment monitoring certain features α of initial or final states.

Since we are interested in the limit of strong coupling to the pointer, we calculate the transition probability from property α_0 to another one, α in lowest order perturbation theory. Starting from

$$|\Psi(0)\rangle = |\alpha_0 E_0\rangle\,|\Phi\rangle, \tag{32}$$

where Φ is the pointer state, the transition probability is

$$P_{\alpha E} = 4 \int dp\,|V_{\alpha E,\alpha_0 E_0}|^2\,|\Phi(p)|^2\,\frac{\sin^2(E - E_0 + \gamma(\alpha)p)t/2}{(E - E_0 + \gamma(\alpha)p)^2} \tag{33}$$

(assuming $\gamma(\alpha_0) = 0$ for simplicity) . This expression shows a familiar resonance factor, but now we have new resonances for each value of p with weight $|\Phi(p)|^2$, shifted from $E = E_0$ to a new value $E = E_0 - \gamma(\alpha)p$. Summing over many states with property α gives

$$P_\alpha \approx 2\pi t \int dE \frac{\sigma_\alpha(E)\,|V_{\alpha E,\alpha_0 E_0}|^2}{\gamma(\alpha)}\left|\Phi\left(\frac{E - E_0}{\gamma(\alpha)}\right)\right|^2 \tag{34}$$

Three limiting cases can be extracted from this expression (see also Fig. 10).

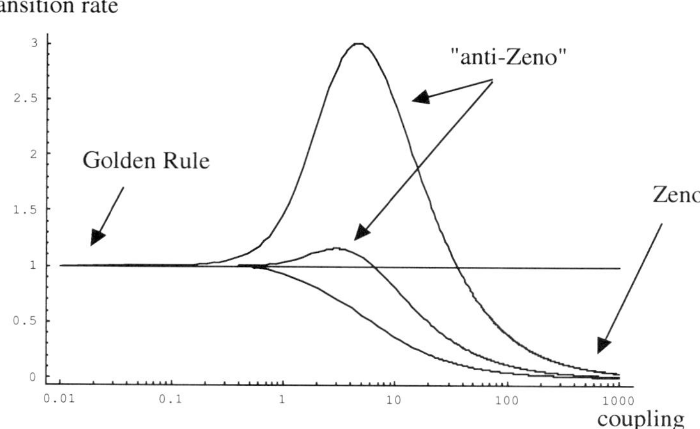

Figure 10. Continuous coupling to a pointer changes the transition rate from an initial state $|\alpha_0 E_0\rangle$ to a group of final states in various ways. For small coupling we find the standard Golden rule result (here normalized to unity). Increasing the coupling to the measuring agent may in some cases increase the transition probability by shifting the effective resonance frequency to regions with higher level density or larger transition matrix elements (anti-Zeno effect). Strong interaction always leads to decreasing transition rates (Zeno effect).

- Case 1: *Zeno limit*: For large coupling $\gamma(\alpha)$ we have

$$P_\alpha \approx \frac{2\pi t}{\gamma(\alpha)} \int dE \; \sigma |V|^2(E) |\Phi(0)|^2 \sim \frac{1}{\gamma(\alpha)} \tag{35}$$

Transitions are suppressed as expected.

- Case 2: *Golden Rule limit*: For small coupling, transition rates become independent of γ and the usual result is recovered,

$$P_\alpha = 2\pi t \sigma_\alpha(E_0) |V(E_0)|^2. \tag{36}$$

- Case 3: *"Anti-Zeno effect"*: If the contributions from each transition are comparable, that is, if $\sigma |V|^2 \approx const.$ in the releveant interval $[E_{\min}, E_{\max}]$ then it is easy to see that we have a smooth transition from the Zeno region to the Golden limit. If this is not the case, it can happen that in the intermediate range transition probabilities are *enhanced* above the Golden rule value. This is occasionally called "anti-Zeno effect".

7 Summary

We have seen that unitary evolution depends decisively on interference between components of the wave function. If phase relations are lost, evolution is hindered. This leads finally to the Zeno freezing of motion. No coherence, no motion.

The destruction of phase relations can be understood as phase *de-localization* arising from *unitary* quantum evolution, if the interaction of a system with its environment is taken into account. In this way, the Zeno effect can be completely understood as a dynamical effect. No collapse of the wave function is required, but only quantum nonlocality.

Many-state systems can escape Zeno freezing. This is important for the properties of our experienced macroscopic world, but also for common "quantum" features such as radioactive decay, which happens whether or not a counter is setup to observe the decay. (In fact, in most cases Nature herself provides the necessary "counters".)

Systems with only a few degrees of freedom are very sensitive to quantum entanglement and can therefore never escape the Zeno effect, if they are interacting with other systems. Zeno freezing can thus be used to delineate the borderline between classical and quantum objects.

References

1. B. Misra and E.C.G. Sudarshan: "The Zeno's paradox in quantum theory." *Journal of Math. Phys.* **18**, 756–763 (1977).
2. E. Joos, H.D. Zeh, C. Kiefer, D. Giulini, J. Kupsch, I.-O. Stamatescu: *Decoherence and the Appearance of a Classical World in Quantum Theory.* 2nd edition (Springer, Berlin, 2003).
3. E. Joos: "Continuous Measurement: Watchdog Effect versus Golden Rule." *Phys. Rev.* **D29**, 1626–1633 (1984).
4. W.M. Itano, D.J. Heinzen, J.J. Bollinger, and D.J. Wineland: "Quantum Zeno effect." *Phys. Rev.* **A41**, 2295–2300 (1990).
5. A.C. Elitzur and L. Vaidman: "Quantum Mechanical Interaction-Free Measurements." *Found. Phys.* **23**, 987–997 (1993).
6. C.H. Bennett: "Night thoughts, dark sight." *Nature* **371**, 479–480 (1994).
7. P.G. Kwiat, A.G. White, J.R. Mitchell, O. Nairz, G. Weihs, H. Weinfurter, and A. Zeilinger: "High-efficiency quantum interrogation measurements via the quantum Zeno effect." *Phys. Rev. Lett.* **83**, 4725–4728 (1999).
8. H.D. Zeh: "On the Interpretation of Measurement in Quantum Theory." *Found. Phys.* **1**, 69–76 (1970).
9. E. Joos and H.D. Zeh: "The Emergence of Classical Properties Through Interaction with the Environment." *Z. Phys.* **B59**, 223–243 (1985).
10. A.O. Caldeira and A.J. Leggett: "Path Integral Approach to Quantum Brownian Motion." *Physica* **121A**, 587–616 (1983).
11. W.H. Zurek: "Reduction of the Wave Packet: How long does it take?" In: *Frontiers in Nonequilibirum Statistical Physics*, ed. by G.T. Moore and M.T. Sculley (Plenum, New York, 1986).

COHERENCE AND THE CLOCK

L. STODOLSKY

*Max-Planck-Institut für Physik (Werner-Heisenberg-Institut), Föhringer Ring 6, 80805
München, Germany
E-mail:les@mppmu.mpg.de*

We discuss the notion of quantum mechanical coherence in its connection with
time evolution and stationarity. The transition from coherence to decoherence is
examined in terms of an equation for the time dependence of the density matrix.
It is explained how the decoherence rate parameter arising in this description is
related to the "unitarity defect" in interactions with the environment as well as
the growth in entropy of the system. Applications to the "Zeno-Watched Pot
Effect" and gravitational interactions are given. Finally, some recent results on
applications to macroscopic coherence with the rf SQUID, where the transition
from quantum to classical behavior could be studied experimentally, are shown.

There is no doubt that the concept of time is absolutely fundamental, be it for
philosophers, physicists or the man and woman on the street. Kant even snubbed
those haughty mathematicians and put time before number, making the intuition
of time the origin of arithmetic [1]. Peculiar to physicists however, and probably
not agreed upon or well understood by these other demographic groups, is that
with the 20th Century lessons of relativity and quantum mechanics behind them,
physicists in their overwhelming majority will agree that time should be viewed as
a physical process. There seems to be no avoiding–even in principle– the need to
understand how the clock works as a physical object.

We are familiar with the slowing down of time in special relativity, perhaps most
easily understood as a transformation of fields due to velocity. The various fields
change in such a way, according to relativity, that any physical process–including
some "clock"– goes slower. Similarly, gravitational fields can affect the clock, as in
the red shift.

While these are interesting enough, there are further and perhaps less familiar
aspects of time as a physical process resulting from quantum mechanics. This will
be the subject of our discussion.

1 Time Evolution as a Coherence Phenomenon

For a clock to work, according to quantum mechanics, it is necessary to have A)
different energy states present, and B) coherence between these states. This might
be easier to understand if we first look at the more familiar situation for space
instead of time. In order to have a highly accurate definition of a point in space we
need, according to quantum mechanics, a well localized wavepacket. This contains a
large spread of momentum states. Indeed if we had only *one* momentum component
the wavefunction would be completely spread out everywhere and in no definite
position at all. So localization in position requires a spread of momentum states.

But this is not enough. In addition these momentum states must be *coherent*. If
one momentum state alone is spread out everywhere, it is only the phases between

the different momentum states that allows them to cancel everywhere but in the vicinity of one point. And so if these states are added *incoherently* the system is still not localized.

And so is it also with energy eigenstates and time. A single energy eigenstate by itself does not tic, it must "beat" or interfere with something else. Typically, with atoms or molecules, we have a second energy eigenstate nearby and we observe the oscillating electromagnetic field resulting from the beating of the two levels. Or we can suppose a K^0-meson clock [2] or neutrino clock using the oscillations in flavor (strangeness, electron-muon number) resulting from the interference of different mass eigenstates. If the two atomic or molecular levels are not coherent, the emitted photons will be "chaotic" and the electromagnetic field has no well-defined phase suitable for a clock; if the two kaon or neutrino states [3] are incoherent there will be no oscillations.

We thus see that in addition to the slowing down of time which is possible by relativistic effects, there is a further aspect to time as physical process: coherence and incoherence.

2 Density Matrix

To deal with such questions of coherence and incoherence in quantum mechanics it is not enough to think about the wavefunction of the system in question. We need a quantity which can describe something like an average over different wavefunctions.

The tool for this in quantum mechanics is the density matrix. The density matrixis easy to understand if we suppose having–in our minds–the complete, big wavefunction of the whole world, or at least that part of it that might interact with what we want to study. What we want to study will in general only be a subsystem of what is described by the big wavefunction and involves only a few of the total number of variables. Let us take the simple case where this subsystem is a two-state system, with basis states ϕ_1, ϕ_2. (One state would be *too* simple.) Then the big wavefunction can be written

$$\Psi = |x_1, x_2.....x_N; 1 > \phi_1 + |x_1, x_2.....x_N; 2 > \phi_2 \, , \tag{1}$$

where the x's are all the variables outside the subsystem, with N a big number like 10^{23}, and the kets $|.... >$ are the wavefunctions for these variables. The "1" or "2" indicate that these in general depend on the state of the subsystem. Now if we only look at the subsystem and not at the x's, that is are only concerned with operators that act on ϕ_1, ϕ_2, then all we need to know is the "reduced" density matrix of the two-state system. This is a 2 x 2 matrix which can be expressed [4] as follows

$$\rho = \begin{pmatrix} < x_1, x_2.....x_N; 1|x_1, x_2.....x_N; 1 > & < x_1, x_2.....x_N; 1|x_1, x_2.....x_N; 2 > \\ < x_1, x_2.....x_N; 2|x_1, x_2.....x_N; 1 > & < x_1, x_2.....x_N; 2|x_1, x_2.....x_N; 2 > \end{pmatrix} \tag{2}$$

where the interpretation of this matrix can be given as:

$$\begin{pmatrix} \text{how much of 1} & \text{coherence between 1 and 2} \\ (\text{coherence between 1 and 2})^* & \text{how much of 2} \end{pmatrix} \, , \tag{3}$$

where of course the "amounts" are probabilities normalized to one: (how much of 1 + how much of $2 = 1$). The justification for calling the off-diagonal elements the degree of coherence between the two states may be seen by considering two extreme cases: let $|x_1, x_2.....x_N; 1 >$ and $|x_1, x_2.....x_N; 2 >$ be either the same or orthogonal. In the first case ϕ can be factored out in Eq. (1) and Ψ is just a product wavefunction; ϕ_1 and ϕ_2 are completely coherent. In the second case no operator working just on the ϕ_1, ϕ_2 subspace can cause the two states to interfere, they are completely incoherent. If we think of the x's as describing an environment in which our subsystem is immersed, we can say that to the degree the subsystem is correlated with the environment, the subsystem is internally incoherent.

Finally, this environment could be a "measuring apparatus". This is something that establishes a correlation–usually a strong one– between the state of the subsystem and the $|x... >$, which now represent the "apparatus". In the limit of very strong correlation we can know what ϕ is just by looking at the $|x... >$, which is of course how a measurement instrument should function. On the other hand, we thus see how the measurement "collapses the wavefunction " of the subsystem: ϕ_1 and ϕ_2 become incoherent [4] when the $|x... >$ states are very different.

3   Thermal State, Stationarity

An easy and limiting case of incoherence to understand is the thermal state. The density matrix operator ρ for a thermal state is proportional to

$$\rho \sim e^{-H/kT} , \tag{4}$$

where H is the Hamilton operator, T the temperature and k the Boltzmann constant. Looking at this expression you might be inclined to see only the usual Boltzmann factor of statistical physics, which it is. However, more important for our discussion is the fact that it is an *operator*. Furthermore this operator is a function of the Hamilton operator only, T and k being just numerical parameters. This means that there is a definite basis in which ρ is diagonal, namely the basis of energy eigenstates $|E >$:

$$\rho_{E,E'} \sim C_E \delta_{E,E'} . \tag{5}$$

This diagonal property has the physical interpretation that different energy eigenstates are *incoherent*. That is, consider the expectation value of some operator $\mathcal{O}$:

$$\overline{\mathcal{O}} = Tr\left[\rho \mathcal{O}\right] \sim \sum_{E,E'} C_E \delta_{E,E'} \mathcal{O}_{E,E'} \sim \sum_E C_E \mathcal{O}_{E,E} , \tag{6}$$

In words, the contributions to $\overline{\mathcal{O}}$ for each energy state are to be added separately and incoherently.

One obvious consequence of this incoherence between energy states is that nothing can be time dependent: as we well know the thermal state is *stationary*. That is, if we consider the time dependence of any operator $\mathcal{O}(t) = e^{iHt}\mathcal{O}e^{-iHt}$ in the thermal state we get

$$\overline{\mathcal{O}}(t) = Tr\left[\rho \, e^{iHt}\mathcal{O}e^{-iHt}\right] = Tr\left[\rho \mathcal{O}\right] = constant , \tag{7}$$

since we can use the fact that e^{iHt} commutes with ρ and the permutation property of the trace to bring the e^{iHt} factors together and cancel them.

Is the thermal state the only state like this? Obviously not, for Eq. (7) we see that the important point is Eq. (5), the diagonality, and not the precise value of the C_E. The above arguments go through for *any* $\rho = f(H)$ and not just Eq. 4. For this reason we wrote a general coefficient C_E and not the Boltzmann factor. So if it is not the thermal state, what is the characterizing feature of the density matrixof Eq. (5)? The property which will be true of any state described by Eq. (5) is that it is *stationary*. Since there is no coherence between different energy states no operator can show a time dependence.

Conversely, if we know on physical grounds that some system is stationary, we know that its energy states are incoherent. An important case of this is the radiation coming from a thermal source. The radiation itself, say neutrinos from the sun, need not follow a Boltzmann distribution. But since the sun (to a very high degree of approximation) is stationary, we know that different energy states are incoherent. Understanding this helped to clear up a long-standing confusion on the treatment of neutrino oscillations [5].

4 Loss of Coherence

Limiting cases like perfect coherence for a pure state, or incoherence of energy states for stationary systems, are easy to understand. However, we are often confronted with the more difficult situation of partial and time-dependent coherence. Typically we imagine the problem of a subsystem in a pure, perfectly coherent, state at $t = 0$; which as time goes on becomes more incoherent as it interacts with the external world. Although this is in principle a perfectly well defined problem: just go ahead, find the total wavefunction at any time and average out the unobserved variables by calculating Eq. (2) —there is no simple way to do this perfectly rigorously for a many-variable system. We must resort to some kinds of plausible approximations or guesses.

A simple and reasonable picture, from which however we can learn a lot, amounts to assuming that the outer world or environment has steady properties with time and there is a loss of coherence at a constant rate for our subsystem. Formally this is like the "golden rule" calculation used in decay problems or scattering theory. For the two state system as described by Eq. (2) we can give a quite complete description of the phenomenology of the problem on this basis and draw a number of interesting conclusions.

5 Parameterization of the Density Matrix ρ

To set up the framework for this approach, we first parameterize the 2x2 matrix for ρ in terms of the familiar Pauli matrices σ:

$$\rho = 1/2 + \mathbf{P} \cdot \sigma , \tag{8}$$

which is appropriately normalized to $Tr\rho = 1$. The two-state subsystem is governed by an internal Hamiltonian which completely describes its behavior in the absence of interaction with the environment. It is also a 2 x 2 matrix which we can parameterize in the same way:

$$H = \mathbf{V} \cdot \sigma , \tag{9}$$

where we have left out a possible constant term which simply represents an overall energy shift and has no interesting effects.

These are just mathematical definitions, but because of the familiar identification of the Pauli matrices with spin, they allow a simple intuitive picture. The 3-vector $\mathbf{P}$ is like a "polarization". According to Eq. (3) its vertical or "z" component represents the relative amounts of the two states "1" or "2". The "transverse" or x-y components, according to Eq. (3) represent the degree of coherence within the subsystem. In generalizing to more than two- state systems, this remains true; the diagonal elements represent the amount of the "quality" in question and the off-diagonal elements the coherence between them. Note that this is necessarily a basis-dependent picture: we must specify which "quality" we are talking about.

6 Isolated Systems

Naturally if the system is isolated and has no interactions with the outside world the situation is very simple. The evolution is governed by the above Hamiltonian and the equation of motion for ρ given by the basic formula

$$i\dot{\rho} = [\rho, H] \tag{10}$$

Using the algebra of the σ this translates into

$$\dot{\mathbf{P}} = \mathbf{P} \times \mathbf{V} . \tag{11}$$

Pictorially, the "polarization" $\mathbf{P}$ precesses around a "pseudo-magnetic field" $\mathbf{V}$. Note that $\mathbf{V}$ can be time dependent if we wish. So far we have not done very much, this is just equivalent to the evolution of a spin in a magnetic field. Observe that with this equation the length of $\mathbf{P}$ cannot change:

$$\frac{dP^2}{dt} = 2\mathbf{P} \cdot (\mathbf{P} \times \mathbf{V}) = 0 . \tag{12}$$

That is, $\mathbf{P}$ rotates without shrinking. This represents the fact that a pure state remains a pure state for an isolated system.

7 Environment

The non-trivial part is now when we try to handle the interaction with the outside world. As already mentioned, if we make the assumption of an outside world with constant properties, interacting repeatedly with our subsystem such that there is a steady loss of coherence, that is a constant rate of decrease of the off-diagonal elements, we can obtain a plausible generalization [6] of Eq. (11). This has the form

$$\dot{\mathbf{P}} = \mathbf{P} \times \mathbf{V} - D\mathbf{P}_T . \tag{13}$$

This involves one new parameter, D. This is the damping or decoherence rate and describes the rate of loss of coherence between the two basis states. $\mathbf{P}_T$ means the

122

"transverse" or (x, y) components of $\mathbf{P}$. If we mentally turn off the first term in the equation, the D term leads to an exponential decay of the overlap or coherence terms in ρ, as in the "golden rule", and this rate is independent of the time when we start. If our subsystem is immersed in a large essentially infinite system it is difficult to imagine any other behavior.

We have made one inessential simplification in writing Eq. (13). In order to concentrate on the interesting interaction between the decoherence and the internal evolution we have left out a possible term $D'P_z$ in which there would be a direct relaxation between the two basis states. For example in the problem of two states of a chiral molecule separated by a barrier [4] we are assuming that the temperature is low enough that there is no direct jumping over the barrier. This is the simple classical relaxation process which can easily be dealt with but is not essentially relevant to the quantum mechanical aspects on which we wish to focus. Similarly, it will be seen that at long times $\mathbf{P}$ in Eq. (13) tends to zero. This may not be exactly correct if the internal Hamiltonian is such that there is a constant energy difference between the two basis states, large compared to an ambient temperature. Then P_z should end up being given by the Boltzmann factor and not be zero. So we implicitly make the assumption that any constant component V_z is small compared to the temperature.

The most important aspect of the presence of the damping term is that now the length of $\mathbf{P}$ can shrink:

$$\frac{dP^2}{dt} = -2D\mathbf{P} \cdot \mathbf{P}_T \neq 0 \,, \tag{14}$$

in general, and so the density matrix can change its "purity". That is, even without explicitly "quality changing" relaxations the length of P will inevitably decrease, if there is a $\mathbf{P}_T$ component. Such a component however, will be produced, in general, by the "rotation" induced by $\mathbf{V}$.

Although we shall explain below how to arrive at a microscopic understanding of D, even the simple phenomenological Eq. (13) allows us to arrive at some interesting conclusions. One of these is the "Zeno Effect".

8   Zeno or 'Watched Pot' Effect

The Zeno or 'Watched Pot' Effect, perhaps first proposed by Alan Turing [4], goes something like this. According to the "Copenhagen School"'s treatment of the "Measurement Problem" (all these things are in "..." because I am not sure they exist), a measurement leaves the state of the system in one particular eigenstate of the measurement. For example after a position measurement, the object is in some definite position–not in a superposition of positions. So we or Turing might say, let us keep rapidly measuring the object's position. It will repeatedly be fixed to the spot and not be able to move at all!

Sometimes this "paradox" is used to say there is something wrong with quantum mechanics or something of the sort. But actually it represents perfectly reasonable quantum mechanical behavior and in fact is just a solution Eq. (13) for the case of large D.

To see this, consider the behavior of Eq. (13) in the case of a constant $\mathbf{V}$, say along the x-axis. If we start $\mathbf{P}$ along the z-axis, it will tend to rotate in a circle in z-y plane: P_z–the probabilities or "amounts" we have defined as represented by P_z will oscillate. Now if we turn on D, the oscillations will be damped and gradually die out. As we make D larger and larger the oscillations will become overdamped and disappear completely. Finally in the limit of very strong damping one finds [4,3]

$$P_z \sim e^{-\frac{V^2}{D}t} . \tag{15}$$

In the limit of large D, P_z is practically frozen and will hardly budge from the spot! According to this formula the time scale, initially set by the oscillation time $1/V$, gets stretched by a factor V/D, which can be enormous when it turns out D is something like the interaction rate in a gas (see formula for D below) and $1/V$ is the tunneling time between two states of a chiral molecule. In this way we gave a second answer to Hund's paradox on the stability of the optical isomers [4] (the first was parity violation [7]). On a practical level this means that in relaxation processes at low temperature , where the classical barrier-hopping type of mechanisms are frozen out and quantum tunneling takes over, we have "anti-intuitive" relaxation, where a bigger D means *slower* relaxation [4].

You might say the repeated "measurements" by the environment have fixed the state a la Zeno-Watched Pot, but in any case it is just a result of a simple solution to Eq. (13) and has nothing mysterious about it.

At this point we should warn of a misunderstanding which sometimes arises. The arguments, or similar arguments leading to Eq. (15) or its equivalent, depend very much on our dealing with a system with two, or in any case a finite number of, levels. They do *not* apply to a continuous density of levels, as for the decay of a quasi-stable state into the continuum. This misunderstanding has occasionally led people to erroneously suggest that the decay of a particle or atomic or nuclear level could be inhibited by observing if it has decayed or not. This is obviously silly. Mathematically the difference between the two cases is that with two or a finite number of levels P_z has a "flat-top" near $t = 0$ in the absence of damping. That is, $P_z \sim 1 - (Vt)^2$ for small times; while for a true decay problem we have $e^{-\Gamma t} \sim 1 - \Gamma t$, a linear behavior. This leads to qualitatively different behavior with respect to "stopping and restarting" the system as we do when we turn on the damping to get the Zeno-Watched Pot behavior.

Another nice way of understanding this (suggested to me by Michael Berry) is to consider the decay process as the tunneling through a barrier, as in the Gamow picture of alpha decay. Now, when there is tunneling through a barrier connecting two *discrete* states, as in the chiral molecule problem [4], the origin of the Zeno-Watched Pot effect may be understood as follows. Tunneling between two discrete states is a very delicate process and depends very much on the quasi-degeneracy of the energy between the two states. With E_{split} the energy splitting , the tunneling probability goes as $\sim (\omega_{tunnel}/E_{split})^2$, with ω_{tunnel} the tunneling energy. Since ω_{tunnel} is typically very small, so must E_{split} be small to have a non-neglible tunneling probability. Now if some external influence is causing the energy levels to move around constantly, even by small amounts, say by some shifting of the two potential wells in the two-state tunneling problem, the degeneracy is constantly being lifted and the tunneling is slowed down. Depending on the ratio of D/ω_{tunnel} it can be practically stopped. This leads to the Zeno-Watched Pot effect. But the situation is entirely different if we have a quasi-bound state decaying into a *continuum* of states, as in the alpha decay situation. Even if the energy of the quasi-bound state is fluctuating, it will always find some continuum state with which it is degenerate. Hence there is no Zeno-Watched Pot effect.

9 Formula for D or the "Unitarity Deficit"

We now come to the microscopic understanding of D, the damping or decoherence rate. Using our general ideas, we can derive a nice formula for this, which furnishes considerable insight into the "decoherence" or "measurement process".

From its definition in Eq. (13) and looking at Eq. (3) we see that D is a rate parameter, one which gives the rate at which the overlap between $|x_1, x_2.....x_N; 2 >$ and $|x_1, x_2.....x_N; 1 >$ is decreasing with time. We take our subsystem to be interacting with the big environment in a constant, steady manner–this is implicit in our assumption of a constant D with no further time dependence. We can model this situation by looking at the subsystem as being bombarded by a constant flux from the environment. This may be thought of as particles, or excitations like phonons or quasiparticles. Our basic idea is that if the two states 1 and 2 of the subsystem scatter this incoming flux in a way which is different for 1 and 2, then the overlap $< x_1, x_2.....x_N; 1|x_1, x_2.....x_N; 2 >$ decreases.

We recall the concept of the S matrix of scattering theory. This is an operator which turns the incoming wavefunction into the outgoing wavefunction:

$$\psi_{out} = S\psi_{in} \ . \tag{16}$$

Now the important point here is that states 1 and 2 may scatter the incoming wavefunction differently, so we have two S matrices, S_1 and S_2 [8]. If these are different, the incoming particle or excitation, which was *uncorrelated* with the state of the subsystem before the scattering, will be correlated with it afterwards. So every scattering decreases the overlap and if we work out what this means for D we get the following formula:

$$D = (flux) \ \mathrm{Im} \ i \ < i|(1 - S_1^\dagger S_2)|i > \ , \tag{17}$$

The expression is proportional to the incoming flux since it is a rate parameter and $|i >$ refers to the incoming state, with a possible average implied if we have many kinds of incoming states. "Im " means imaginary part. There is also a significance to the real part, which is an energy shift of the subsystem induced by the environment [4].

Since usually we have $S^\dagger S = 1$ for the S matrix, the formula says that the decoherence rate is related to the "unitarity deficit" produced by the fact that the different components of the subsystem don't interact the same way with the outside world.

Eq. (17) has two illuminating limits:

$$S_1 = S_2 \qquad\qquad D = 0 \ , \ \ no \ decoherence \tag{18}$$

If both states of the subsystem interact equally with the outer world, $D = 0$, there is no decoherence. This is in accord with our feelings about "measurement". If the "outside" does not respond to the differences in the subsystem there is no "measurement" and no "reduction" of the state. However, we do not need to use this language, the equation stands by itself. Note an important point: interaction with the environment is possible without decoherence.

The other interesting limit occurs if one state, say 1, doesn't interact so $S_1 = 1$ then

$$S_1 = 1 \qquad\qquad D = 1/2 \times (scattering\ rate\ on\ state\ 2) \qquad (19)$$

This result follows from an application of the optical theorem of scattering theory $(cross\ section) \sim Im\ S$ and $Rate = (flux) \times (cross\ section)$. This corresponds to the naive expectation that the decoherence rate is the scattering rate. This is just one limit however, and there can be subtleties such as effects of phases on D [9].

10   The Unitarity Deficit and Entropy Production

There is a classical conundrum which says, in apparent contradiction to common sense, that the entropy of a isolated system cannot increase. This is particularly simple to show in quantum mechanics where the definition of entropy is $-Tr[\rho\,ln\rho]$. Since for an isolated system the evolution of ρ is governed by a Hamiltonian, we get for the time dependence of the entropy, just as in Eq. (7), $Tr\left[e^{iHt}\rho\,ln\rho e^{-iHt}\right] = Tr\left[\rho\,ln\rho\right] = constant$. This argument says that as long as the time dependence of ρ is governed by a Hamiltonian, the entropy is constant.

This constancy would also apply to the entropy $-Tr[\rho\,ln\rho]$ of our two-state system where ρ is given by Eq. (8) if it were isolated –not coupled to the environment. Or as we learn from the argument, it could even be coupled to the environment; but in such a way that the evolution of ρ is given by a Hamiltonian. However we see from Eq. (17) that the coupling to the environmnet is not governed by a single Hamiltonian but rather by two Hamiltonians, giving the two S matrices. If we had one Hamiltonian we would have $S_1 = S_2$, in which case $D = 0$, and there is no decoherence.

Hence there is a connection between D and the entropy increase of the two state system. In fact diagonalizing Eq. (8) and taking the trace, we find for the entropy $-Tr\left[\rho\,ln\rho\right] = ln2 - \frac{1}{2}\big((1+P)ln(1+P) + (1-P)ln(1-P)\big)$. Taking the time derivative, we find for the entropy change of the two state system

$$\frac{d(Entropy)}{dt} = -\dot{P}P\frac{1}{2P}ln\frac{(1+P)}{(1-P)} = (D\mathbf{P}\cdot\mathbf{P_T})\frac{1}{2P}ln\frac{(1+P)}{(1-P)} \approx D\mathbf{P}\cdot\mathbf{P_T}\ , \quad (20)$$

where we used Eq. (??). The $\approx$ refers to the limit of small P. It seems intuitively clear that the rate of entropy increase and the decoherence rate should be closely related and Eq. (20) expresses this quantitatively. By appropriate generalization of the various quantities, this could be extended to systems larger than just two states. Note $\mathbf{P}\cdot\mathbf{P_T} = \mathbf{P}_T^2$ is necessarily positive. Furthermore, in thermal equilibrium where there is no coherence between states of different energy i.e $\mathbf{P}_T = 0$, there is no entropy production.

11   Decoherence in Mesoscopic Devices

In recent years our subject has moved from the theoretical- philosophical to the domain of the almost practical with the realization of quantum behavior for large, essentially macroscopic devices–"mesoscopic systems". This has been given additional impetus in view of the possible use of such devices for the implementation of the "quantum computer". Furthermore the subject is interesting in connection with the idea–to my mind wrong–that there might be some limit where large objects don't obey the rules of quantum mechanics. Decoherence is of course the main question for the observability of such effects and their possible practical use.

One of the devices that has been studied in detail is the rf SQUID, where by suitable adjustment of parameters it is possible to bring the system into the configuration of the familiar double potential well separated by a tunneling barrier. The "x" coordinate stands for the flux in the superconducting ring, and the system obeys–in the absence of decoherence– a Schroedinger equation in this variable. The states localized in one of the potential wells represent the supercurrent flowing in a given sense around the ring, with a very large number (microamps) of electrons reversing direction when we go from one well to the other. Creation of quantum linear combinations of these states, which can occur by tunneling, would certainly be impressive evidence for the general applicabilty of quantum mechanics. Some beautiful experiments [10] [11] using microwave technique have seen evidence for such combinations in SQUID systems.

We have suggested [12] a particularly simple way to both see the quantum linear combination of the two states and to measure the decoherence time of the system, hopefully directly sensitive only to the true quantum decoherence. It involves the idea of "adiabatic inversion". This occurs when a slowly varying external field can cause a quantum system to reverse its state, as when a spin "follows" a rotating magnetic field and goes from up to down. This phenomenon is also often refered to as a "level crossing". It is an intrinsically quantum mechanical phenomenon and, –important for us– is hindered when the decoherence time is short compared to the time in which the inversion takes place.

We propose to produce such an inversion in the SQUID by sweeping an external field and then observing the reversal of the direction of the flux in the SQUID. Note that the system need not be "observed" until the procedure is over—our method is "noninvasive". When the sweep is faster than the decoherence time the inversion should take place, and when it is slower it should be inhibited. We are witnessing the transition from quantum (tunneling allowed) to classical (tunneling forbidden) behavior as the decoherence is increased.

Going from fast to slow sweeps, the sweep time where the inversion begins to become unsuccesful thus gives a determination of the decoherence time. A possible difficulty here is that the sweep cannot be too fast, otherwise the procedure becomes non-adiabatic. However, our estimates indicate that a region of SQUID parameters and temperature should exist where fast-enough sweeps are possible without violating adiabaticity.

In order to study these points in more detail we have developed numerical simulations of such systems, both for the study of the adiabatic inversion (the logical NOT) as well as for a two-SQUID system operating as a CNOT quantum logic gate [13]. In Fig 1 we show the results of a simulation [14] for the one-SQUID adiabatic inversion. The decoherence time $1/D$ was chosen to be about 39 000 units and simulated as a random flux noise. The SQUID parameters were $\beta = 1.19, L = 400\,pH, C = 0.1pF$, giving a time unit of $6.3 \times 10^{-12}s$ and so $1/D = 0.25\mu s$. This decoherence time would correspond to about $T = 50\,mK$ in the estimate $D = T/(e^2 R)$, with $R = 5M\Omega$ [12]. The simulation included the first 8 quantum levels of the SQUID so that possible effects of non- adiabaticity are taken into account.

The vertical axis in Fig 1 shows the probability for finding the flux in the ring reversed after the sweep. We see that while the inversion is essentially always successful for sweep times less than decoherence time, it becomes progressively less so for longer sweeps. Hence we are seeing the transition from quantum towards classical behavior, and a measuremnt of when this takes place furnishes a determination of the decoherence time. The gradual fall-off seen after the completion of the longest sweep is indicative of another difficulty, relaxation. Our final states will in general not be distributed according to thermal equilibrium, and the final detection should take place quickly on the time scale for relaxation.

12   Decoherence and Gravity

Although the above ideas are rather simple and phenomenological, they can be applied to a wide variety of interesting problems, where of course the various quantities like $\mathbf{V}$ and S_1, S_2 must be adapted to each particular case. These range from understanding the permanence of optical or "chiral" isomers (Hund's paradox) [4], to the study of neutrino oscillations [3], to understanding decoherence in mesoscopic devices [9], [12], and the design of quantum logic gates [13].

Here, however, I would like to conclude in a more speculative vein, coming back to the question of the need for coherence between energy states to define time. Is the loss of coherence always a more or less accidental happening, depending on the particulars of the situation? Or is there something universal about it?

There is of course one kind of interaction which universally and inevitably couples to energy states and which in principle "measures" or distinguishes them: gravity [15]. Gravity couples to mass/energy and so states of different mass/energy interact differently. Indeed one can derive the gravitational redshift by considering the K^0 meson "clock" [2].

For this clock, we have two mass/energy eigenstates and interference effects between them tell time. By the same token this must imply some degree of decoherence between the different mass/energy states due to gravitational interactions. There ought to be an imaginary or dispersive part to the redshift. Naturally because the coupling, Newton's constant, is so small we expect the effects to be negligible under everyday conditions. It is nevertheless amusing to see what the formulas look like. We shall incidentally find that there is in fact an "everyday" application, involving the passage of oscillating neutrinos near galaxies.

Qualitatively, we expect in regions of large and rapidly varying gravitational fields that different energy states become incoherent. Since gravity couples univer-

128

sally, this cannot be dismissed as an incidental effect. It will affect all energy states, and thus time equally for all processes. If the effects become strong, time in some sense becomes ill-defined or stands still.

We try to model the decoherence due to gravity by calculating D for the K^0 clock. There are two interfering mass eigenstates and an environment interacting gravitationally with these two states. If we take this environment to be a flux of particles, we can use our formula Eq. (17). The calculation of the S matrices is (non relativistically) the same as for Coulomb scattering, where the "charge" is the mass. The two components of the clock have different masses M_1, M_2 and so S_1, S_2 are different and thus there is a contribution to D. In an impact parameter representation we have

$$D = (flux) \int 2\pi b \, db \, Im \, i(1 - S^\dagger_{M_1}(b) S_{M_2}(b)) \tag{21}$$

with $S(b) = e^{2i\delta(b)}$ and

$$\delta(b) = 2 \int_0^{l_{max}} \alpha \frac{dl}{\sqrt{l^2 + b^2}} \, , \tag{22}$$

where $\alpha = GEM/v$, with G the Newton constant and E the energy of the incoming particle. As is usual in Coulomb problems there is a divergence at large distances, and so we introduce a large distance cutoff l_{max} which must be interpreted according to the physical conditions. Taking the two S's and the imaginary part we finally get

$$D \approx (flux) l_{max}^2 (\Delta\alpha)^2 \, , \tag{23}$$

with $\Delta\alpha = G(\Delta M)E/v$, which we have taken to be small and expanded to lowest order. ΔM is $(M_1 - M_2)$. We can try to estimate this under thermal conditions, as in the early universe, where all dimensional quantities are related to the temperature T. We end up with

$$D \approx T^3(G\Delta M)^2 = T^3 \left(\frac{\Delta M}{M_{planck}^2} \right)^2 . \tag{24}$$

As would have been expected, T must be at least on the order of the Planck scale for this to be significant. And then the decoherence rate is $\Delta M(\Delta M/M_{planck})$, which is still small, unless we are considering ΔM also on the Planck scale. The result Eq. (24) is of order $1/M_{planck}^4$ since decoherence, as explained, is a unitarity effect and so comes in to second order in the coupling. Although conceptually interesting this seems very remote from any potentially observable effects.

Nevertheless there is, as promised, a "present-day" application. We now know there are neutrino oscillations. So possibly if the time scale for the release of the neutrinos in the Big Bang were short compared to their oscillation time (then), they would still be oscillating in step today. Instead of the K^0 clock we have a neutrino clock. A beautiful clock–if we could ever read it–a precise timepiece going back to the very first minutes of the Big Bang. But there's a difficulty: in traveling to us the neutrinos will pass near various mass and gravitational field inhomogeneities, like galaxies, and these will tend to mix up the phases. Because of the huge mass of a galaxy, $\Delta\alpha$ will no longer be small. One finds [15], with typical mass parameters for neutrinos and galaxies, that essentially any neutrino traveling through or near a galaxy will be decohered. Unfortunately, there will be no oscillations to read. So much for the Big Bang Clock, but at least there's a "practical" application of decoherence theory involving gravity.

Note that these ideas, based on the different interaction of different mass/energy states, do not contradict the classical equivalence principle according to which different masses move in the same way. The interaction is so cleverly constructed that while different masses follow the same path classically, they get different quantum phases [2], and these are essentially what concern us here. The interplay between the classical and quantum is subtle and interesting. The last has certainly not been said on these questions.

References

1. *What is Mathematics Really?* R. Hersh, Oxford (1997).
2. Matter and Light Wave Interferometry in Gravitational Fields, L.Stodolsky, Gen. Rel. Grav. **11**, 391 (1979).
3. On the Treatment of Neutrino Oscillations in a Thermal Environment, L.Stodolsky, Phys. Rev. D 36(1987)2273 and chapter 9 of G.G. Raffelt, *Stars as Laboratories for Fundamental Physics* (Univ. Chicago Press, 1996).
4. Two Level Systems in Media and 'Turing's Paradox', R.A. Harris and L.Stodolsky, Phys. Let. B 116(1982)464. Quantum Damping and Its Paradoxes, L.Stodolsky, in *Quantum Coherence*, J. S. Anandan ed. World Scientific, Singapore (1990).
5. When the Wavepacket is Unnecessary, L.Stodolsky, Phys. Rev. **D58** 036006, 1998 and www.arxiv.org, hep-ph/9802387.
6. See ref 4, and for a more formal derivation under certain assumptions, focused however on neutrino applications: "Non-Abelian Boltzmann Equation for Mixing and Decoherence" G. Raffelt, G. Sigl and L.Stodolsky, Phys Rev Let. 70(1993)2363.
7. Quantum Beats in Optical Activity and Weak Interactions, R.A. Harris and L.Stodolsky, Phys Let B78 (1978) 313.
8. The sophisticated reader will recognized that although a very simple and intuitive picture arises through the use of the S matrix with distinct scattering events, one could reach a similar result by using a hamiltonian with a continuous time evolution and depending on a quantum number with values "1" and "2".
9. Measurement Process In a Variable-Barrier System, L.Stodolsky, Phys. Lett. **B459** 193, (1999).
10. J. R. Friedman, V. Patel, W. Chen, S. K. Tolpygo, and J. E. Lukens, Quantum Superposition of Distinct Macroscopic States, Nature **406** 43, (2000).
11. I. Chiorescu, Y. Nakamura, C. J. P. M. Harmans, and J. E. Mooij, Coherent Dynamics of a Superconducting Flux Qubit, Science **299** 186 (2003).
12. Study of Macroscopic Coherence and Decoherence in the SQUID by Adiabatic Inversion, Paolo Silvestrini and L.Stodolsky, Physics Letters **A280**

17-22 (2001);www.arxiv.org, cond-mat/0004472. Also Adiabatic Inversion in the SQUID, Macroscopic Coherence and Decoherence, Paolo Silvestrini and L.Stodolsky, *Macroscopic Quantum Coherence and Quantum Computing*, pg.271, Eds. D. Averin, B. Ruggiero and P. Silvestrini, Kluwer Academic/Plenum, New York (2001) www.arxiv.org, cond-mat/0010129.

13. Design of Adiabatic Logic for a Quantum CNOT Gate, Valentina Corato, Paolo Silvestrini, L.Stodolsky, and Jacek Wosiek, www.arxiv.org, cond-mat/0205514, Physics Letters **A309** 206 (2003). Adiabatic Evolution of a Coupled- Qubit Hamiltonian, V.Corato, P. Silvestrini, L.Stodolsky, and J. Wosiek cond-mat/0310386; Physical Review **B 68**, 224508 (2003).

14. Valentina Corato, Paolo Silvestrini, L.Stodolsky, and Jacek Wosiek, to be published. We use a C program developed from the work of J. Wosiek by summer students at the Max-Planck-Institute, A.T. Goerlich and P. Korcyl of the Jagellonian University.

15. Decoherence Rate of Mass Superpositions, L.Stodolsky, Acta Physica Polonica **B27**,1915(1996).

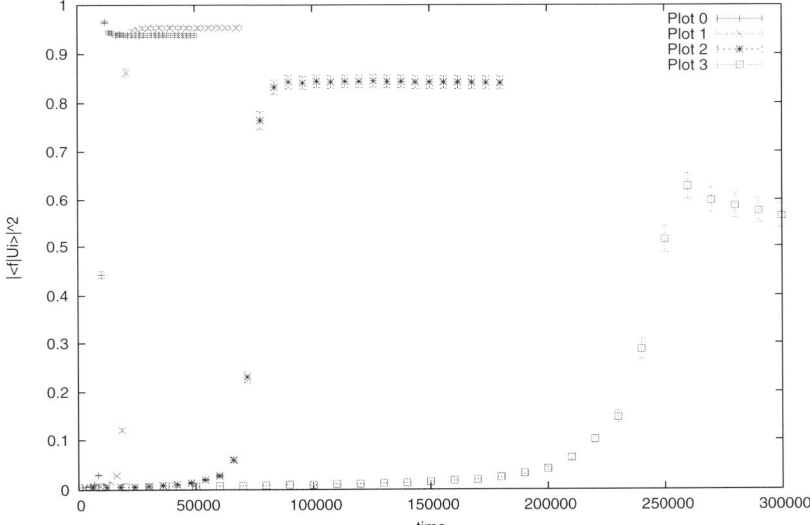

Figure 1. Simulations of adiabatic inversion in the rf SQUID for sweep times long and short compared to the decoherence time. The vertical axis is the probability for finding the flux in the SQUID reversed. The simluation parameters have been chosen so that decoherence time is about $1/D = 39\,000$. For sweep times less than $1/D$ the inversion is essentially completely succesful, while for longer sweep times it becomes less and less so . An experiment of this type would show the transition from quantum to classical behavior and permit a determination of the decoherence time. For the SQUID parameters given in the text the time interval corresponds to $6.3 \times 10^{-12}s$, so the decoherence time is about $0.25\,\mu s$.

DECOHERENCE, CHAOS AND THE SECOND LAW

W.H. ZUREK

Theory Division, LANL, Mail Stop B210,Los Alamos, New Mexico 87545, U.S.A.
E-mail: whz@santafe.edu

There is no write-up of this talk. However the reader can find very similar material discussed in: W.H. Zurek, *Rev. of Mod. Physics* **75** (2003) 715.

WHAT COULD WE HAVE BEEN MISSING WHILE PAULI'S THEOREM WAS IN FORCE?

ERIC A. GALAPON

Theoretical Physics Group, National Institute of Physics
University of the Philippines, Diliman, Quezon City, 1101 Philippines
E-mail: egalapon@nip.upd.edu.ph

Pauli's theorem asserts that the canonical commutation relation $[T, H] = iI$ only admits Hilbert space solutions that form a system of imprimitivities on the real line, so that only non-self-adjoint time operators exist in single Hilbert quantum mechanics. This, however, is contrary to the fact that there is a large class of solutions to $[T, H] = iI$, including self-adjoint time operator solutions for semi-bounded and discrete Hamiltonians. Consequently the theorem has brushed aside and downplayed the rest of the solution set of the time-energy canonical commutation relation.

1 Introduction

Time is widely recognized as a parameter in quantum mechanics, and no one emphatically asserts this conviction in recent times more than Sakurai's assertion in his well known textbook—"The first important point we should keep in mind is that time is just a parameter in quantum mechanics, not an operator. In particular, time is not an observable. It is nonsensical to talk about the time operator in the same sense as we talk about the position operator."[1] On the other hand, the scalar status of time has been seen as a weakness of quantum theory. Von Neumann has much earlier categorically expressed this view—"First of all we must admit that this objection [time being just a number] points at an essential weakness which is, in fact, the chief weakness of quantum mechanics. In fact, while all other quantities are represented by operators, there corresponds to time an ordinary number-parameter t, just as in classical mechanics."[2]

But how did *time* acquire its notorious parametric label? If time was an observable represented by a self-adjoint operator, then this operator would be canonically conjugate with the Hamiltonian, in accordance with quantum dynamics. But as early as 1933, Pauli[3] has "shown" that the existence of a self-adjoint time operator canonically conjugate with a Hamiltonian implies that the time operator and the Hamiltonian have completely continuous spectra spanning the entire real line, a result widely known as Pauli's theorem. Thus for a semibounded or discrete Hamiltonian, no self-adjoint time operator would exist. Since physical systems are assumed to have a stable ground state, the Hamiltonian will generally be semibounded. Pauli then concluded that any "attempt of introducing time as an operator in quantum mechanics must be fundamentally abandoned, and that the time t in quantum mechanics has to be regarded as an ordinary number." From then on, it has been believed that no self-adjoint time operator exists in quantum mechanics,[1,3,4,5,6,7,8,9,10,11,12,13,14,15] and has been widely recognized that time is a parameter merely serving to mark the evolution of a quantum system.[1,17]

However, it is likewise widely recognized that time undoubtedly acquires dynamical significance in questions involving the occurrence of an event,[4,13,16] e.g. when a nucleon decays,[18] or when a particle arrives at a given spatial point,[19,20] or when a particle emerges from a potential barrier.[21] Moreover, there is the time-energy uncertainty principle begging for an interpretation, a reasonable interpretation of which requires more than a parametric treatment of time.[4,22,23,24,25,26,27,28,29] It has then become part of the physicist's common sense that if one acknowledges the legitimacy of these problems and attempts to find quantum mechanical solutions to them, then one must solve them without contradicting Pauli's theorem. One recourse is to abandon the standard framework of quantum mechanics, and build a framework that can support the introduction of time operators or that can accommodate the temporal aspects of quantum mechanics.[18,30,31,32]

Another recourse is to stick with the single Hilbert space formulation; but it is acknowledged that this cannot be done without a compromise: *If one imposes self-adjointness as a desirable requirement for any observable, then one necessarily has to abandon the requirement that time operator be conjugate to the Hamiltonian; if, on the contrary, one decides that any proper time operator must be strictly conjugate to the Hamiltonian, then one has to renounce the search for a self-adjoint operator.*[33] In recent years, the problem of introducing time in quantum mechanics has taken the later route, abandoning self-adjointness in favor of the canonical commutation relation. And this specifically calls for extending quantum observables to maximally symmetric but not necessarily self-adjoint operators; in which case, quantum observables are generally positive operator valued measures.[5,9,10]

However, we have shown recently that Pauli's theorem does not hold in single Hilbert spaces, and thus there is no a priori reason to exclude the existence of self-adjoint operators canonically conjugate to a semibounded Hamiltonian, contrary to the claim of Pauli.[34] Moreover, we have explicitly proved that to every discrete, semibounded Hamiltonian with compact inverse there exists a characteristic self-adjoint time operator conjugate to the Hamiltonian in a dense subspace of the system Hilbert space.[35] Furthermore, we have likewise shown that the non-self-adjoint free time of arrival operator in unbounded space, generally considered as an explicit demonstration of Pauli's theorem, defines a class of bounded, self-adjoint, and canonical time operator for a spatially confined particle.[36]

These results inevitably force us to reassess our opinions on the role of time in single Hilbert space quantum mechanics and what constitutes a quantum time operator. To this end, we tackle in this paper two related issues that have been grossly neglected while Pauli's theorem was in force. First, we point out the existence of a large class of solutions to the canonical commutation relation in a Hilbert space, and argue that each solution can be identified as solution to a specific physical problem, say, to one facet of the quantum time problem. Second, we demonstrate the existence of multiple self-adjoint time operator solutions to the time-energy canonical commutation relation for a given discrete, semibounded Hamiltonian. Our discussion is focused on the interplay between these two in explicating self-adjoint time operators in single Hilbert space quantum mechanics.

The paper is then organized as follows. In Sect.2 we revisit Pauli's theorem. In Sect.3 we discuss the basic properties of the Hilbert space solutions to the canonical commutation relation (CCR). In Sect.4 we address the question on the physical relevance of the different solutions to the CCR. In Sect.5 we illustrate that there may be more than one self-adjoint time operator corresponding to a given Hamiltonian. Finally, in Sect.6 we synthesize our discussion on the previous two sections, and address the issue of POVM-time observables.

2 Pauli's Theorem: It's Traditional and Modern Readings

Pauli's argument goes as follows. Assume that there exists a self-adjoint time operator T, i.e an operator canonically conjugate with the Hamiltonian H, $[T, H] = iI$. Since T is self-adjoint, the operator $U_\epsilon = \exp(-i\epsilon T)$ is unitary for all real numbers ϵ. Now let φ_E be an eigenvector of H with eigenvalue E, then, according to Pauli, we have the implication

$$[T, H] = iI \longrightarrow U_\epsilon \varphi_E = \varphi_{E+\epsilon}, \tag{1}$$

which further implies that H has a continuous spectrum spanning the entire real line because ϵ is an arbitrary real number. Hence, Pauli concluded, if the Hamiltonian is semibounded or discrete, no self-adjoint time operator T will exist to satisfy $[T, H] = iI$, otherwise, the operator U_ϵ will map the discrete or semibounded spectrum of H into the entire real line, which is not possible for unitary U_ϵ.

A modern interpretation of the theorem is that if P_E is the spectral decomposition of the (self-adjoint) Hamiltonian, then we have the implication

$$[T, H] = iI \longrightarrow U_\epsilon P_E U_\beta^\dagger = P_{E+\epsilon}. \tag{2}$$

By reversing the roles of the Hamiltonian and the time operator in Pauli's argument, one gets similar conclusion about the properties of the time operator. Specifically, if a self-adjoint time operator exists such that its spectral decomposition is P_T, then we have the similar implication

$$[T, H] = iI \longrightarrow V_t P_E V_t^\dagger = P_{T+t}, \tag{3}$$

where $V_t = \exp(-itH)$, for all real t. The right-hand sides of the last two alleged implications identify the Hamiltonian and the time operator pair as forming a system of imprimitivities over the real line. Of course it is well-known that a system of such pairs has the property that the operators have continuous spectra taking values in the entire real line. If one upholds the validity of the above implications, then one recovers Pauli's original conclusion for semibounded or discrete Hamiltonians—no self-adjoint time operator exists.

These readings of Pauli's theorem imply that self-adjointness and canonicality of a time operator cannot be imposed simultaneously: If canonicality is required, self-adjointness has to be renounced; on the other hand, if self-adjointness is required, canonicality has to be renounced. Moreover, they give the impression that any time operator is not only canonically conjugate with the Hamiltonian but *must* also be a generator of energy shifts.

However, we have recently shown and demonstrated that Eq.(1) does not hold in Hilbert space, and have argued that there are no Eqs.(2) and (3)[34]. These explicitly belie the belief that an operator canonically conjugate with a semibounded Hamiltonian is necessarily a generator of energy shifts and cannot be self-adjoint.

They inevitably put into question the traditional reading of what is a time operator in quantum mechanics. If there were no Implications-1, 2, and 3, then is there any justification in requiring the time operator and Hamiltonian pair to form a system of imprimitivies, as has been since Pauli, or is it just enough to require that time operators be canonically conjugate with the Hamiltonian?

3 Canonical Pairs in Hilbert Spaces

We cannot answer the above questions without a clear understanding of the properties of a canonical pair in a Hilbert space. It has been the lack of understanding of these properties that led many to numerous false conclusions and unwarranted generalizations concerning the existence and non-existence of self-adjoint time operators. To the physicist, a canonical pair is a pair of operators (Q, P) satisfying the canonical commutation relation, $[Q, P] = i\mathsf{I}$, (CCR). Much of the inferred properties of Q and P have been derived from formal manipulations and have been assumed to hold in Hilbert spaces. Unfortunately, these inferences are generally valid only under some strict, unstated conditions, which may exclude the assumed range of validity of the inferred properties. So the first step to a better perspective on the quantum time problem is to understand the properties of the CCR in Hilbert spaces. We stick with the basics.

If we seek a pair of Hilbert space operators, Q and P,[a] satisfying the CCR, then two things must be borne in mind. First, no pair (Q, P) exists to satisfy the CCR in the entire $\mathcal{H}$; that is, there are no Q and P such that $[Q, P]\varphi = i\varphi$ for all φ in $\mathcal{H}$, or $[Q, P] = iI_{\mathcal{H}}$, where $I_{\mathcal{H}}$ is the identity in $\mathcal{H}$. Had such a pair existed, the operators Q and P would have to be everywhere defined or bounded; and the CCR would lead to $[Q, P^n] = inP^{n-1}$, which, upon taking the norm of both sides, yields the inequality $2 \|Q\| \|P\| \geq n$ for all $n > 1$, which is a contradiction for bounded Q and P, and for arbitrarily large integer n. This is Weidlant's theorem. A pair (Q, P) then can at most satisfy the CCR in a—proper—subspace, $\mathcal{D}_c$, of $\mathcal{H}$; that is, the relation $[Q, P]\varphi = i\varphi$ holds only for all those φ in $\mathcal{D}_c$, where $\mathcal{D}_c$ is always smaller than $\mathcal{H}$. Thus a canonical-pair in a Hilbert space is a triple $\mathcal{C}(Q, P; \mathcal{D}_c)$—a pair of Hilbert space operators, Q and P, together with a non-trivial, proper subspace $\mathcal{D}_c$ of $\mathcal{H}$, which we shall hereafter refer to as the canonical domain.

Second, there are canonical pairs in the *same* Hilbert space that are not unitarily equivalent, or pairs with distinct properties, say, spectral properties. This can be best appreciated by giving an example. Let $\mathcal{H} = L^2(-\infty, \infty)$, the Hilbert space of square integrable complex valued functions in the real line. The pair of operators

$$(Q_1\varphi)(q) = q\varphi(q), \quad (P_1\varphi)(q) = -i\varphi'(q),$$

together with the dense subspace $\mathcal{D}_c^1 \subset \mathcal{H}$, consisting of all infinitely differentiable complex valued functions with compact support, forms the canonical pair $\mathcal{C}_1(Q_1, P_1; \mathcal{D}_c^1)$; moreover, Q_1 and P_1 are essentially self-adjoint in $\mathcal{D}_c^1$. Also the pair of operators[40]

$$(Q_2\varphi)(q) = q\varphi(q) + \varphi(q + i\sqrt{2\pi}), \quad (P_2\varphi)(q) = -i\varphi'(q) + e^{-\sqrt{2\pi}q}\varphi(q),$$

---

[a] A Hilbert space operator A is more accurately denoted by $A : \mathcal{D}_A \subseteq \mathcal{H} \mapsto \mathcal{H}$, where $\mathcal{D}_A$ is the domain of A in $\mathcal{H}$.

together with the dense subspace $\mathcal{D}_c^2 \subset \mathcal{H}$, consisting of the linear span of $q^n e^{-rq^2+cq}$, with $n = 0, 1, 2, \ldots$, $r > 0$, and c a complex number, forms the canonical pair $\mathcal{C}_2(Q_2, P_2; \mathcal{D}_c^2)$; moreover, Q_2 and P_2 are likewise essentially self-adjoint in $\mathcal{D}_c^2$. Now the pairs $\mathcal{C}_1$ and $\mathcal{C}_2$ reside in the same Hilbert space $\mathcal{H}$, yet they do not share the same properties. For one, the self-adjoint extensions of Q_1 and P_1, $\bar{Q}_1$ and $\bar{P}_1$, satisfy the Weyl relation $U(s)V(t) = e^{ist}V(t)U(s)$, for all real numbers s and t, where $V(t) = \exp(it\bar{Q}_1)$ and $U(s) = \exp(is\bar{P}_1)$. On the other hand, the self-adjoint extensions of Q_2 and P_2 do not satisfy the same relation. This means that $\mathcal{C}_1$ and $\mathcal{C}_2$ are two distinct canonical pairs in $\mathcal{H}$.

Clearly there could be numerous distinct canonical pairs in a given Hilbert space $\mathcal{H}$. We shall refer to each pair as a Hilbert space *solution*, or simply a solution, to the CCR. Generally solutions split into two major *categories*, according to whether the canonical domain $\mathcal{D}_c$ is dense or closed. We will call a canonical pair of *dense-category* if the corresponding canonical domain is dense; otherwise, of *closed-category* if the corresponding canonical domain is closed. Solutions under these categories further split into distinct *classes* of unitary equivalent pairs, and each class will have each own set of properties. Under such categorization of solutions the CCR in a given Hilbert space $\mathcal{H}$, assumes the form $[Q, P] \subset i\mathcal{P}_c$, where $\mathcal{P}_c$ is the projection operator onto the closure $\bar{\mathcal{D}}_c$ of the canonical domain $\mathcal{D}_c$. If the pair $\mathcal{C}$ is of dense category, then the closure of $\mathcal{D}_c$ is just the entire $\mathcal{H}$, so that $\mathcal{P}_c$ is the identity $I_{\mathcal{H}}$ of $\mathcal{H}$. One should immediately recognize that we are considering a more general solution set to the CCR than has been considered so far. The traditional reading of the CCR in $\mathcal{H}$ is the form $[Q, P] \subset iI_{\mathcal{H}}$, which is just the dense category.

The pair $\mathcal{C}_1$ above and all its unitary equivalents are then canonical pairs of dense categories. These pairs satisfy the Weyl relation and are unbounded with completely continuous spectrum taking values in the entire real line. On the other hand, the pair $\mathcal{C}_2$ and all its unitary equivalents are canonical pairs of dense categories as well, but they do not satisfy the Weyl relation. These later pairs have different spectral properties from the former. Clearly, these sets of pairs belong to different classes. They are not unitarily equivalent, and they represent two distinct classes of solutions of dense categories to the canonical commutation relation. Later we will give an example of a canonical pair of closed category in relation to the quantum time problem.

A question immediately arises—Is there a preferred solution to the CCR? That is, should we accept only solutions of dense or closed category of a specific class?

4 An Answer from the Position-Momentum Canonical Pairs

Let us refer to the well-known position and momentum operators in three different configuration spaces: The entire real line, $\Omega_1 = (-\infty, \infty)$; the bounded segment of the real line, $\Omega_2 = (0, 1)$; and the half line $\Omega_3 = (0, \infty)$. Quantum mechanics in each of these happens in the Hilbert spaces $\mathcal{H}_1 = L^2(\Omega_1)$, $\mathcal{H}_2 = L^2(\Omega_2)$, $\mathcal{H}_3 = L^2(\Omega_3)$, respectively. The position operators, Q_j, in $\mathcal{H}_j$, for all $j = 1, 2, 3$, arise from the fundamental axiom of quantum mechanics that the propositions for the location of an elementary particle in different volume elements of Ω_j are compatible (see

Jauch[37] for a detailed discussion for Ω_1, which can be extended to Ω_2 and Ω_3). They are self-adjoint and are given by the operators $(Q_j\varphi)(q) = q\varphi(q)$ for all φ in the domain $\mathcal{D}_{Q_j} = \{\varphi \in \mathcal{H}_j : Q_j\varphi \in \mathcal{H}_j\}$. Note that Q_1 and Q_3 are both unbounded, while Q_2 is bounded.

Now each of the configuration spaces, Ω_1, Ω_2, and Ω_3, has an identifying property. Ω_1 is fundamentally homogeneous—points in there are physically indistinguishable. On the other hand, Ω_2 and Ω_3 are not homogeneous, the boundaries being the distinguishing factor. However, their inhomogeneities are not the same: Ω_2 has two boundaries, while Ω_3 has one. These properties can be expressed mathematically in terms of the respective representation of translation in each of these configuration spaces. Translation in Ω_1 is isomorphic to the additive group of real numbers; in Ω_2, to the group of rotations of the circle[b]; in Ω_3, to the semigroup of additive positive numbers. Thus in $\mathcal{H}_1$ and $\mathcal{H}_2$ there are one parameter unitary operators $\mathcal{U}_1(s)$, $\mathcal{U}_2(s)$ representing translations in $\mathcal{H}_1$ and $\mathcal{H}_2$, respectively. And in $\mathcal{H}_3$ there is a completely one-parameter semigroup $\mathcal{U}_3(s)$ representing translations.

The properties of the three configuration spaces can now be explicitly stated in the following respective forms

$$\Omega_1 : \quad \left(\mathcal{U}_1^\dagger(s)\varphi\right)(q) = \varphi(q - s), \quad \text{for all } s \in \Re,$$

$$\Omega_2 : \quad \left(\mathcal{U}_2^\dagger(s)\varphi\right)(q) = \begin{cases} \varphi(q - s) & : \quad \text{for } 1 > q > s > 0 \\ \varphi(1 + (q - s)) & : \quad \text{for } 1 > s > q > 0 \end{cases},$$

$$\Omega_3 : \quad \left(\mathcal{U}_3^\dagger(s)\varphi\right)(q) = \begin{cases} \varphi(q - s) & : \quad \text{for } q > s \\ 0 & : \quad \text{for } q < s \end{cases}.$$

If we define the momentum operator as the generator of translation in the configuration space, then the momentum operator in $\mathcal{H}_j$ is the operator P_j defined on all vectors φ for which the limit $\lim_{s\to 0}(is)^{-1}(\mathcal{U}_j(s) - \mathcal{I}_\mathcal{H})\varphi = P_j\varphi$ exists. Explicitly, it is given by $(P_j\varphi)(q) = -i\varphi'(q)$.

In each $\mathcal{H}_j$, there exists a dense common subspace $\mathcal{D}_j$ of Q_j and P_j, which is invariant under Q_j and P_j, for which we have the canonical pair $\mathcal{C}_j(Q_j, P_j; \mathcal{D}_j)$. The $\mathcal{C}_j$'s are of the same dense category, but they belong to different classes: Q_1 and P_1 are both self-adjoint, having absolutely continuous spectra spanning the entire real line $\Re$ and forming a system of imprimitivities in $\Re$, and their restrictions in $\mathcal{D}_1$ are essentially self-adjoint. Q_2 is self-adjoint with an absolutely continuous spectra in (a, b), and its restriction in $\mathcal{D}_2$ is essentially self-adjoint; P_2 is self-adjoint with a pure point spectrum, but its restriction in $\mathcal{D}_2$ is not essentially self-adjoint. Q_3 is self-adjoint with an absolutely continuous spectrum in $(0, \infty)$, and its restriction in $\mathcal{D}_3$ is essentially self-adjoint; P_3 is maximally symmetric and non-self-adjoint, thus without any self-adjoint extension. These varied properties of the position and momentum canonical pairs are obviously the consequences of the underlying properties of their respective configuration spaces.

---

[b]The proper treatment of Ω_2 is more elaborate than our treatment here. Our treatment is sufficient though for our present purposes.

Now we can go back to the question we have posed in the previous section. *Is there a preferred solution to the CCR?* Recall that there is only one separable Hilbert space; that is, all separable Hilbert spaces are isomorphically equivalent to one other, so that there are unitary operations transforming one Hilbert space into another. The three Hilbert spaces, $\mathcal{H}_1$, $\mathcal{H}_2$, and $\mathcal{H}_3$, are separable, and hence can be transformed to a common Hilbert space $\mathcal{H}_C$, together with all the operators in them, including their respective position and momentum operators. The canonical pairs, $\{\mathcal{C}_1,\, \mathcal{C}_2,\, \mathcal{C}_3\}$, are then solutions of the CCR in the same Hilbert space $\mathcal{H}_C$. And we have seen that they are of dense category solutions, but of different classes. If we look at the diverse properties of the above $\mathcal{C}_j$'s, we can see that these properties are reflections of the fundamental properties of the underlying configuration spaces. It is then misguided to prefer one solution of the CCR over the rest or to require a priori a particular category of a specific class of a solution without a proper consideration of the physical context against which the solution is sought. For example, if we insist that only canonical pairs forming a system of imprimitivities over the real line are acceptable, then, within the context of position-momentum pairs, we are insisting homogeneity in all configuration spaces, a preposterous proposition. Why impose the homogeneity of, say, Ω_1 in intrinsically inhomogeneous configuration spaces like Ω_2 and Ω_3?

The example of the position and the momentum operators makes it clear that the set of properties of a specific solution to the CCR is consequent to a set of underlying fundamental properties of the system under consideration, or to the basic definitions of the operators involved, or to some fundamental axioms of the theory, or to some postulated properties of the physical universe. That is to say that a specific solution to the CCR is canonical in some *sense*, i.e. of a particular category and of a particular class. It is conceivable to impose that a given pair be canonical as a priori requirement based, say, on its classical counterpart, but not the sense the pair is canonical without a deeper insight into the underlying properties of the system. In other words, we don't impose in what *sense* a pair is canonical if we don't know much, we derive in what *sense* instead. Furthermore, if a given pair is known to be canonical in some *sense*, then we can learn more about the system or the pair by studying the structure of the *sense* the pair is canonical.

5 Self-adjoint Time Operators

Pauli's theorem has made the impression that an operator canonically conjugate with a Hamiltonian, i.e. a time operator, is necessarily a generator of energy shifts or an operator of an imprimitivity pair in the real line, and thus cannot be self-adjoint. With Implications-1, 2 and 3 exhibited not to hold in Hilbert spaces, there thus exists no a priori reason for the non-existence of self-adjoint time operators for semibounded Hamiltonians. However, with the conjugacy of the time operator with the Hamiltonian and imprimitivity of the time-operator-Hamiltonian-pair taken synonymous for a long time, one now wonders what a self-adjoint time operator is without the imprimitivity requirement. It is in this context that the need for the appreciation of the different solutions to the CCR becomes necessary. In this section, we illustrate how self-adjoint time operators can arise without satisfying the

140

imprimitivity requirement, and demonstrate that a given Hamiltonian can in fact form a canonical pair with two self-adjoint time operators of different categories, one dense and another closed.

Let us consider a particle confined between two points with length $2l$, subject to no force in between the boundaries. We attach the Hilbert space $\mathcal{H} = L^2[-l, l]$ to the system. The position operator is unique and is given by the bounded operator q, $(\mathsf{q}\varphi)(q) = q\varphi(q)$ for all φ in $\mathcal{H}$. On the other hand, the momentum operator and the Hamiltonian are not unique, and have to be considered carefully. We assume the system to be conservative and we require that the evolution of the system be generated by a purely kinetic Hamiltonian. The former requires a self-adjoint Hamiltonian to ensure that time evolution is unitary. The latter requires a self-adjoint momentum operator commuting with the Hamiltonian. These requirements are only satisfied by the following choice of the momentum operator. For every $|\gamma| < \pi$, define the self-adjoint momentum operator $(\mathsf{p}_\gamma \phi)(q) = -i\phi'(q)$, with domain $\mathcal{D}_{p_\gamma}$ consisting of those vectors ϕ in $\mathcal{H}$ such that $\int |\phi'(q)|^2 \, dq < \infty$, and satisfying the boundary condition $\phi(-l) = e^{-2i\gamma}\phi(l)$. With p_γ self-adjoint, the Hamiltonian is purely kinetic, i.e. $\mathsf{H}_\gamma = (2\mu)^{-1}\mathsf{p}_\gamma^2$. The momentum and the Hamiltonian then commute and have the common set of eigenvectors $\phi_k^{(\gamma)}(q) = (2l)^{-\frac{1}{2}} \exp\!\big(i\,(\gamma + k\pi)\tfrac{q}{l}\big)$, with respective eigenvalues $p_{k,\gamma} = (\gamma + k\pi)l^{-1}$, $E_k = p_{k,\gamma}^2 (2\mu)^{-1}$, for all $k = 0, \pm 1, \pm 2 \cdots$. In the following, we give two time operators of different categories for the Hamiltonian H_γ.

5.1 A Self-adjoint Time Operator of Dense Category

There exists a compact and self-adjoint operator T_c^γ such that T_c^γ and H_γ form a canonical pair of dense category. This operator has the integral representation

$$(\mathsf{T}_c^\gamma \varphi)(q) = \int_{-l}^{l} \langle q | \, \mathsf{T}_c^\gamma \, | q' \rangle \, \varphi(q') \, dq',$$

whose kernel is given by

$$\langle q | \, \mathsf{T}_c^\gamma \, | q' \rangle = i \sum_{k,k'}{}' \frac{\varphi_k^{(\gamma)}(q)\varphi_{k'}^{(\gamma)}(q')^*}{E_k - E_{k'}},$$

where the primed sum indicates that $k = k'$ is excluded from the summation. That is, the pair T_c^γ and $\mathcal{H}_\gamma$ satisfies the canonical commutation relation in some dense subspace $\mathcal{D}_c^1$ of $\mathcal{H}$,

$$([\mathsf{T}_c^\gamma, \mathsf{H}_\gamma]\varphi)(q) = i\varphi(q), \quad \text{for all } \varphi(q) \in \mathcal{D}_c^1$$

$$\mathcal{D}_c^1 = \left\{ \varphi(q) = \sum_k a_k \varphi_k^{(\gamma)}(q), \ \sum_k |a_k|^2 < \infty, \ \sum_k a_k = 0 \right\}.$$

Since the canonical domain is dense, i.e. orthogonal only to the zero vector, the canonical pair $\mathcal{C}(\mathsf{T}_c^\gamma, \mathsf{H}_\gamma; \mathcal{D}_c^1)$ is of dense category. While T_c^γ and H_γ form a canonical pair, they do not form a system of imprimitivies over the real line since T_c^γ is compact.

But what is T_c^γ? In a separate work we have explicitly shown that to every discrete, semibounded Hamiltonian with constant degeneracy and with compact inverse there exists a time operator characteristic of the Hamiltonian.[35] The Hamiltonian H_γ satisfies all these conditions and the operator T_c^γ is the corresponding characteristic time operator to H_γ. But what is the physical content of the characteristic time operator? It is sufficient at this moment to say that its any two dimensional projection can serve as a quantum clock. Let us consider the general case. Given k and l, consider the closed subspace spanned by the eigenstates φ_k and φ_l of the Hamiltonian. Denote this subspace by $\mathcal{H}_{kl}$. Obviously the state $\varphi_{kl} = 2^{-\frac{1}{2}}(\varphi_k - \varphi_l)$ in $\mathcal{D}_c$ belongs to $\mathcal{H}_{kl}$. Let P_{kl} be the projection operator onto $\mathcal{H}_{kl}$. Let $\mathsf{T}_{kl} = \mathsf{P}_{kl}\mathsf{T}\mathsf{P}_{kl} = i\omega_{kl}^{-1}\left(|\varphi_k\rangle\langle\varphi_l| - |\varphi_l\rangle\langle\varphi_k|\right)$, where $\omega_{kl} = (E_k - E_l)$; and let $\mathsf{H}_{kl} = \mathsf{P}_{kl}\mathsf{H}\mathsf{P}_{kl} = E_k|\varphi_k\rangle\langle\varphi_k| + E_l|\varphi_l\rangle\langle\varphi_l|$. Since T and H are canonically conjugate in $\mathcal{D}_c$, we expect that T_{kl} and H_{kl} continue to satisfy the canonical commutation relation in the subspace of $\mathcal{H}_{kl}$ spanned by φ_{kl}, i.e. $(\mathsf{T}_{kl}\mathsf{H}_{kl} - \mathsf{H}_{kl}\mathsf{T}_{kl})\varphi_{kl} = i\varphi_{kl}$. The state φ_{kl} evolves according to $\varphi_{kl}(t) = 2^{-\frac{1}{2}}\left(e^{-iE_k t}\varphi_k - e^{-iE_l t}\varphi_l\right)$. Now T_{kl} has the expectation value and variance

$$\langle\varphi_{kl}(t)|\,\mathsf{T}_{kl}\varphi_{kl}(t)\rangle = \frac{1}{\omega}\,\mathrm{Sin}\,\omega_{kl}t, \quad \Delta T_{\varphi_{kl}(t)}\,\Delta H_{\varphi_{kl}(t)} = |\mathrm{Cos}\,\omega_{kl}t|\,\frac{1}{2}.$$

Here we see that the observable T_{kl} and the entire $\mathcal{D}_c$ wrap the entire time axis into the circle. Thus for a given t there is a positive integer k and a time interval τ such that $t = \tau + 2\pi n\,\omega_{kl}$. For a given n and for small τ's, equations (5.1) reduce to

$$\langle\varphi_{kl}(t(\tau))|\,\mathsf{T}_{kl}\varphi_{kl}(t(\tau))\rangle = \tau, \quad \Delta T_{\varphi_{kl}(t(\tau))}\,\Delta H_{\varphi_{kl}(t(\tau))} = \frac{1}{2}.$$

The operator T then is a quantum clock wrapping the entire time axis into the circle and saturating the time energy uncertainty relation in every neighborhood of $|t - 2\pi n\omega_{kl}|$. We shall give a more detailed analysis of characteristic time operators elsewhere.

5.2 A self-adjoint Time Operator of Closed Category

There exists a compact and self-adjoint operator T^γ, $\gamma \neq 0$, such that T^γ and H_γ form a canonical pair of closed category. This operator has the integral representation

$$(\mathsf{T}^\gamma\varphi)(q) = \int_{-l}^{l} \langle q|\,\mathsf{T}^\gamma\,|q'\rangle\,\varphi(q')\,dq',$$

whose kernel is given by

$$\langle q|\,\mathsf{T}^\gamma\,|q'\rangle = \frac{\mu}{4\,\mathrm{sin}\gamma}(q+q')\left(e^{i\gamma}\,\mathrm{H}(q-q') + e^{-i\gamma}\,\mathrm{H}(q'-q)\right)$$

where $\mathrm{H}(q-q')$ is the Heaviside function. That is, the pair H_γ and T^γ satisfies the canonical commutation relation in a closed subspace of $\mathcal{H}$,

$$([\mathsf{T}^\gamma, \mathsf{H}_\gamma]\varphi)(q) = i\,\varphi(q) \quad \text{for all} \quad \varphi(q) \in \mathcal{D}_c^\gamma \tag{4}$$

$$\mathcal{D}_c^\gamma = \left\{ \int_{-l}^l \varphi(q)\, dq = 0,\ \varphi(\partial) = 0,\ \varphi'(\partial) = 0 \right\}.$$

Since the canonical domain $\mathcal{D}_c^\gamma$ is closed, i.e. orthogonal to the one-dimensional subspace $\mathcal{D}^{(\gamma)\perp} = \{\phi = c,\ c \in \mathcal{C}\}$ the canonical pair $\mathcal{C}(\mathsf{T}^\gamma, \mathsf{H}_\gamma; \mathcal{D}_c^\gamma)$ is of closed category. Also, while T^γ and H_γ form a canonical pair, they do not form a system of imprimitivies over the real line since T^γ is compact.

But what is T^γ? This operator is the quantization of the classical passage time $T = \mu q/p$ in the Hilbert space $\mathcal{H} = L^2(-l, l)$, subject to the requirement that the quantum Hamiltonian is the quantization of the purely kinetic classical Hamiltonian of the freely evolving particle between the boundaries. This condition leads to the momentum operator p_γ and the Hamiltonian H_γ given above. For a given $\gamma \neq 0$, we have the following correspondences

$$\mathcal{Q} : H = \frac{p^2}{2\mu} \quad \mapsto \quad \mathsf{H}_\gamma = \frac{\mathsf{p}_\gamma^2}{2\mu},$$

$$\mathcal{Q} : T = \mu \frac{q}{p} \quad \mapsto \quad \mathsf{T}^\gamma = \frac{\mu}{2}\left(\mathsf{q}\mathsf{p}_\gamma^{-1} + \mathsf{p}_\gamma^{-1}\mathsf{q}\right),$$

$$\mathcal{Q} : \{T, H\} = 1 \quad \mapsto \quad [\mathsf{T}^\gamma, \mathsf{H}_\gamma] \subset i\mathcal{I}_\gamma, \tag{5}$$

where $\mathcal{Q}$ is a quantization and $\mathcal{I}_\gamma$ is the identity in the closure of the canonical domain $\mathcal{D}_c^\gamma$. Notice that Eq.(5), the right hand side of which is just Eq.(4), is Dirac's Poisson-bracket-commutator correspondence at work. It is interesting to note though that Dirac's correspondence principle holds in a closed subspace for the $(\mathsf{T}^\gamma, \mathsf{H}_\gamma)$ pair, not in a dense subspace, as expected in the theory of quantization. Now T^γ is best appreciated by observing that the operator $\mathsf{T}_\gamma = -\mathsf{T}^\gamma$ is identifiable as the time of arrival operator at the origin for the spatially confined particle. We have referred to T_γ as the confined, non-periodic time of arrival operator for a given $|\gamma| < \pi$.[36] T_γ possesses the expected set of symmetry of a time of arrival operator and its eigenfunctions are identifiably time of arrival states. That is positive eigenvalue eigenfunctions evolve to symmetrically collapse at the origin, while negative eigenvalue eigenfunctions evolve to symmetrically collapse at the origin in the time reversed direction. We refer the reader to our earlier work[36] for a fuller account of the confined time of arrival operators.

6 Discussion

For a long time, Pauli's theorem has led most, if not all, to believe that the canonical commutation relation $[T, H] = iI$ only admits Hilbert space solutions that form a system of imprimitivities on the real line, contrary to the fact that there is a large class of solutions to the CCR, as we have discussed above. Consequently the theorem has brushed aside and downplayed the rest of the solution set of the canonical commutation relation for a given Hamiltonian. Our example clearly demonstrates that there are self-adjoint time operator solutions to $[T, H] = iI$ for semibounded Hamiltonians. It further demonstrates that for a given Hamiltonian there are possibly more than one class of solutions to the canonical commutation relation. In our example, the Hamiltonian is conjugate with two self-adjoint time operators belonging to two different categories. That these operators belong to distinct categories can be traced from the fact that they have distinct physical origins. For the time operator forming a dense-category with the Hamiltonian, it is characteristic of the system—it can be taken for an inherent quantum clock. For the time operator forming a closed category with the Hamiltonian, it is problem-specific—it is a direct result of quantization of the classical first passage time. As we have asserted in Sect.4, a specific solution to the CCR is canonical in some *sense*, i.e. of a particular category and of a particular class, and its *sense* is consequent to a set of underlying fundamental properties of the system under consideration, or to the basic definitions of the operators involved. Our example clearly illustrates this assertion. Obviously the condition of imprimitivity is not necessary for a time operator to satisfy in order to be physically meaningful.

But how about the current prevailing claim that time operators—in order to be meaningful—must be characterized as positive-operator-valued-measure (POVM) observables that transform covariantly under time translations?[4,5,7,13,15,24,38,39] Such a claim would require non-self-adjoint time operators for semibounded Hamiltonians. However, this claim has been introduced under the understanding that Pauli's theorem is the statement that *an operator canonically conjugate to a Hamiltonian is a generator of energy shifts*. But now we know that this traditional reading of Pauli's theorem, together with its modern rendering, is not correct in single Hilbert spaces. Should we then rule these covariant non-self-adjoint time operators misplaced?

It is misplaced to require covariance if one does so under the assumption that a time operator must *necessarily* be a generator of energy shifts. However, if one upholds the legitimacy of POVM to accommodate non-self-adjoint observables, as one should, then one may still be justified in requiring covariance as long as the required covariance is anchored on physical grounds. One must only acknowledge that covariant non-self-adjoint time operators are just one class and not the only class of solutions to the canonical commutation relation for a given Hamiltonian. Covariance can then be seen as a specific property of one class of solutions to the canonical commutation relation that can be anchored on specific problems. One must note though that over requiring covariance can lead to non-normalizable positive operator valued measures[13] which results in missing probabilities. On this instance, a solution to the canonical commutation relation is in conflict with the axioms of quantum mechanics. Thought is required to consider whether they are acceptable or not, an acceptance of which requires further revision of the axioms of quantum mechanics.

References

1. J.J. Sakurai *Modern Quantum Mechanics* (Addison-Wesley Publishing company, 1985).
2. J. von Nuemann *Mathematische Grundlagen der Quantenmechanik* (Springer Berlin, 1955).
3. W. Pauli, *Hanbuch der Physik* vol V/1 ed. S Flugge (Springer-Verlag, Berlin,1926) 60.

144

4. P. Busch *et. al. Phys. Lett. A* **191**, 357 (1994).
5. P. Busch, M. Grabowski and P. Lahti *Operational Quantum Physics* (Springer, Berlin, 1995).
6. V. Delgado and J.G. Muga *Phys. Rev. A* **56**, 3425 (1997).
7. R. Giannitrapani *Int. Jour. Theor. Phys.* **36**, 1575 (1997).
8. Gottfried, K. 1966 *Quantum Mechanics* (Benjamin/Cummings, Reading) vol. 1, 248.
9. A.S. Holevo *Rep. Math. Phys.* **13**, 379 (1978).
10. A.S. Holevo *Probabilistic and Statistical Aspects of Quantum Theory* (North-Holland, Amsterdam, 1982).
11. Cohen-Tannoudji, C. 1977 *Quantum Mechanics* Vol.1 (John Willey and Sons).
12. D. Park *Fundamental Questions in Quantum Mechanics* eds. L Roth and A Inomata (Gordon and Breach Science Publishers, 1984).
13. M.D. Srinivas and R. Vijayalakshmi *Pramana* **16**, 173 (1981).
14. V.S. Olhovsky V S and E. Recami *Nuovo Cimento* **22** 263 (1974).
15. M. Toller *Phys. Rev. A* **59**, 960 (1999).
16. J.G. Muga, R. Sala Mayato, I.L. Egusquiza (eds.) *Time in Quantum Mechanics* (Springer, Berlin, 2002).
17. R. Omnes, *The Interpretation of Quantum Mechanics* ed. P. W. Anderson et al (Princeton University Press, Harvard, 1994).
18. E. Eisenberg and L.P. Horwitz *Ad. Chem. Phys.* **XCIX** 245 (1997).
19. J.G. Muga, J. Palao and P. Sala *Superlatices and Microstructures* **24**, 23 (1998).
20. N. Grot, C. Rovelli, and R.S. Tate, R. S. *Phys. Rev. A* **54**, 4676 (1996).
21. R. Landuaer *Rev. Mod. Phys.* **66**, 217 (1994).
22. Y. Aharonov and D. Bohm Phys. Rev. **10**, 1127 (1969).
23. M. Bauer and P.A. Mello *Ann. Phys.* **111**, 38 (1978).
24. P. Busch *et. al. Annals of Phys.* **237**, 1 (1995).
25. H. Fick and F. Engelmann *Zeitschrift fur Physik* **178**, 551 (1964).
26. V. Fock and Kirilov *J. Phys.* (USSR) **11**, 112 (1974).
27. W. Heisenberg *Z. Phys.* **43**, 172 (1927).
28. J. Hilgevoord, *Am. J. Phys.* **64**, 1451 (1996).
29. J. Hilgevoord *Am. J. Phys.* **66**, 396 (1998).
30. P.R. Holland, *The Quantum Theory of Motion* (Press Syndicate, Cambridge University Press, 1993).
31. J.J.Halliwell and E. Zafiris *Phys. rev. D* **57**, 3351 (1997).
32. Ph. Blanchard and A. Jadczyk *Helv. Phys. Acta* **69**, 613 (1996).
33. D. Atkitson, http://www-th.phys.rug.nl/ atkinson/time.html.
34. E.A. Galapon, *Proc. R. Soc. Lond. A* **487**, 451 (2002).
35. E. A. Galapon *Proc. R. Soc. Lond. A* **487**, 2671 (2002).
36. E. A. Galapon and R. T. Bahague Jr, *Confined Quantum Time of Arrival Operators*, arXiv:quant-ph/0302036 (2003).
37. J.M. Jauch *Foundations of Quantum Mechanics* (Addison-wesley, 1968).
38. H. Atmanspacher and A. Amann, *Int. J. Theo. Phys.*, 629 (1998).
39. I. Egusquiza and J.G. Muga *Phys. Rev. A* **61**, 012104 (1999).
40. B. Fuglede *Math. Scand.* **20**, 79 (1958).

SIMULTANEITY AND THE CONCEPT OF 'PARTICLE'

CARL E. DOLBY

Department of Theoretical Physics, 1 Keble Rd, Oxford OX1 3RH, U.K.
E-mail: dolby@thphys.ox.ac.uk

The history of the particle concept is briefly reviewed, with particular emphasis on the 'foliation dependence' of many particle creation models, and the possible connection between our notion of particle and our notion of simultaneity. It is argued that the concept of 'radar time' (originally introduced by Sir Hermann Bondi in his work on k-calculus) provides a satisfactory concept of 'simultaneity' for observers in curved spacetimes. This is used to propose an observer-dependent particle interpretation, applicable to an arbitrary observer, depending solely on that observers motion and not on a choice of coordinates or gauge. This definition is illustrated with application to non-inertial observers and simple cosmologies, demonstrating its generality and its consistency with known cases.

1 Introduction

In this conference we have heard illuminating discussions of various aspects of the role of time in physics, and the conceptual tension that often surrounds it. One well-known tension is between the 'effectively absolute' role that time plays in quantum mechanics, and the role it plays in general relativity, where it is just one coordinate in a covariant theory. My contribution to these proceeding will discuss the problem of particle creation in gravitational backgrounds, and in accelerating reference frames. In so doing I hope to shed some light on the aforementioned tension, and also to describe a fascinating connection that exists between our concept of simultaneity, and our concepts of 'particle' and 'vacuum'.

The first prediction of particle creation in gravitational backgrounds came in 1939 when Schrödinger[1] predicted that if the universe is expanding then "it would mean production of matter merely by its expansion". This prediction was readressed in detail by Parker[2,3,4] in the late 60's. However, gravitational particle creation first hit the headlines in 1975, with the discovery of Hawking radiation from black holes [5]. Perhaps an even more intriguing discovery was made later that year, by Unruh[6] and independently by Davies[7]. They showed that an observer who accelerates uniformly through flat empty space will also observe a thermal bath of particles, at a temperature given by their acceleration. This means that a state which is empty according to an inertial observer will not be empty according to an accelerating observer, and hence it demonstrates that the concept of 'vacuum' (and hence of 'particle') must be observer-dependent. To see how these predictions could arise, consider a globally hyperbolic spacetime, and for definiteness, consider massive Dirac fermions. Then we have a field operator $\hat{\psi}(x)$ satisfying[8,9]:

$$(i\gamma^\mu \nabla_\mu - m)\hat{\psi}(x) = 0 \tag{1}$$

where $\{\gamma^\mu, \gamma^\nu\} = 2g^{\mu\nu}$, $\nabla_\mu \psi \equiv \partial_\mu \psi + \Gamma_\mu \psi$ and $\Gamma_\mu = \frac{1}{4}\gamma_\nu \nabla_\mu \gamma^\nu$ is the spin connection. Since no interactions are present, we can expand $\hat{\psi}(x)$ in terms of a complete set of normal modes as:

$$\implies \hat{\psi}(x) = \sum_i \{u_i(x)a_i + v_i(x)b_i^\dagger\} \tag{2}$$

However, these modes are no longer simple plane waves, so it is no longer obvious which modes should be put with the a_i operators and interpreted as particle modes, and which should be put with the $b_i^\dagger$ operators and interpreted as antiparticle modes. Two choices must be made. The 'in' modes $\{u_{i,\text{in}}, v_{i,\text{in}}\}$, chosen to represent particle/antiparticle modes at early times, determine the 'in' vacuum $|0_\text{in}\rangle$ by the requirement:

$$a_{i,\text{in}}|0_\text{in}\rangle = 0 = b_{i,\text{in}}|0_\text{in}\rangle$$

The 'out' modes $\{u_{i,\text{out}}, v_{i,\text{out}}\}$ determine the 'out' number operator

$$\hat{N}_\text{out} = \sum_i \{a_{i,\text{out}}^\dagger a_{i,\text{out}} + b_{i,\text{out}}^\dagger b_{i,\text{out}}\}$$

By expanding the 'out' modes in in terms of the 'in' modes we obtain:

$$a_{i,\text{out}} = \sum_j \{\alpha_{ij}a_{j,\text{in}} + \beta_{ij}b_{j,\text{in}}^\dagger\} \tag{3}$$

$$b_{i,\text{out}} = \sum_j \{\gamma_{ij}^* a_{j,\text{in}}^\dagger + \epsilon_{ij}^* b_{j,\text{in}}\} \tag{4}$$

The number of 'out' particles in the 'in' vacuum is then given by:

$$\langle 0_\text{in}|\hat{N}_\text{out}|0_\text{in}\rangle = Trace(\beta\beta^\dagger + \gamma\gamma^\dagger) \tag{5}$$

hence describing particle creation. The task of describing particle creation then boils down to the question: *How do we choose the 'in' and 'out' modes?*

There are a large variety of methods proposed for this choice (see for instance the common texts[10,11] and the references therein), based on adiabatic expansions, conformal symmetry, killing vectors, the diagonalisation of a suitable Hamiltonian, or many other methods. Broadly speaking these methods are limited by one of two drawbacks. Either they require the spacetime to possess certain desirable symmetries (deSitter, Killing vectors, conformal symmetries etc), or they give results which depend on an arbitrary foliation of spacetime into 'space' and 'time'. Meanwhile, although a choice of observer often motivates the choice of foliation (such as in the Unruh effect), there is no systematic prescription for linking the chosen observer to the chosen foliation.

Many of these drawbacks can be avoided by introducing a model particle detector[6,10,12,13]. This provides an operational particle concept, which directly incorporates the observers motion. However, it can not be used to define the particle/antiparticle modes, for a number of reasons. Firstly, because a detector only counts particles on its trajectory, so could not for instance categorize the emptiness of a state. More importantly, it would be circular. Provided a particle detector is anything that detects particles, a particle cannot also be "anything detected by a particle detector". Even if we stick only to 'tried and tested' detector models[6,14], then the question arises "what were they tested against?" - we must have in mind a concept of particle before fashioning a concept of detector. (In the case of fermions there are also technical difficulties[12,13,15], meaning that the predictions of current detector models are not always proportional to the number of particles present, even for inertial detectors in electro-magnetic fields.)

In this article we offer a resolution to these difficulties[16,17,18] which builds on the so-called 'Hamiltonian diagonalisation' prescription[19,20,21,22]; a method criticized in the past[23] for its reliance on an arbitrarily chosen foliation of spacetime (time coordinate). Our resolution lies in using the concept of 'radar time[a]' (originally introduced by Sir Hermann Bondi[24,25,26] in his work on k-calculus) to uniquely assign a foliation of spacetime to any given observer. The result is a particle interpretation which depends *only* on the motion of the observer, and on the background present, and which generalizes Gibbons' definition[27] to arbitrary observers and non-stationary spacetimes. It also facilitates the definition of a number density operator, allowing us to calculate not just the total asymptotic particle creation, but also to say (with definable precision) where and when these particles were 'created'.

Given the central role that radar time will play in this particle interpretation, the next Section is devoted to describing radar time, while Section 3 describes the application of radar time to an arbitrary observer in 1+1 Dimensional Minkowski Space. The observer-dependent particle interpretation is defined and discussed in Section 4. In Section 5 we return to 1+1 Dimensional Minkowski space, and describe the massless Dirac Vacuum as seen by an arbitrarily moving observer. Conclusions are presented in Section 6.

2 Radar Time

Consider an observer traveling on path $\gamma : x^\mu = x^\mu(\tau)$ with proper time τ, and define:

$\tau^+(x) \equiv$ (earliest possible) proper time at which a null geodesic leaving

point x could intercept γ.

$\tau^-(x) \equiv$ (latest possible) proper time at which a null geodesic could

leave γ, and still reach point x.

$$\tau(x) \equiv \tfrac{1}{2}(\tau^+(x) + \tau^-(x)) \qquad = \text{ 'radar time'.}$$

$$\rho(x) \equiv \tfrac{1}{2}(\tau^+(x) - \tau^-(x)) \qquad = \text{ 'radar distance'.}$$

$$\Sigma_{\tau_0} \equiv \{x : \tau(x) = \tau_0\} = \text{ observer's 'hypersurface of simultaneity at time } \tau_0\text{'.}$$

This is a simple generalization of the definition made popular by Bondi in his work on special relativity and k-calculus[24,25,28]. It can be applied to any observer in any spacetime. We can also define the 'time-translation' vector field:

$$k_\mu(x) \equiv \frac{\frac{\partial \tau}{\partial x^\mu}}{g^{\sigma\nu}\frac{\partial \tau}{\partial x^\sigma}\frac{\partial \tau}{\partial x^\nu}} \tag{6}$$

This represents the perpendicular distance between neighboring hypersurfaces of simultaneity, since it is normal to these hypersurfaces and it satisfies $k^\mu(x)\frac{\partial \tau}{\partial x^\mu} = 1$.

---

[a]Also known[29] as "Märzke-Wheeler coordinates".

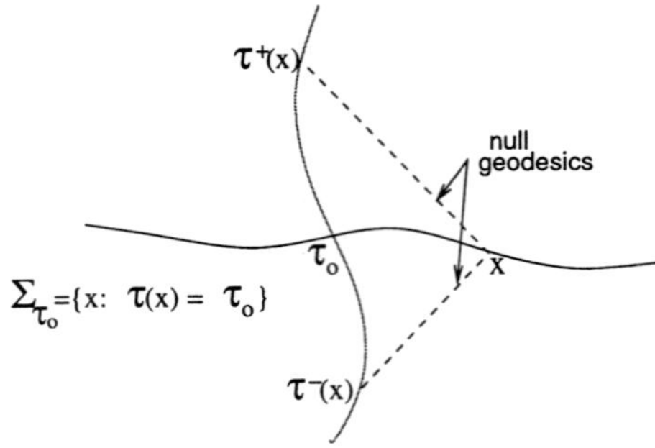

Figure 1. Schematic of the definition of 'radar time' $\tau(x)$.

Radar time is independent of the choice of coordinates, and is single valued in the observers *causal envelope* (the set of all spacetime points with which the observer can both send and receive signals). An affine reparametrisation of the observers worldline leads only to a relabeling of the same foliation, such that the radar time always agrees with proper time on the observer's path. It is invariant under 'time-reversal' - that is, under reversal of the sign of the observer's proper time.

We now illustrate these properties with a simple class of examples; observers in 1+1 Dimensional Minkowski space. Some simple cosmological examples are presented elsewhere[18].

3 Arbitrary Observer in 1 + 1 Dimensions

Let the observer's worldline be described by

$$x^\pm \equiv t \pm x = x_\lambda^\pm(\tau_\lambda) = \int^{\tau_\lambda} e^{\pm\alpha(\tau)}\mathrm{d}\tau$$

where τ_λ is the observers proper time, and $\alpha(\tau_\lambda)$ is the observers 'rapidity' at time τ_λ; $e^{\alpha(\tau_\lambda)}$ is the obvious time-dependent generalization of the 'k' of Bondi's k-calculus[24,25,28]. The observers acceleration is $a(\tau_\lambda) = \frac{d\alpha}{d\tau_\lambda}$. The observers worldline is completely specified by the choice of origin (i.e. $x^\mu(0)$) and the rapidity function $\alpha(\tau_\lambda)$, or by the choice of origin, the initial velocity, and the function $a(\tau_\lambda)$.

It is straightforward to show that the coordinates $\tau^\pm = \tau \pm \rho$ are given by:

$$x^\pm = x_\lambda^\pm(\tau^\pm)$$

while the metric in these coordinates[b] is:

$$\mathrm{ds}^2 = e^{(\alpha(\tau^+)-\alpha(\tau^-))}(\mathrm{d}\tau^2 - \mathrm{d}\rho^2)$$

We see that the radar coordinates are obtained from the Minkowski coordinates simply by rescaling along the null axes. The 'time-translation vector field' [18,17] is simply $k^\mu \frac{\partial}{\partial x^\mu} = \frac{\partial}{\partial\tau}$, while the hypersurfaces Σ_τ are hypersurfaces of constant τ.

---

[b]For convenience we have reversed the role of τ^+ and τ^- to the observers left, so that ρ plays the role of a spatial (rather than radial) coordinate, being negative to the observers left - the radar time is unchanged by this.

As a useful consistency check, consider an inertial observer with a velocity v relative to our original frame. Then α is constant, and $x_\lambda^\pm(\tau_\lambda) = e^{\pm\alpha}\tau_\lambda$. The coordinates $\tau^\pm$ are hence given by $\tau^\pm = e^{\mp\alpha}x^\pm$ so:

$$\tau = \tfrac{1}{2}(e^{-\alpha}x^+ + e^{\alpha}x^-) = \frac{t - vx}{\sqrt{1 - v^2}}, \qquad \rho = \tfrac{1}{2}(e^{-\alpha}x^+ - e^{\alpha}x^-) = \frac{x - vt}{\sqrt{1 - v^2}}$$

The radar coordinates of an inertial observer are just the coordinates of their rest frame, as expected.

3.1 Constant Acceleration

The simplest nontrivial case is constant acceleration. In this case $\alpha(\tau) = a\tau$, and we have $x_\lambda^\pm(\tau_\lambda) = \pm a^{-1}e^{\pm a\tau_\lambda}$ which gives:

$$\tau = \frac{1}{2a}\log\left(\frac{x+t}{x-t}\right) \qquad\qquad \rho = \frac{1}{2a}\log\left(a^2(x^2 - t^2)\right) \tag{7}$$

$$ds^2 = e^{2a\rho}(d\tau^2 - d\rho^2) \tag{8}$$

These are Rindler coordinates, which cover only region U of Figure 2, as expected. The hypersurfaces of constant τ are given by $t_{\tau_0}(x) = x\tanh(a\tau_0)$.

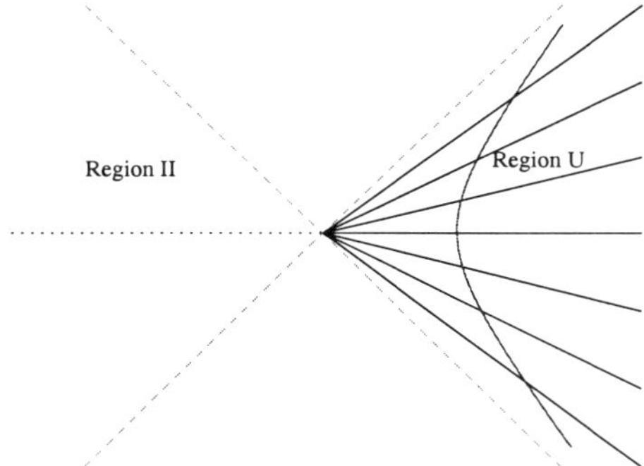

Figure 2. Hypersurfaces of simultaneity of a uniformly accelerating observer.

3.2 Gradual Turnaround Cases

Consider now an observer (Barbara say) who accelerates uniformly for $|\tau_\lambda| < \tau_c$, but is otherwise inertial. Then $\alpha(\tau_\lambda) = a\tau_\lambda$ for $|\tau_\lambda| < \tau_c$ and $= \pm a\tau_c$ for $\tau_\lambda > \tau_c$ or $< -\tau_c$ respectively.

The hypersurfaces of simultaneity for this observer are shown[c] in Figure 3. For comparison we have included the standard 'instantaneous rest frame' in Figure 4. The instantaneous rest frame suffers from being triple valued to the observer's left. It is also sensitively dependent on the small-scale details of Barbara's trajectory. Consider for instance a small deviation of Barbara's trajectory, like the small dotted line in the turnaround point of Figs. 3 and 4. In Figure 4 this has serious effects - Barbara now assigns five times to events far to her left, and three to events far to

---

[c]This is also described elsewhere[26], in the context of the well-known relativistic twin "paradox".

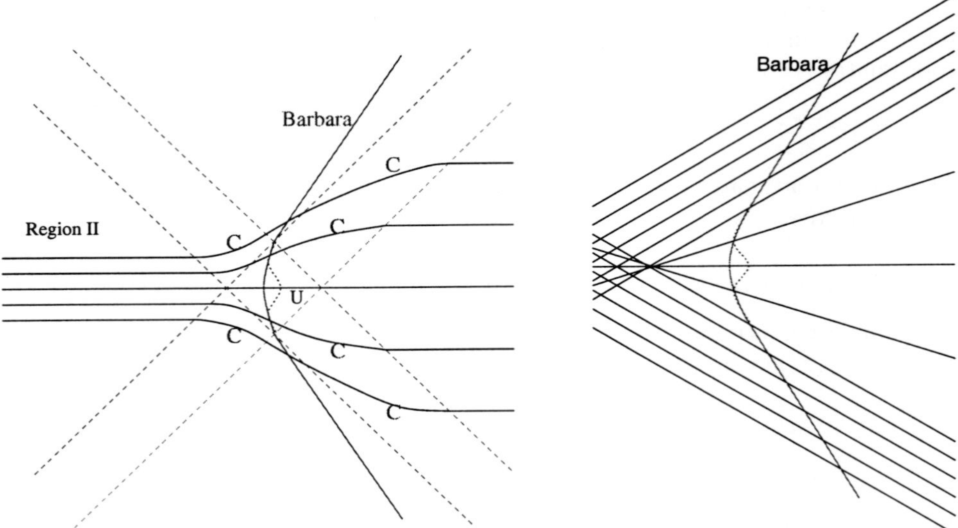

Figure 3. Barbara's hypersurfaces of constant τ.

Figure 4. Barbara's instantaneous rest frames.

her right! In Figure 3 however, this change causes only a small change in the times assigned to events in the vicinity of the points marked C.

A similar example is that of an observer with trajectory given by:

$$x_\lambda^\pm(\tau_\lambda) = \int_0^{\tau_\lambda} e^{\pm a\tau_c \tanh(\tau/\tau_c)} \mathrm{d}\tau$$

This observer has acceleration $a(\tau) = a\cosh^{-2}(\tau/\tau_c)$, so is uniformly accelerating for $|\tau| << \tau_c$ and inertial for $|\tau| >> \tau_c$. We will return to this example shortly.

4   An Observer-Dependent Particle Interpretation

Consider again the field operator:

$$\hat{\psi}(x) = \sum_i \{u_{i,\mathrm{in}}(x)a_{i,\mathrm{in}} + v_{i,\mathrm{in}}(x)b^\dagger_{i,\mathrm{in}}\} \tag{9}$$

and the state $|\mathrm{in}\rangle$ defined by $a_{i,\mathrm{in}}|\mathrm{in}\rangle = 0 = b_{i,\mathrm{in}}|\mathrm{in}\rangle$. We will consider the time-dependent particle content of this state, as measured by an observer O. We mentioned in the introduction that this definition stems from the diagonalisation of a suitable Hamiltonian. The Hamiltonian in question is:

$$\hat{H}(\tau) = \int_{\Sigma_\tau} \sqrt{-g}\, k^\mu T_{\mu\nu}(\hat{\psi}, \hat{\psi})\mathrm{d}\Sigma^\nu \tag{10}$$

where $T_{\mu\nu}(\psi,\phi) = \frac{1}{2}i[\bar{\psi}\gamma_{(\mu}\nabla_{\nu)}\phi - (\nabla_{(\mu}\bar{\psi})\gamma_{\nu)}\phi]$ $\tag{11}$

. $T_{\mu\nu}(\hat{\psi}, \hat{\psi})$ is the (unregularised) stress-energy tensor for Dirac fermions[10]. Diagonalising this Hamiltonian[22] entails expanding $\hat{\psi}$ as:

$$\hat{\psi}(x) = \sum_i u_{i,\tau_0}(x)a_{i,\tau_0} + \sum_i v_{i,\tau_0}(x)b^{\dagger}_{i,\tau_0} + \sum_i w_{i,\tau_0}(x)c_{i,\tau_0} \tag{12}$$

$$= \hat{\psi}^{(+)}_{\tau_0}(x) + \hat{\psi}^{(-)}_{\tau_0}(x) + \hat{\psi}^{(0)}_{\tau_0}(x) \tag{13}$$

and choosing these modes such that the Hamiltonian becomes:

$$\hat{H}(\tau_0) = \sum_{ij} h^{(+)}_{ij}(\tau_0)a^{\dagger}_{i,\tau_0}a_{j,\tau_0} - \sum_{ij} h^{(-)}_{ij}(\tau_0)b_{i,\tau_0}b^{\dagger}_{j,\tau_0} \tag{14}$$

where the matrices $h^{(\pm)}$ are positive definite. To consider the content of this requirement, it is convenient to define the '1st quantized Hamiltonian' $\hat{H}_1(\tau)$ (on the space of finite-norm solutions of the Dirac equation) by:

$$\langle \psi | \hat{H}_1(\tau) | \phi \rangle = \int_{\Sigma_\tau} \sqrt{-g} \; k^\mu T_{\mu\nu}(\psi, \phi) \mathrm{d}\Sigma^\nu$$

Then equation (14) requires that $\{u_{i,\tau_0}(x)\}$ span the positive spectrum of $\hat{H}_1(\tau_0)$, $\{v_{i,\tau_0}(x)\}$ span the negative spectrum of $\hat{H}_1(\tau_0)$, and $\{w_{i,\tau_0}(x)\}$ span the null space of $\hat{H}_1(\tau_0)$. The $w_{i,\tau_0}(x)$ will generally be states of compact support outside the causal envelope of the observer[d]. Having defined $\hat{\psi}^{(\pm)}(x)$ and $\hat{\psi}^{(0)}(x)$ by this requirement, we can now define the particle number operator on Σ_{τ_0}, $\hat{N}^+_{\tau_0}$ by:

$$\hat{N}^+_{\tau_0} = \int_{\Sigma_{\tau_0}} \sqrt{-g} \; \bar{\hat{\psi}}^{(+)}_{\tau_0} \gamma^\mu \hat{\psi}^{(+)}_{\tau_0} \mathrm{d}\Sigma_\mu \quad = \int_{\Sigma_{\tau_0}} \sqrt{-g} \; \hat{J}^{\mu(+)} \mathrm{d}\Sigma_\mu \tag{15}$$

$$\text{where } \hat{J}^{\mu(+)} = \bar{\hat{\psi}}^{(+)}_{\tau(x)} \gamma^\mu \hat{\psi}^{(+)}_{\tau(x)} \tag{16}$$

For any state and any chosen observer, the field $\langle J^{\mu(+)} \rangle$ is a covariant vector field, which can be interpreted as describing the 'flow of particles' as seen by this observer. $\hat{J}^{\mu(+)} \mathrm{d}\Sigma_\mu$ represents the number of particles in $\mathrm{d}\Sigma_\mu$. Similarly, the antiparticle number operator is given by:

$$\hat{N}^-_{\tau_0} = - : \int_{\Sigma_{\tau_0}} \sqrt{-g} \; \bar{\hat{\psi}}^{(-)}_{\tau_0} \gamma^\mu \hat{\psi}^{(-)}_{\tau_0} \mathrm{d}\Sigma_\mu : \quad = \int_{\Sigma_{\tau_0}} \sqrt{-g} \; \hat{J}^{\mu(-)} \mathrm{d}\Sigma_\mu \tag{17}$$

$$\text{where } \hat{J}^{\mu(-)} = - : \bar{\hat{\psi}}^{(-)}_{\tau(x)} \gamma^\mu \hat{\psi}^{(-)}_{\tau(x)} : \tag{18}$$

The normal-ordering is with respect to the observer's particle interpretation at the time of measurement (i.e. the $b_{i,\tau}$). These operators allow the observer to calculate the total number of particles/antiparticles on Σ_τ for all τ, and to determine how this particle content is distributed throughout Σ_τ. Although the total number

---

[d]However, even for inertial observers in electromagnetic backgrounds, there exist topologically non-trivial backgrounds for which zero energy eigenstates exist, leading to the existence of fractional charge[30]. Although such situations are straightforward to describe within the present approach, we will not discuss them further here.

operator $\hat{N}_\tau = \hat{N}_\tau^+ + \hat{N}_\tau^-$ is necessarily non-local (no local operator could possibly be consistent with the Unruh effect) it will generally be effectively local[16,17] on scales larger than the Compton length $\lambda_c = \frac{h}{mc}$ of the particle concerned. Equating expressions (9) and (12) for $\hat{\psi}(x)$ gives:

$$a_{i,\tau_0} = \sum_j \{\langle u_{i,\tau_0}|u_{j,\text{in}}\rangle a_{j,\text{in}} + \langle u_{i,\tau_0}|v_{j,\text{in}}\rangle b_{j,\text{in}}^\dagger\} \tag{19}$$

$$b_{i,\tau_0} = \sum_j \{\langle v_{i,\tau_0}|u_{j,\text{in}}\rangle^* a_{j,\text{in}} + \langle v_{i,\tau_0}|v_{j,\text{in}}\rangle^* b_{j,\text{in}}^\dagger\} \tag{20}$$

which allows us to deduce for instance:

$$\langle\text{in}|\hat{N}_{\tau_0}^+|\text{in}\rangle = \text{Trace}(\beta\beta^\dagger) \tag{21}$$

$$\langle\text{in}|\hat{N}_{\tau_0}^-|\text{in}\rangle = \text{Trace}(\gamma\gamma^\dagger) \tag{22}$$

$$\text{where } \beta_{ij}(\tau_0) \equiv \langle u_{i,\tau_0}|v_{j,\text{in}}\rangle \qquad \text{and} \qquad \gamma_{ij}(\tau_0) \equiv \langle v_{i,\tau_0}|u_{j,\text{in}}\rangle \tag{23}$$

as in equation (5). Note that, in the presence of horizons, the observer O cannot define a unique 'vacuum state' at any time τ_0. All he can say is that "a state $|0,\tau_0\rangle$ is vacuum throughout Σ_{τ_0}" if:

$$a_{i,\tau_0}|0,\tau_0\rangle = 0 = b_{i,\tau_0}|0,\tau_0\rangle$$

for all i. This condition is not unique, since we have said nothing about $c_{i,\tau_0}|0,\tau_0\rangle$. This is a natural limitation however; since O cannot communicate with points outside his causal envelope, we can't expect him to be able to determine particle content in such regions.

Although we have specified the 'out' modes $\{u_{i,\tau}, v_{i,\tau}\}$ for all possible 'out times', we have not yet discussed the choice of 'in' modes $\{u_{i,\text{in}}, v_{i,\text{in}}\}$. This choice is largely a question of convenience, and depends on what state we wish to consider the properties of. In the absence of particle horizons (when $\hat{\psi}^{(0)} = 0$) we may wish the 'in' state to be our observer's 'in-vacuum' $|0,\tau_{\text{in}}\rangle$ prepared at some 'in' time τ_{in}. Alternatively, we may wish that the state $|\text{in}\rangle$ be prepared by someone other than the observer. This will be the case shortly, where the content of the inertial vacuum will be studied by an accelerating observer. Or we may wish (as is common in cosmological applications) to consider a state $|\text{in}\rangle$ which is never considered 'empty' by any observer, but is instead justified by symmetry considerations[10].

5   The Massless Dirac Vacuum in 1+1 Dimensions

As a concrete example of these definitions, consider now the massless Dirac vacuum $|0_M\rangle$ of flat 1+1 dimensional Minkowski space, as measured by an arbitrarily moving observer (more detail is presented elsewhere[31]). Then the 'in' modes are the plane wave states, which can be written in the massless case as:

$$u_{p,\pm,\text{in}}(x) = e^{-ipx^\mp}\phi_\pm, \qquad v_{p,\pm,\text{in}}(x) = e^{ipx^\mp}\phi_\pm \qquad \text{for } p > 0 \tag{24}$$

where the subscript $\pm$ denotes forward/backward moving modes, and the basis spinors $\phi_\pm$ satisfy $\bar{\gamma}_1\bar{\gamma}_0\phi_\pm = \pm\phi_\pm$ where $\bar{\gamma}_\mu$ are the flat space Dirac matrices in

1+1 Dimensions. It can be shown[31] that the modes:

$$u_{\omega,\pm,O} = e^{\pm\frac{1}{2}\alpha(\tau^{\mp})}e^{-i\omega\tau^{\mp}}\phi_{\pm} \qquad v_{\omega,\pm,O} = e^{\pm\frac{1}{2}\alpha(\tau^{\mp})}e^{i\omega\tau^{\mp}}\phi_{\pm} \qquad \text{for } \omega > 0 \tag{25}$$

diagonalise $\hat{H}(\tau)$ for all τ. Substituting (24) and (25) into (23) and calculating the integral over $p > 0$ that is implicit in the Trace, gives:

$$(\beta\beta^{\dagger})_{\omega\omega',\pm} = 2\int_{-\infty}^{\infty} d\tau_a e^{-i\omega_d\tau_a} \int_0^{\infty} d\tau_d \sin(\omega_a\tau_d)g_{\pm}(\tau_a,\tau_d) \quad = (\gamma\gamma^{\dagger})^*_{\omega\omega',\pm} \tag{26}$$

$$\text{where } g_{\pm}(\tau_a,\tau_d) = \frac{1}{\tau_d} - \frac{\exp\left(\frac{\mp 1}{2}(\alpha(\tau_a + \tau_d/2) + \alpha(\tau_a - \tau_d/2))\right)}{\int_{-\tau_d/2}^{\tau_d/2}\exp(\mp\alpha(\tau_a + \tau))d\tau} \tag{27}$$

$$\omega_a = \tfrac{1}{2}(\omega + \omega') \qquad \text{and} \qquad \omega_d = \omega' - \omega \tag{28}$$

From these we can deduce that[31] the distribution $n_F^+(x)$ of forward moving particles exactly matches the distribution of forward moving antiparticles, and is given by:

$$n_F(\tau^-) = \int_{-\infty}^{\infty} \frac{d\tau}{2\pi\tau}g_+(\tau^- + \tau, \tau) \quad = \int_0^{\infty} \frac{d\omega_a}{2\pi}n_{F,\omega}(\tau^-) \tag{29}$$

$$\text{where } n_{F,\omega}(\tau^-) = \int_{-\infty}^{\infty} d\tau_a \frac{\sin[\omega(\tau^- - \tau_a)]}{\pi(\tau^- - \tau_a)} \int_0^{\infty} d\tau_d \sin(\omega\tau_d)g_+(\tau_a,\tau_d) \tag{30}$$

This is a function only of $\tau^- = \tau - \rho$, as would be expected for forward-moving massless particles. It is defined such that $n_F(\tau^-)d\rho$ gives the number of particles within $d\rho$ of the point (τ, ρ). The function $n_{F,\omega}(\tau^-)$ can be interpreted as the frequency distribution of forward moving particles/antiparticles at the point τ^-. Equation (30), together with (27),expresses this distribution anywhere in the space-time, directly in terms of the observer's rapidity. Similarly, the spatial distribution of backward-moving particles matches that of backward moving antiparticles, and can be defined by $n_B(\tau^+) = \int_0^{\infty} \frac{d\omega}{2\pi}n_{B,\omega}(\tau^+)$ where the expressions for $n_{B,\omega}(\tau^+)$ and $n_B(\tau^+)$ are as above, but with g_+ replaced with g_- and τ^- replaced with τ^+. Notice that if the observer's worldline is time-symmetric about $\tau_\lambda = 0$ then $\alpha(-\tau_\lambda) = -\alpha(\tau_\lambda)$ which gives $n_{F,\omega}(\tau') = n_{B,\omega}(-\tau')$ for all τ'. On the hypersurface $\tau = 0$ for instance, where $\tau^{\pm} = \pm\rho$, this implies that the distribution of forward moving particles exactly matches that of backward moving particles. On the observers worldline on the other hand, the distribution of forward moving particles is the time-reverse of that for backward moving particles.

As a consistency check, consider again an inertial observer. Then α is constant, so $g_{\pm}(\tau_a,\tau_d) = 0$, and the particle content is everywhere zero, as expected. We now consider other examples.

5.1 Constant Acceleration

For a uniformly accelerating observer we have:

$$g_+(\tau_a,\tau_d) = g_-(\tau_a,\tau_d) = \frac{1}{\tau_d} - \frac{a}{2\sinh\left(\frac{a\tau_d}{2}\right)}$$

which is independent of τ_a. Hence the forward and backward moving particles are each distributed uniformly in ρ for all τ, and the frequency distribution is everywhere given by:

$$n_{F,\omega} = n_{B,\omega} = 2 \int_0^\infty d\tau_d \; \sin(\omega\tau_d) \left(\frac{1}{\tau_d} - \frac{a}{2\sinh\left(\frac{a\tau_d}{2}\right)} \right) \qquad (31)$$

$$= \frac{2\pi}{1 + e^{\frac{2\pi\omega}{a}}} \qquad (32)$$

which is a thermal spectrum at temperature $T = \frac{a}{2\pi k_B}$, as expected.

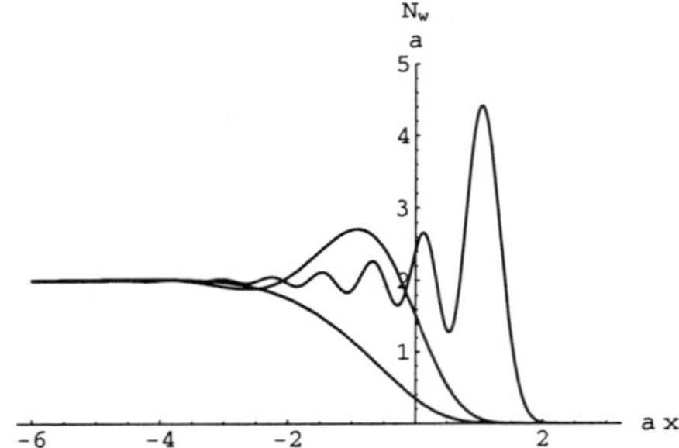

Figure 5. $n_\omega(\rho)/a$ as a function of $a\rho$, for $m = a$ and $\omega = a/4$ (lowest curve), a, and $4a$ (most oscillatory curve).

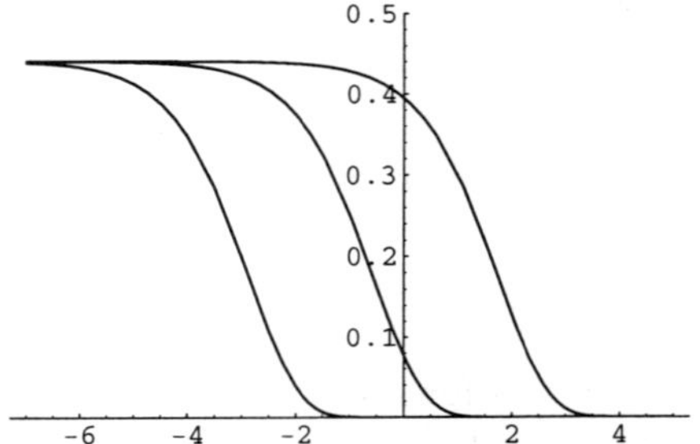

Figure 6. $n(\rho)/a$ as a function of $a\rho$, for $m = a/10$ (right curve), a and $10a$ (left curve).

For comparison, briefly consider the case of massive fermions in 1+1 dimensions (which no longer decompose into forward/backward moving modes). In this case the spatially averaged frequency distribution[8,12,32,18] is as in (32), but the massive particles are no longer distributed uniformly in ρ (although the distribution is completely independent of τ). Nor is the spatial distribution independent of ω. Figs. 5 and 6 show the spatial distribution of Rindler particles in this case[18]. Figure 5 shows $n_\omega(\rho)/a$ as a function of $a\rho$ for $m = a$ and $\omega = a/4$ (lowest curve), a, and $4a$ (most oscillatory curve), while Figure 6 shows $n(\rho)/a$ as a function of $a\rho$, for $m = a/10$ (right curve), a and $10a$ (left curve). These can be understood by considering that these particles see an 'effective mass gap' of $2me^{a\rho}$. Each frequency penetrates to a value of ρ for which $me^{a\rho} \sim \omega$. Changing the ratio m/a is equivalent to a translation in ρ. We see that in general the particle number density is uniform to the observer's left and negligible to the observer's right, with the transition happening at $\rho \sim \frac{1}{a}\log\left(\frac{a}{m}\right)$. As $m \to 0$ this transition point goes to ∞, reproducing the spatial uniformity of the massless limit. However, for non-zero m and realistic accelerations, the particle density is small even at low ρ (where it is $\propto a$), while the transition to a negligible density occurs far to the observer's left.

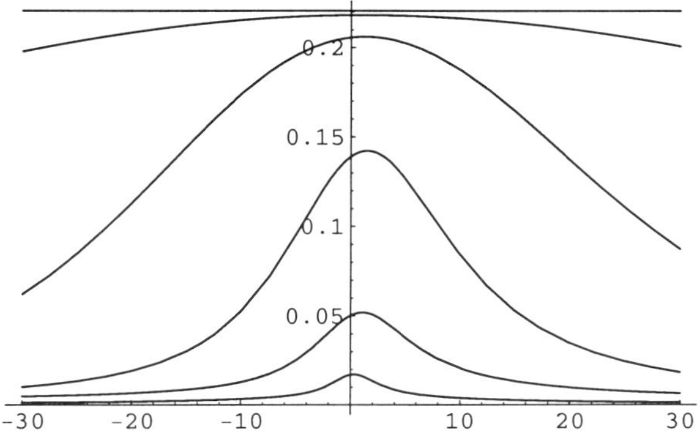

Figure 7. $n_{F,B}(\rho)/a$ as a function of $a\rho$ for $\tau = 0$ and $a\tau_c = 1$ (bottom curve), $3, 10, 30, 100$ and ∞ (top curve).

5.2 Gradual Turnaround Observer

Returning to the massless case, consider now the observer with acceleration

$$a(\tau_\lambda) = a \cosh^{-2}\left(\frac{\tau_\lambda}{\tau_c}\right)$$

Their rapidity is $\alpha(\tau_\lambda) = a\tau_c \tanh(\tau/\tau_c)$. They are accelerating uniformly for $|\tau| << \tau_c$, but are inertial at asymptotically early and late times (with velocity $\pm \tanh(a\tau_c)$). There are no particle horizons in this case; the observer's causal envelope covers the whole spacetime. By substituting the rapidity into equation (27) we immediately obtain the spatial distribution of forward or backward moving particles. At time $\tau = 0$ these distributions are equal. They are shown in Figure 7, as a function of $a\rho$ for $a\tau_c = 1$ (bottom curve), $3, 10, 30, 100$ and ∞ (top line). As τ_c increases the particle density increases, and approaches the spatial uniformity of the $\tau_c \to \infty$ limit.

In Figs. 8 and 9 we have shown the frequency distribution $n_{F,\omega}(\rho)$ of forward-moving particles, as a function of ω/a for $\tau = 0$. In Figure 8 we have $\rho = 0$ and $a\tau_c = 3, 10, 30$ and ∞. We can clearly see that the distribution approaches thermal as τ_c is increased. In Figure 9 $a\tau_c = 10$ and $a\rho = \pm 10$. We have also included a plot corresponding to a thermal spectrum appropriate to a constant acceleration of $a \cosh^{-2}(1)$. The difference between the actual spectrum and the thermal spectrum is more significant here. Since $n_{F,B}$ depend only on $\tau^{\pm}$ respectively, then Figure 9 also represents the distribution of forward/backward moving particles on the observer's worldline, at $\tau_\lambda = 10/a$. We see that the observer sees a different number of forward moving particles than backward moving particles. The forward/backward moving distributions are the reverse at $\tau = -10/a$.

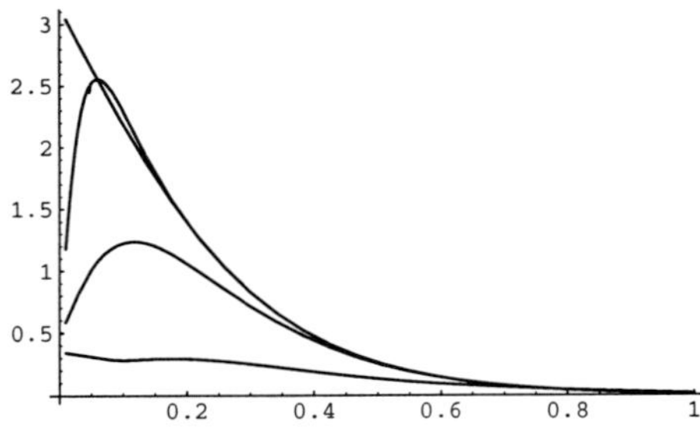

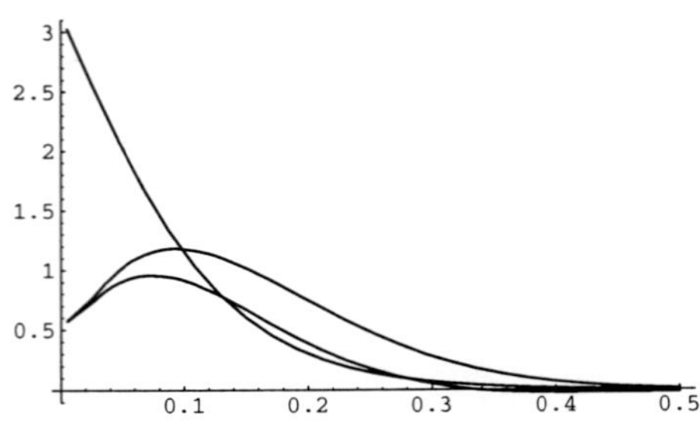

Figure 8. $n_{F,\omega}(\rho)$ as a function of ω/a for $\rho = 0 = \tau$, and $a\tau_c = 3, 10$ and 30.

Figure 9. $n_{L,R,\omega}(\rho)$ as a function of ω/a for τ, $a\tau_c = 10$ and $a\rho = \pm 10$.

6 Conclusion

Particle creation has been discussed, as seen by non-inertial observers in gravitational backgrounds. The observer-dependence of the particle interpretation has been emphasised, and the problem of foliation dependence discussed. Bondi's[24,25,28] *radar time* has been introduced, which provides an observer-dependent foliation of spacetime, depending only on the observer's motion, and not an any choice of coordinates. We have argued that this observer-dependent foliation resolves the problem of foliation dependence, by uniquely connecting it to the known observer-dependence of the particle concept (demonstrated by effects such as the Unruh[6,7] effect). The result is a particle interpretation which depends *only* on the motion of the observer, and on the background present, and which generalizes Gibbons' definition[27] to arbitrary observers and non-stationary spacetimes. It also facilitates the definition of a number density operator, allowing us to calculate not just the total asymptotic particle creation, but also to say (with definable precision) where and when these particles were 'created'. By incorporating the motion of the observer/detector, it links the 'Bogoliubov coefficient' approach to particle creation with that provided by operational 'detector' models, and provides a concrete answer to the question "what do particle detectors detect?" Concrete applications of these definitions have been presented, to non-inertial observers in 1+1D Minkowski spacetime (other examples are presented elsewhere[18]). We have shown how the thermal spectrum associated with a uniformly accelerating observer emerges as the limit of a class of 'smooth turn-around' observers, none of whom have acceleration horizons.

This conference, on "Time and Matter", has fueled much successful discussion of the role of time in physics, and the conceptual tensions that often surround it. In this contribution I have described what I believe to be quite a deep connection between 'time' and 'matter'. That is, between our concept of 'simultaneity', and our concepts of 'particle' and 'vacuum'. It is also hoped that some light may have been shed on the well-known conceptual tension between the 'effectively absolute' role that time plays in quantum mechanics, and the role it plays in general relativity, where it is just one coordinate in a covariant theory. While the relevance and faintness of this light is for the reader to decide, the availability of radar time appears to suggest that there need not be any inconsistency between the foliation dependence of quantum mechanics, and the coordinate covariance of general relativity, provided the role of the observer is properly considered.

References

1. E. Schrödinger, *Physica.* **VI(9)**, 899 (1939).
2. L. Parker, *Phys. Rev.* **183(5)**, 1057 (1969).
3. L. Parker, *Phys. Rev. D* **3(2)**, 346 (1971).
4. L. Parker, *Phys. Rev. Lett.* **21(8)**, 562 (1969).
5. S.W. Hawking, *Nature* **248**, 30 (1974).
6. W.G. Unruh, *Phys. Rev. D* **14**, 870 (1976).
7. P.C.W. Davies, *J. Phys. A.* **8(4)**, 609 (1975).
8. W. Greiner, B. Müller and J. Rafelski, *Quantum Electrodynamics of Strong Fields* (Springer, 1985).
9. M. Kaku, *Quantum Field Theory* (Oxford University Press, 1993).
10. N.D. Birrell and P.C.W. Davies, *Quantum Fields in Curved Spacetime* (Cambridge University Press, 1982).
11. S.A. Fulling, *Aspects of Quantum Field Theory in Curved Space-Time* (Cambridge University Press, 1989).
12. S. Takagi, *Prog. Theo. Phys.* Supplement No. **86** (1986).
13. L. Sriramkumar and T. Padmanabhan, *Int. J. Mod. Phys. D* **11**, 1 (2002).
14. B.S. DeWitt, in *General Relativity,* eds. S.W. Hawking and W. Isreal (Cambridge University Press, 1979).
15. L. Sriramkumar, *Mod. Phys. Lett. A.* **14**, 1869 (1999).
16. C.E. Dolby, PhD Thesis. Available from http://www.mrao.cam.ac.uk/~clifford/publications/abstracts/carl_diss.html
17. C.E. Dolby and S.F. Gull, *Annals. Phys.* **293**, 189 (2001).
18. C.E. Dolby and S.F. Gull, gr-qc/0207046.
19. A.A. Grib and S.G. Mamaev, *Sov. J. Nuc. Phys.* **10(6)**, 722 (1970).
20. A.A. Grib and S.G. Mamaev, *Sov. J. Nuc. Phys.* **14(4)**, 450 (1972).
21. S.G. Mamaev, et. al., *Sov. Phys. JETP* **43(5)**, 823 (1976).
22. A.A. Grib, et. al., *J. Phys. A: Math. Gen.* **13**, 2057 (1980).
23. S.A. Fulling, *Gen. Rel. and Grav.* **10(10)**, 807 (1979).
24. H. Bondi, *Assumption and Myth in Physical Theory* (Cambridge University Press, 1967).
25. D. Bohm, *The Special Theory of Relativity* (W. A. Benjamin, 1965).
26. C.E. Dolby and S.F. Gull, *Am. J. Phys.* **69**, 1257 (2001).
27. G.W. Gibbons, *Comm. Math. Phys.* **44**, 245 (1975).
28. R. D'Inverno, *Introducing Einsteins Relativity* (Oxford University Press, 1992).
29. M. Pauri and M. Vallisneri, *Found. Phys. Lett.* **13(5)**, 401 (2000).
30. R. Jackiw, Dirac Prize Lecture, Trieste, 1999. Available at hep-th/9903255.
31. M.D. Goodsell, C.E. Dolby and S.F. Gull. In preparation.
32. M. Soffel, B. Müller and W. Greiner, *Phys. Rev. D* **22**, 1935 (1980).

SECTION IV
CP & T VIOLATION

CP and T Violation in the Kaon System
  E. Blucher

Time-Reversal Non-Invariance
  D. Zavrtanik

'Per Aspera ad Astra' — A Short Essay on the Long Quest for CP Violation
  I. I. Bigi

CP AND T VIOLATION IN THE KAON SYSTEM

E. BLUCHER

Department of Physics and The Enrico Fermi Institute,
The University of Chicago,
5640 S. Ellis Ave., Chicago, IL 60637, USA
E-mail: blucher@hep.uchicago.edu

Experimental tests of CP, CPT, and T symmetry in the neutral kaon system are described. In particular, experiments contributing to the observation of direct CP violation will be reviewed.

1 Introduction

The origin of the observed matter-antimatter asymmetry in the Universe is one of the fundamental questions of physics. This asymmetry implies that the laws of nature somehow distinguish between particles and antiparticles. Although the matter-antimatter asymmetry appears maximal today, it corresponds to a tiny imbalance of matter over antimatter 1 μsec after the big bang – about 1 billion plus one particles per billion antiparticles. It is the origin of this tiny asymmetry that we hope to understand.

Until recently, the only other evidence for this type of asymmetry was found in the neutral kaon system, which is the topic of this paper. The discussion begins with a brief review of the neutral kaon system that has proven to be a unique laboratory for testing discrete symmetries in weak interactions. The next sections describe experimental studies of CP violation in the kaon system. The final sections discuss tests of CPT invariance and T violation.

2 Discrete Symmetries and the Neutral Kaon System

This paper involves the testing of three discrete symmetries: C (charge conjugation), P (parity), and T (time reversal). Until the 1950s, all three were thought to be separately conserved, although only the combination of all three symmetries (CPT) had been shown to be a property of any relativistic local quantum field theory.[1] Lee and Yang, in response to the "τ-θ" puzzle (involving decays of the particle now known as the charged kaon), pointed out that the parity symmetry had not been tested in weak interactions.[2] Following their suggestion, a series of experiments[3] demonstrated that parity was in fact violated maximally in weak interactions. These same experiments showed that the CP symmetry appeared to be conserved (*i.e.*, both C and P are violated such that CP remains a good symmetry).[a] A few years later, studies of neutral kaon decays revealed that the CP symmetry was also violated at a small level.[5]

The neutral kaon was first observed by Rochester and Butler using a cloud chamber in 1947.[6] Figure 1 shows a photograph of the first detected "V" particle.[7]

---

[a]This possibility had been pointed out by Landau[4] after the paper of Lee and Yang, but before these experiments were performed.

162

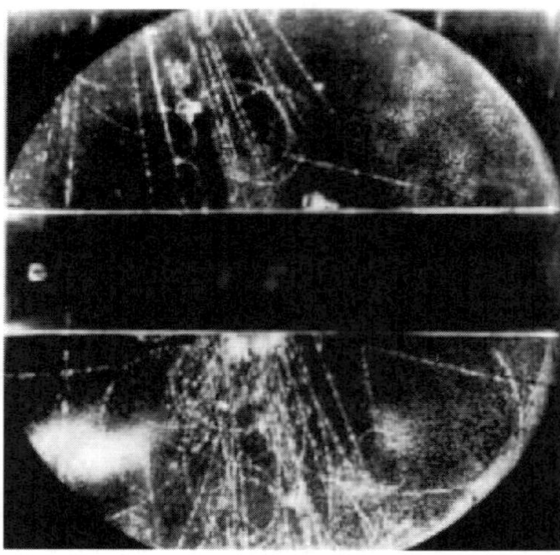

Figure 1. The first observed "V" particle. Rochester and Butler identified the inverted V in the lower right quadrant as two oppositely-charged particles coming from the decay of a previously unknown neutral particle.

We now know that this image shows the decay of a neutral kaon to $\pi^+\pi^-$, but we cannot tell whether it was a K^0 $(\bar{s}d)$ or $\overline{K}^0$ $(s\bar{d})$. This ambiguity is the key to kaon's unique properties. Since the K^0 and $\overline{K}^0$ have common decay modes, they can mix into each other through second-order weak processes ($\Delta s = 2$). The physical states that decay will therefore be mixtures of the K^0 and $\overline{K}^0$. Assuming CP invariance, these states are

$$
\begin{aligned}
K_{even} &\sim K^0 + \overline{K}^0 \\
K_{odd} &\sim K^0 - \overline{K}^0,
\end{aligned}
\tag{1}
$$

where $CP|K_{even}> = +|K_{even}>$ and $CP|K_{odd}> = -|K_{odd}>$.

The CP-even state almost always decays to pairs of pions. Since the CP-odd state cannot decay to the two-pion final state, it should have a longer lifetime than the CP-even state. Gell-Mann and Pais[8] predicted the existence of the long-lived neutral kaon, which had not yet been observed, based on this argument.[b] Lederman and collaborators[9] quickly searched for and observed the predicted particle with a lifetime 580 times longer than that of the short-lived neutral kaon.

The large lifetime difference between the long and short-lived neutral kaons makes it easy to produce a beam of the long-lived, CP-odd state: an experiment may be placed far from a target so that all of the short-lived state has decayed away. Such an experiment was performed by Christenson, Cronin, Fitch, and Turlay in 1963 to investigate an anomalous result on regeneration of K_{even} mesons. They also planned to test CP invariance by obtaining a better limit on $K_{odd} \to \pi^+\pi^-$. To their surprise, they observed about 45 decays of the long-lived neutral kaon to

---

[b]Their argument assumed C invariance rather than CP invariance.

$\pi^+\pi^-$, establishing the existence of CP violation.[5]

3 Direct and Indirect CP violation

Much of the experimental work in kaon physics since the discovery of CP violation has been devoted to distinguishing between two possible mechanisms contributing to the observed $K_L \to \pi\pi$ decays: indirect CP violation, resulting from an asymmetry between the rates for $K^0 \to \overline{K}^0$ and $\overline{K}^0 \to K^0$ oscillations, and direct CP violation, resulting from CP violation in decay amplitudes.[c]

Indirect CP violation results in physical states that are not the pure eigenstates given in Eq. 1, but are mixtures of the even and odd eigenstates:

$$
\begin{aligned}
K_S &\sim (1+\epsilon)K^0 + (1-\epsilon)\overline{K}^0 \\
&\sim K_{\text{even}} + \epsilon K_{\text{odd}} \\
K_L &\sim (1+\epsilon)K^0 + (1-\epsilon)\overline{K}^0 \\
&\sim K_{\text{even}} + \epsilon K_{\text{odd}},
\end{aligned}
\tag{2}
$$

where ϵ quantifies the CP impurity of the K_S and K_L states. Experiments done shortly after the Cronin-Fitch experiment showed that the observed $K_L \to \pi^+\pi^-$ events resulted mainly from indirect CP violation with $|\epsilon| = 2.28 \times 10^{-3}$.

The clearest measurement of indirect CP violation is the so-called charge asymmetry in semileptonic kaon decays. Semileptonic decays obey the $\Delta S = \Delta Q$ rule, so K^0 decays to a positive lepton and $\overline{K}^0$ decays to a negative lepton: $K^0 \to \pi^- \ell^+ \nu$ and $\overline{K}^0 \to \pi^+ \ell^- \overline{\nu}$. Using these decays, one can measure the K^0 and $\overline{K}^0$ content of the K_L and determine ϵ. The charge asymmetry is defined as

$$
\delta_\ell = \frac{N(\pi^- e^+ \nu) - N(\pi^+ e^- \overline{\nu})}{N(\pi^- e^+ \nu) + N(\pi^+ e^- \overline{\nu})} = 2Re(\epsilon).
\tag{3}
$$

The world average value[10,11] $\delta_\ell = (3.30 \pm 0.07) \times 10^{-3}$ corresponds to $|\epsilon| = (2.27 \pm 0.05) \times 10^{-3}$, This result is consistent with the value extracted from $K_L \to \pi^+\pi^-$ assuming only indirect CP violation (see below), showing that direct CP violation cannot be very large.

Direct CP violation can be detected by comparing the level of CP violation in different decay modes. The parameters ϵ and ϵ' are related to the ratio of CP violating to CP conserving decay amplitudes for $K_L \to \pi^+\pi^-$ and $K_L \to \pi^0\pi^0$:

$$
\begin{aligned}
\eta_{+-} &\equiv \frac{A\left(K_L \to \pi^+\pi^-\right)}{A(K_S \to \pi^+\pi^-)} = \epsilon + \epsilon', \\
\eta_{00} &\equiv \frac{A\left(K_L \to \pi^0\pi^0\right)}{A\left(K_S \to \pi^0\pi^0\right)} = \epsilon - 2\epsilon'.
\end{aligned}
\tag{4}
$$

As discussed above, ϵ is a measure of indirect CP violation, which is common to all decay modes. The quantity ϵ' is a measure of direct CP violation, which contributes differently to the $\pi^+\pi^-$ and $\pi^0\pi^0$ decay modes, and is proportional to the difference

---

[c]Note that it is direct CP violation, which results in different partial decay rates for particles and antiparticles, that is required to produce a matter-antimatter asymmetry in almost all models of baryogenesis.

between the decay amplitudes for $K^0 \to \pi^+\pi^- (\pi^0\pi^0)$ and $\overline{K^0} \to \pi^+\pi^- (\pi^0\pi^0)$. Experimentally, $Re(\epsilon'/\epsilon)$ is determined from the double ratio of the two pion decay rates of K_L and K_S:

$$\frac{\Gamma(K_L \to \pi^+\pi^-)/\Gamma(K_S \to \pi^+\pi^-)}{\Gamma(K_L \to \pi^0\pi^0)/\Gamma(K_S \to \pi^0\pi^0)} = \left|\frac{\eta_{+-}}{\eta_{00}}\right|^2 \approx 1 + 6Re(\epsilon'/\epsilon). \tag{5}$$

The Standard Model, in which CP violation arises from a single complex phase in the CKM matrix, accommodates both direct and indirect CP violation.[12,13,14] Unfortunately, there are large hadronic uncertainties in the calculation of $Re(\epsilon'/\epsilon)$. Most recent Standard Model predictions for $Re(\epsilon'/\epsilon)$ are less than 30×10^{-4}.[15] The superweak model,[16] proposed shortly after the discovery of $K_L \to \pi^+\pi^-$, includes only indirect CP violation.

Following the discovery of CP violation in 1964, a series of experiments with increasing precision were performed to search for direct CP violation. In the early 1990s, experiments E731 at Fermilab[17] and NA31 at CERN[18] reported the following results:

$$Re(\epsilon'/\epsilon) = (7.4 \pm 5.9) \times 10^{-4} \text{ (E731)}$$
$$Re(\epsilon'/\epsilon) = (23 \pm 6.5) \times 10^{-4} \text{ (NA31)}.$$

The CERN result was 3.5 standard deviations from zero, while the Fermilab result was only 1 sigma from zero. To clarify the experimental situation and definitively resolve the question of whether direct CP violation occurs, new experiments were built at Fermilab (KTeV), CERN (NA48), and Frascati (KLOE) to measure $Re(\epsilon'/\epsilon)$ at the $(1-2) \times 10^{-4}$ level. KTeV and NA48 are rather similar fixed-target experiments. As will be described below, they differ mainly in the method used to produce K_S, and in the technique used to correct for the difference in detector acceptance for K_S and K_L decays resulting from the large $K_S - K_L$ lifetime difference. The KLOE experiment uses an $e^+e^- \to \phi$ collider to produce pairs of kaons; unfortunately, KLOE has not yet collected enough data to attempt a measurement of $Re(\epsilon'/\epsilon)$.

4   The Search for Direct CP Violation

The experimental challenge in measuring $Re(\epsilon'/\epsilon)$ with a sensitivity of 10^{-4} is to collect millions of each of the four decay modes in Eq. 5, and to understand relative acceptances between different modes at better than the 10^{-3} level. To achieve this level of statistical and systematic uncertainty, the KTeV (Fig. 4) and NA48 (Fig. 5) experiments use rather similar approaches. Both experiments collect all 4 decay modes simultaneously using two beams – one for K_L decays and one for K_S decays. Each detector includes a long, evacuated decay region, followed by a charged spectrometer and a very precise electromagnetic calorimeter. The KTeV calorimeter uses pure CsI crystals and NA48 uses liquid krypton. Both calorimeters have excellent energy and position resolution; the average energy resolution is better than 1% and the average position resolution is about 1 mm (transverse to the beam direction) for both experiments. The performance of these calorimeters is crucial to the success of the experiments because the reconstructed position of decays

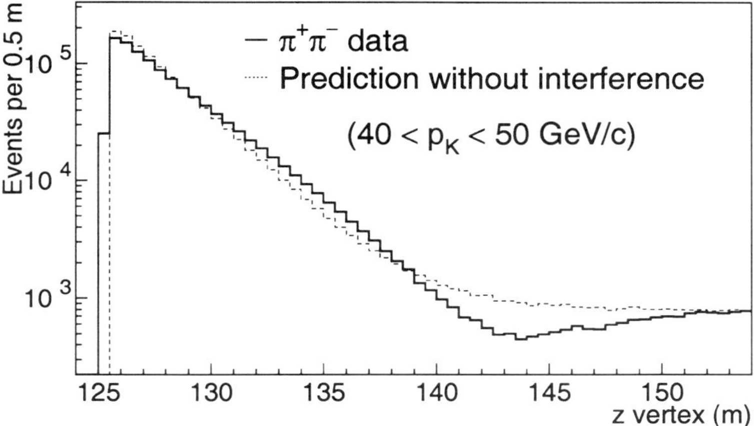

Figure 2. z decay distribution of $K_L \to \pi^+\pi^-$ decays downstream of the KTeV regenerator for the restricted momentum range 40-50 GeV/c.

along the beamline depends directly on the energy scale of the calorimeter. The excellent energy resolution also reduces background for both the $\pi^+\pi^-$ and $\pi^0\pi^0$ decay modes.

The principal difference between KTeV and NA48 is the method used to produce K_S decays. KTeV uses a regenerator in one of the two beams to produce a K_S component through coherent regeneration. The KTeV regenerator is fully active to reduce the background from inelastic interactions. Figure 2 illustrates the interference of K_L and K_S downstream of the regenerator. NA48 uses a bent crystal to transport a small fraction of protons that do not interact in the primary (K_L) target to a secondary (K_S) target close to the experiment. A time coincidence between the detector (*e.g.*, the calorimeter for the $K \to 2\pi^0$ decay mode) and a counter placed in the proton beam upstream of the K_S target is used to identify K_S decays. Figure 3 shows the time difference between the tagging counter and the detector for charged and neutral decays.

The difference between the K_L and K_S lifetimes means that the distribution of decay positions along the beam (z) direction will be very different for the K_L and K_S decays which must be compared to extract ϵ'/ϵ. Figure 6 shows z distributions from KTeV for the 4 decay modes. KTeV corrects for the variation in detector acceptance as a function of z with a Monte Carlo simulation. The quality of the simulation is studied using distributions from both the 2π decays, as well as higher statistics $K_L \to 3\pi^0$, and $K_L \to \pi e\nu$ decays. NA48 greatly reduces the necessary acceptance correction by reweighting K_L decays to have the same z distribution as K_S decays (see Fig. 8). The drawback of the reweighting procedure is that it increases the statistical uncertainty in the result by about a factor of 2.

The results of the experiments are[19,20]

$$\mathrm{Re}(\epsilon'/\epsilon) = (20.7 \pm 2.8) \times 10^{-4} \ (\mathrm{KTeV})$$

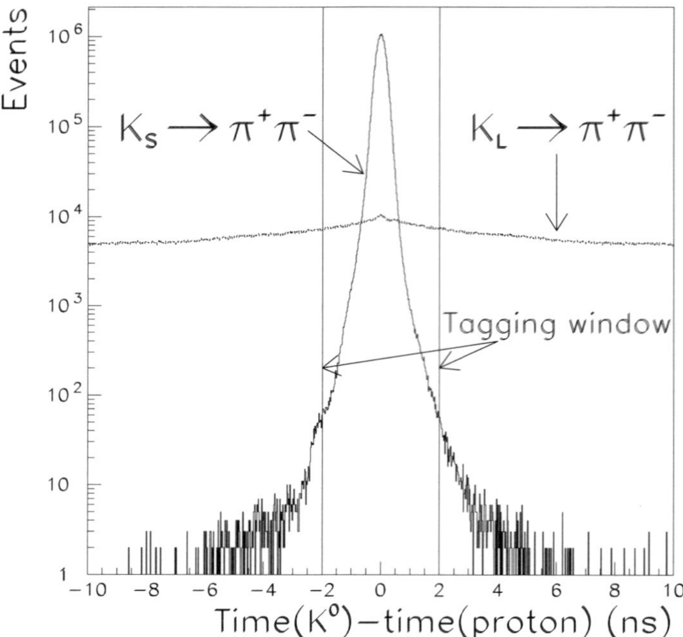

Figure 3. Time difference between NA48 K_S tagging counter and detector for $K_S \rightarrow \pi^+\pi^-$ and $K_L \rightarrow \pi^+\pi^-$ events, identified by the reconstructed vertex.

$$\mathrm{Re}(\epsilon'/\epsilon) = (14.7 \pm 2.2) \times 10^{-4} \text{ (NA48)},$$

clearly establishing the existence of direct CP violation.[d] Figure 9 shows a comparison of these measurements of $Re(\epsilon'/\epsilon)$ with earlier results from E731 and NA31. A weighted average of all measurements gives $Re(\epsilon'/\epsilon) = (16.6 \pm 1.6) \times 10^{-4}$ with a confidence level of 10%.

Although these experiments were designed to measure ϵ'/ϵ, they also have made greatly improved measurements of many parameters of the neutral kaon system. For example, both KTeV and NA48 made measurements of the K_S lifetime that are more precise than the average of all previous measurements (see Fig. 10). KTeV also has measured the interference pattern downstream of the regenerator (see Fig. 2) to make precise measurements of the $K_L - K_S$ mass difference, and to measure the relative phases between the CP violating and CP conserving decay amplitudes for $K_L \rightarrow \pi^+\pi^-$ (ϕ_{+-}) and for $K_L \rightarrow \pi^0\pi^0$ (ϕ_{00}).

It is interesting to note that the single most precise measurement of ϕ_{+-} comes from CPLEAR,[21] another experiment that was originally proposed to measure ϵ'/ϵ. CPLEAR uses low energy antiprotons stopped in a gaseous hydrogen target to produce neutral kaons: $p\bar{p} \rightarrow K^-\pi^+K^0$ or $K^+\pi^-\overline{K}^0$. The charged kaon is used

---

[d]The NA48 result is based on their full data set, while the KTeV result uses only half of the full data sample.

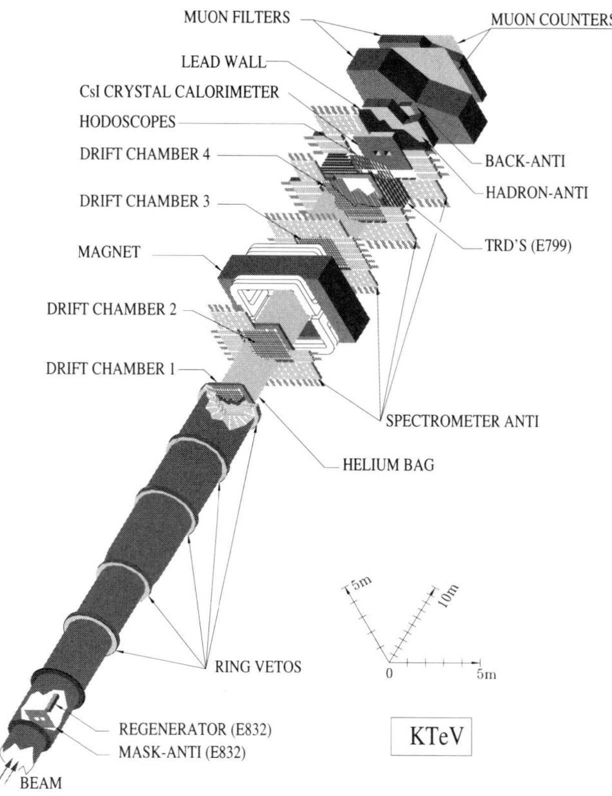

Figure 4. Diagram of the KTeV detector.

to tag the strangeness of the neutral kaon at production (t=0). They then extract ϕ_{+-} from a fit to the asymmetry between the $K \to \pi^+\pi^-$ decay time distributions for K^0 and $\overline{K}^0$ shown in Fig. 11. This asymmetry is very similar to those studied in the B-meson system.[22,23]

5 CPT Invariance

Measurements of the relative phases between the CP violating and CP conserving decay amplitudes, described above, can be used to search for CPT violation in K^0-$\overline{K}^0$ mixing (indirect CPT violation) and/or in decay amplitudes (direct CPT violation). It is most interesting to consider tests of CPT symmetry in K^0-$\overline{K}^0$ mixing where there is significant CP violation. Without the constraint of CPT

168

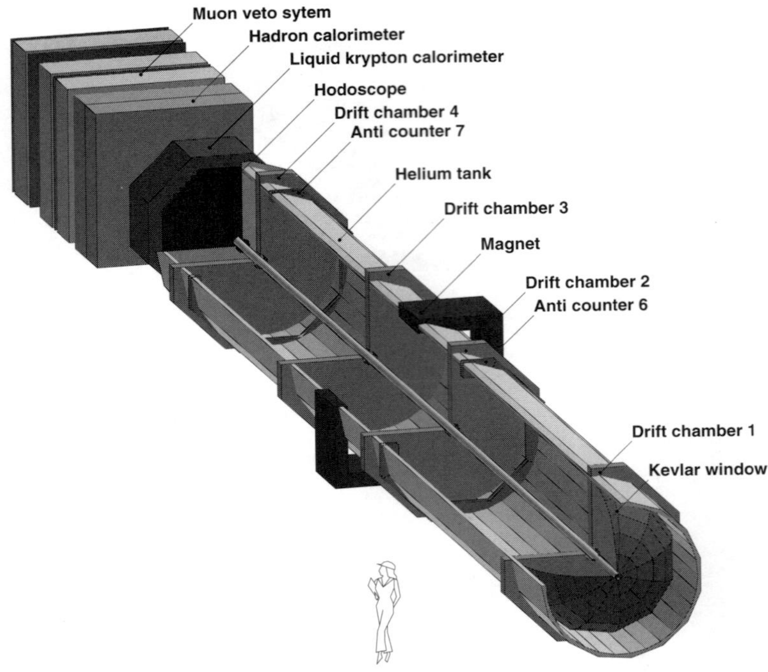

Figure 5. Diagram of the NA48 detector.

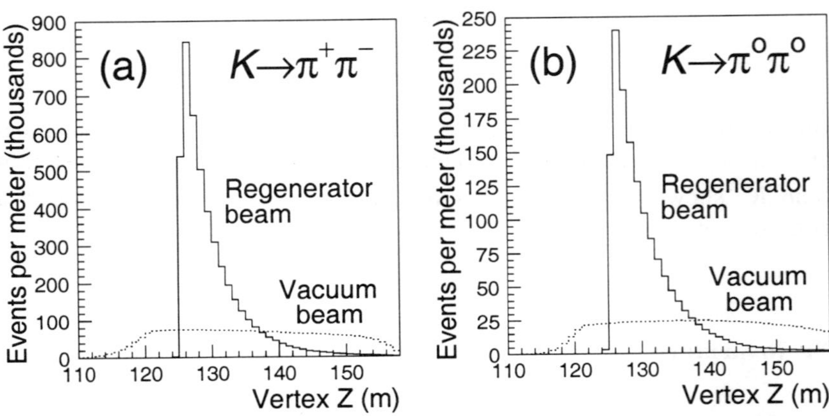

Figure 6. Decay vertex distributions from KTeV for (a) $K \to \pi^+\pi^-$ and (b) $K \to \pi^0\pi^0$ decays, showing the difference between the "regenerator" (K_S) and "vacuum" (K_L) beams.

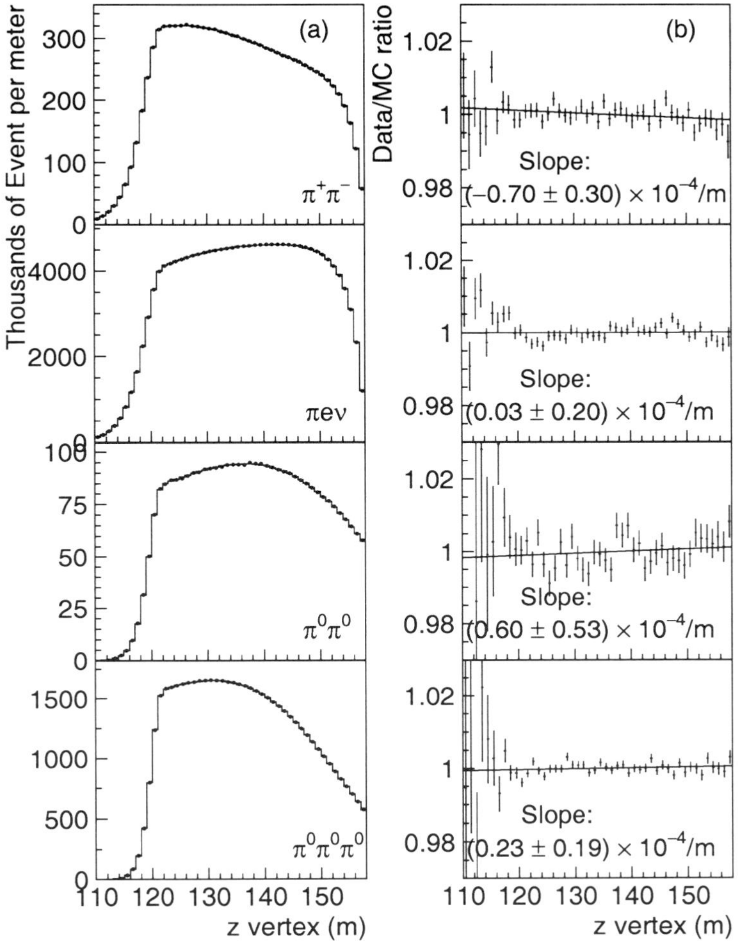

Figure 7. KTeV comparison of the vacuum beam z distributions for data (dots) and MC (histogram). The data-to-MC ratios on the right are fit to a line, and the slopes are shown.

symmetry, the K_S and K_L states may be written as

$$
\begin{aligned}
K_S &\sim (1 + \epsilon_S)K^0 + (1 - \epsilon_S)\overline{K}^0 \\
&\sim K_{\mathrm{even}} + \epsilon_S K_{\mathrm{odd}} \\
K_L &\sim (1 + \epsilon_L)K^0 + (1 - \epsilon_L)\overline{K}^0 \\
&\sim K_{\mathrm{even}} + \epsilon_L K_{\mathrm{odd}},
\end{aligned}
\tag{6}
$$

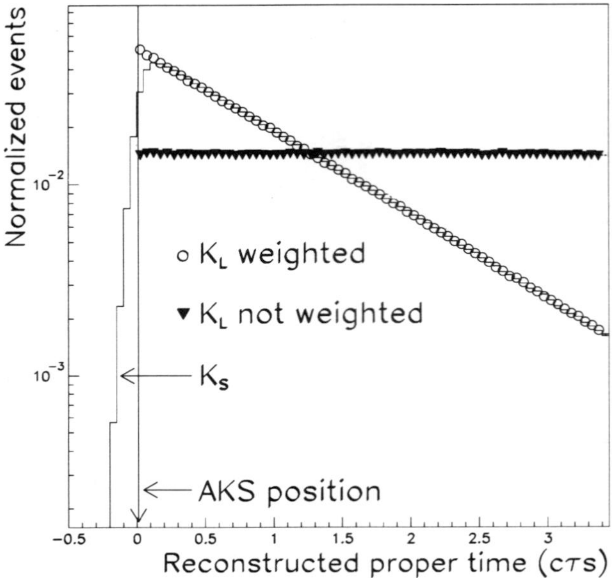

Figure 8. Decay vertex distributions from NA48 for $K \to \pi\pi$ before and after lifetime reweighting.

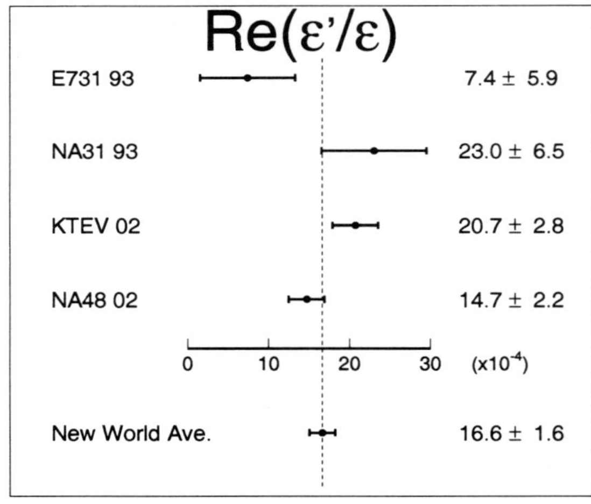

Figure 9. Summary of measurements of $Re(\epsilon'/\epsilon)$ in units of 10^{-4}.

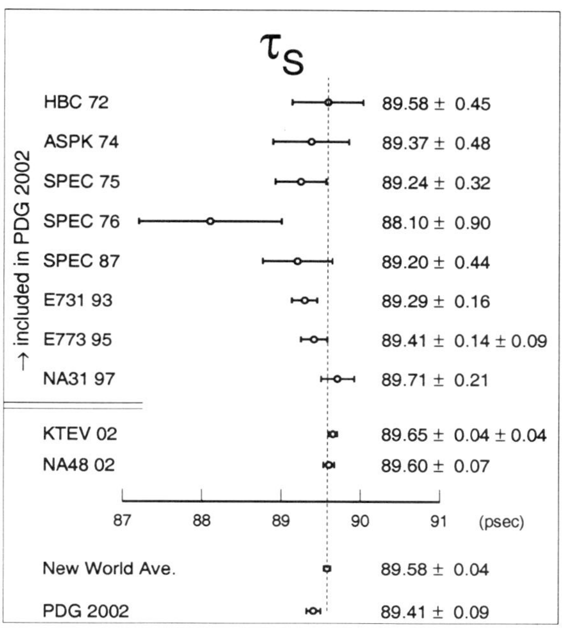

Figure 10. Summary of measurements of the K_S lifetime.

where $\epsilon_S = \epsilon + \Delta$ and $\epsilon_L = \epsilon - \Delta$. Here,

$$\begin{aligned}
\epsilon &= \frac{1}{2}\frac{<\overline{K}^0|H|K^0> - <K^0|H|\overline{K}^0>}{(m_L - m_s) + \frac{1}{2}i(\Gamma_S - \Gamma_L)} \\
\Delta &= \frac{1}{2}\frac{<\overline{K}^0|H|\overline{K}^0> - <K^0|H|K^0>}{(m_L - m_s) + \frac{1}{2}i(\Gamma_S - \Gamma_L)},
\end{aligned} \tag{7}$$

where m_L (m_S) and Γ_L (Γ_S) are the K_L (K_S) mass and width, respectively. If CPT is conserved, $\Delta = 0$ and $\epsilon_S = \epsilon_L = \epsilon$.

The phase of $A(K_L \to \pi\pi)/A(K_S \to \pi\pi)$ may be calculated using Eq. 7 and unitarity.[24] If CPT is conserved and T is violated (i.e., $\Delta = 0$), the phase of ϵ is equal to the "superweak" phase:

$$\phi_{\pi\pi} = \phi_{SW} = \tan^{-1}\frac{2(m_L - m_S)}{\Gamma_S - \Gamma_L} = (43.5 \pm 0.1)^\circ. \tag{8}$$

If instead, CPT is violated and T is conserved,

$$\phi_{\pi\pi} = \tan^{-1}\frac{-(\Gamma_S - \Gamma_L)}{2(m_L - m_S)} = (133.5 \pm 0.1)^\circ. \tag{9}$$

Experimentally, $\phi_{+-} = (43.4 \pm 0.7)^\circ$,[10] consistent with CPT conservation and T violation. Assuming no other sources of CPT violation, the difference between the

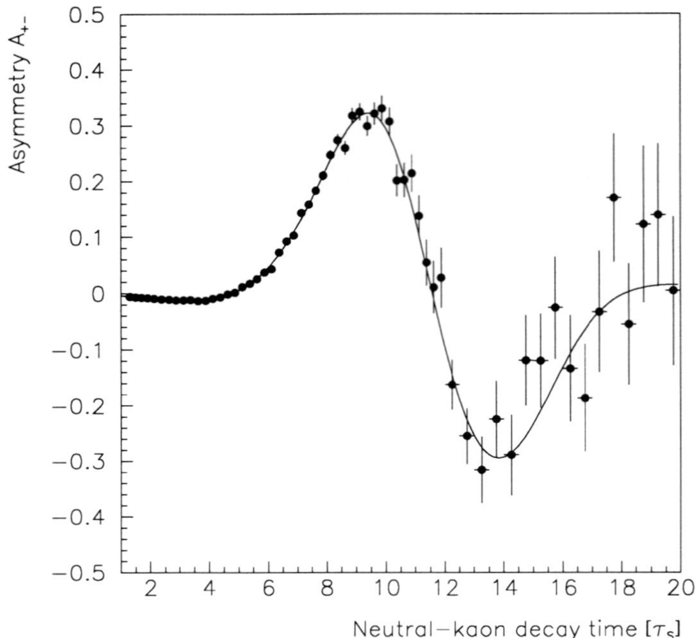

Figure 11. CPLEAR measurement of $A_{+-}(t) = (\overline{N}(t) - N(t))/(\overline{N}(t) + N(t))$, where $\overline{N}(t)$ and $N(t)$ are the number of $\pi^+\pi^-$ decays at time t for events tagged as K^0 and $\overline{K}^0$, respectively, at $t = 0$.

measured $\phi_{\pi\pi}$ and ϕ_{SW} may be converted to a limit on $m_{\overline{K}^0} - m_{K^0}$:

$$|m_{\overline{K}^0} - m_{K^0}| = 2(m_L - m_S)\,|\eta|\,|\phi_{\pi\pi} - \phi_{SW}|/\sin\phi_{SW} < 5 \times 10^{-19} \text{ GeV}/c^2. \quad (10)$$

This limit is often quoted in terms of a fractional mass difference: $|m_{\overline{K}^0} - m_{K^0}|/m_K < 10^{-18}$.

As stated above, it is also possible to search for CPT violation in decay amplitudes (direct CPT violation). Measurements of $\pi\pi$ phase shifts[26] show that, in the absence of CPT violation, the phase of ϵ' is approximately equal to that of ϵ. Therefore, $Re(\epsilon'/\epsilon)$ is a measure of direct CP violation and $Im(\epsilon'/\epsilon)$ is a measure of CPT violation. For small $|\epsilon'/\epsilon|$, $Im(\epsilon'/\epsilon)$ is related to the phases of η_{+-} and η_{00} by

$$\Delta\phi \equiv \phi_{00} - \phi_{+-} \approx -3Im(\epsilon'/\epsilon), \quad (11)$$

so $\Delta\phi$ should be very close to zero if CPT is not violated in decay amplitudes. The world average value of $\Delta\phi$ is $(0.22 \pm 0.43)^\circ$, coming mainly from the recent KTeV measurement.[19] This phase difference corresponds to $Re(B_2)/Re(A_2) = (-1 \pm 2) \times 10^{-4}$, where A_2 and B_2 are the $I = 2$ CP conserving and violating amplitudes, respectively.[25] Figure 12 shows two sigma allowed contours in $Im(\epsilon'/\epsilon)$ vs. $Re(\epsilon'/\epsilon)$, along with predictions of CPT invariance. Additional tests of CPT violation in K decays are described in Refs. 10 and 25.

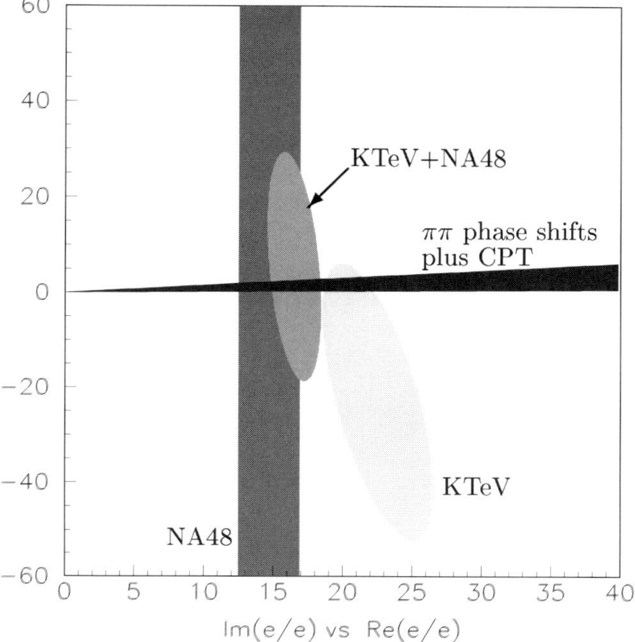

Figure 12. Two sigma constraints on $Im(\epsilon'/\epsilon)$ vs. $Re(\epsilon'/\epsilon)$ from NA48, KTeV, and the combination of NA48 and KTeV. The allowed region based on measured $\pi\pi$ phase shifts and CPT invariance is shown also.

6   T Violation

Although the measurement of ϕ_{+-} demonstrates that the observed CP violation in $K_L \to \pi\pi$ is accompanied by T violation, experiments also have searched for direct signatures of T violation. The clearest of these measurements was the so-called Kabir test performed by the CPLEAR collaboration,[27] in which they compared $K^0 \to \overline{K}^0$ with $\overline{K}^0 \to K^0$ transitions. As described earlier, CPLEAR uses the charge of the accompanying charged kaon to tag the flavor of the neutral kaon at production $(t = 0)$. Semileptonic decays are then used to determine the strangeness of the neutral kaon at the moment it decays $(t = t')$. They measure the following asymmetry as a function of decay time:

$$A_T = \frac{\Gamma[\overline{K}^0(t=0) \to K^0(t=t')] - \Gamma[(K^0(t=0) \to \overline{K}^0(t=t')]}{\Gamma[\overline{K}^0(t=0) \to K^0(t=t')] + \Gamma[(K^0(t=0) \to \overline{K}^0(t=t')]} = 4Re(\epsilon). \quad (12)$$

This asymmetry is plotted in Fig. 13. The average decay rate asymmetry is $<A_T> = (6.6 \pm 1.3 \pm 1.0) \times 10^{-3}$, showing clear evidence for T violation; note that this result is consistent with $4Re(\epsilon) = (6.63 \pm 0.04) \times 10^{-3}$ from measurements of $|\eta_{+-}|$.

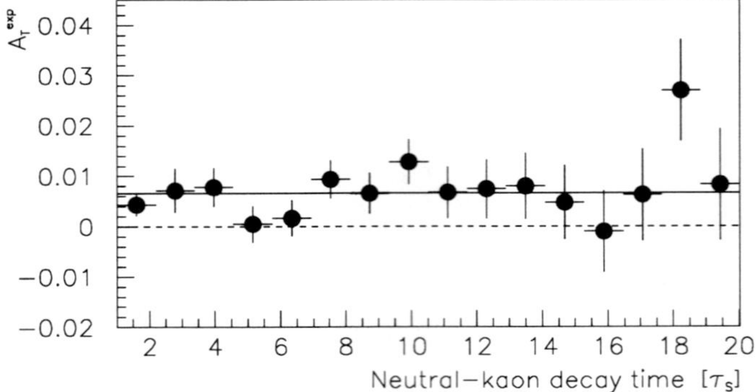

Figure 13. $\overline{K}^0 \to K^0 - K^0 \to \overline{K}^0$ asymmetry as a function of decay time from CPLEAR.

7 Conclusions

The neutral kaon has been studied for more than 50 years and continues to be a unique source of information about basic symmetries of nature. After more than 30 years of effort, direct CP violation has been established, This observation gives strong support to the CKM model and rules out the superweak model as the sole source of CP violation. In addition, all measurements in the kaon system are consistent with CPT invariance, meaning that CP violation is accompanied by T violation.

Following the observation of direct CP violation, several questions remained.

1. Is CP violation unique to the kaon system?

2. Is the Standard "CKM" Model the correct description of CP violation?

3. Are there CP/T violating effects in neutrino oscillations?

4. What is the connection (if any) between CP violation in elementary particles and the matter-antimatter asymmetry in the Universe?

The first of these questions was recently answered by the BaBar and Belle experiments: the CP symmetry also is violated in B mesons.[22,23,28] The second and third questions are the motivation for much of the future experimental work in flavor physics. The last question remains the most intriguing and least well understood; it will continue to motivate great interest in a deeper understanding of symmetry violation.

Acknowledgements

I wish to thank I. Bigi, A. Glazov, R. Kessler, H. Nguyen, V. Prasad, J. Steinberger, and B. Winstein for many useful discussions. Thanks also to I. Bigi and M. Faessler for organizing such an interesting meeting.

References

1. G. Luders, Ann. Phys. **2**, 1 (1957).
2. T.D. Lee and C.N. Yang, Phys. Rev. **105**, 1413 (1957).
3. C.S. Wu *et al.*, Phys. Rev. **105**, 1413 (1957); R. Garwin *et al.*, Phys. Rev. **105**, 1415 (1057); J.I. Friedman and V.L. Telegdi, Phys. Rev. **105**, 1681 (1957).
4. L. Landau, Nucl. Phys. **3**, 127 (1957).
5. J. H. Christenson, J. W. Cronin, V. L. Fitch, and R. Turlay, *Phys. Rev. Lett.* **13**, 138 (1964).
6. Rochester and Butler, Nature **160**, 855 (1947).
7. P. Galison, *Image and Logic* (The University of Chicago Press, Chicago 1997), p. 126.
8. M. Gell-Mann and A. Pais, Phys. Rev. **97**, 1387 (1955).
9. K. Lande, E.T. Booth, J. Empeduglia, L.M. Lederman, and W. Chinowsky, Phys. Rev. **103**, 1901 (1956).
10. Particle Data Group *Phys. Rev.* D **66**, 1 (2002).
11. A. Alavi-Harati *et al.* [KTeV Collaboration], *Phys. Rev. Lett.* **88**, 181601 (2000).
12. M. Kobayashi and T. Maskawa. Prog. Theor. Phys. **49**, 652 (1973).
13. J. Ellis, M. K. Gaillard, and D. V. Nanopoulos. Nucl. Phys. **B109**, 213 (1976).
14. F. J. Gilman and M. B. Wise. Phys. Lett. **83B**, 83 (1979).
15. For example, see references 6-16 in Ref. 20.
16. L. Wolfenstein. Phys. Rev. Lett. **13**, 562 (1964).
17. L.K. Gibbons *et al.* (E731 Collaboration), Phys. Rev. Lett. **70**, 1203 (1993).
18. G.D. Barr *et al.* (NA31 Collaboration), Phys. Lett. **B317**, 233 (1993).
19. A. Alavi-Harati *et al.* [KTeV Collaboration], *Phys. Rev.* D **67**, 012005 (2003).
20. J. R. Batley *et al.* [NA48 Collaboration], *Phys. Lett.* B **544**, 97 (2002).
21. A. Apostolakis *et al.* [CPLEAR Collaboration], Phys. Lett. B **458**, 545 (1999).
22. B. Aubert *et al.* [BABAR Collaboration], Phys. Rev. Lett. **87**, 091801 (2001)
23. K. Abe *et al.* [Belle Collaboration], Phys. Rev. Lett. **87**, 091802 (2001).
24. This discussion follows the treatment in J. Steinberger, "Experimental Status of CP Violation," in Proceedings of CP Violation in Particle Physics and Astrophysics, Blois, France, 55 (1990) based on J.S. Bell and J. Steinberger, "Weak Interactions of Kaons'," in Proceedings of the Oxford Int. Conference on Elementary Particles, Rutherford Lab., Chilton, England, 195 (1965).
25. V.V. Barmin *et al.*, *Phys. Lett.* B **247**, 293 (1984).
26. W. Ochs. πN Newsletter **3**, 25 (1991).
27. D. Zavrtanik, these proceedings; A. Angelopoulos *et al.* [CPLEAR Collaboration], *Phys. Lett.* B **444**, 43 (1998).
28. I. Bigi, these proceedings.

TIME-REVERSAL NON-INVARIANCE

D. ZAVRTANIK

*Nova Gorica Polytechnic, Vipavska 13, POB 301, SI-5000 Nova Gorica, Slovenia and
J. Stefan Institute, Jamova 39, SI-1000 Ljubljana, Slovenia
E-mail: danilo.zavrtanik@p-ng.si*

The arrow of time is one of the great mysteries in elementary particle physics. While the arrow of time is well defined and experimentally confirmed in the macrocosm, there was no experimental evidence for time-reversal violation in the subatomic world till recently. The CPLEAR experiment at CERN has tested the time-reversal invariance in the neutral kaon system with high precision. An asymmetry between $\bar{K}^0 \rightarrow K^0$ and $K^0 \rightarrow \bar{K}^0$ rates was observed leading to the first experimental observation of T violation in the microcosm. Within the experimental accuracy the amount of T violation was found to be commensurate with the amount of CP violation as required by CPT invariance.

1 Introduction

The concept of time and its implications in physics and cosmology are discussed in many textbooks. A "brief history of time" has been told in a book by S. Hawking [1] and his basic ideas on the arrow of time are briefly summarized here as an introduction.

Everyday experience tells us that it is impossible to travel backwards in time. This fact led to at least three different definitions of the arrow of time in the macrocosm. Currently most physicists associate the irreversibility of time with the production of entropy or disorder in the warm macroscopic world. The consequence of this irreversibility is often called the thermodynamic arrow of time which points towards the direction in which entropy increases. Then, there is the psychological arrow of time, the direction in which we feel time passes. This definition is connected to the fact that human beings remember the past but not the future. Finally, we have the cosmological arrow of time. This is the direction of time in which the Universe is expanding. All the three arrows of time point in the same direction. One can argue that the psychological arrow of time is determined by the thermodynamic arrow. Thus, the two definitions are not independent and they necessarily point to the same direction. However, thermodynamic and cosmological arrows of time are well defined. They point in the same direction although it may not be so for the whole history of our Universe.

Although there is a big difference between the forward and backward directions of time in ordinary life, the fundamental classical laws of physics discovered by Galilei, Newton and Einstein do not distinguish between the past and the future. In other words, they appear to be symmetric in time. Thus, one might expect that the thermodynamic arrow of time would be absent in the cold microcosm of elementary particles. Therefore, understanding the nature of time-reversal symmetry and its violation appears essential for a full understanding of the concept of time.

Where to search for time asymmetric processes in the microcosm? To the present knowledge, we are living in a CPT conserving world. On the other hand, CP violation in weak interactions was discovered almost 40 years ago. As a consequence we expect weak processes to be time asymmetric.

2 Historical overview of experimental efforts

2.1 Indirect experimental tests of time-reversal invariance

In order for particles to have electric dipole moments (EDM), the forces concerned in their structure must violate both space parity (P) and time-reversal (T) symmetries [2]. There are many competing theories that attempt to predict the electric dipole moment of elementary particles. Experimental measurements of particle EDMs are providing strong constraints on these theories. The most sensitive are probably the searches for an EDM of the neutron and the electron. Present limits on EDM for the neutron [3] $|d_n| < 6.3 \times 10^{-26} e\ cm$ and the electron [4] $|d_e| = (0.07 \pm 0.07) \times 10^{-26} e\ cm$ are not precise enough to give an answer on possible time-reversal violation.

An indirect demonstration of the time-reversal violation was made by Schubert et al. in 1970 at CERN using data on the decay of long- and short-lived kaons into two neutral pions. Their result $\mathcal{R}e(\epsilon) = (1.68 \pm 0.3) \times 10^{-3}$ was obtained by assuming unitarity. In other words the authors assumed that those kaons that disappeared had decayed into observable states. More details on the analysis can be found elsewhere [5].

A few years ago, the KTeV Collaboration reported results [6] from an experimental study of rare decays of neutral kaons into electron-positron and charged-pion pairs. A T-odd asymmetry was detected through a large asymmetry in the angular distribution between the decay planes of the e^+e^- and $\pi^+\pi^-$ pairs. Since incoming and outgoing states are not exchanged in KTeV experiment, it cannot provide direct evidence for time-reversal violation. However, the rich structure of the measured decay mode may provide new opportunities for the study of novel CP- and T-violation effects.

2.2 Direct experimental tests of time-reversal invariance

In the late sixties, there were some attempts to test time-reversal invariance through experimental verification of the principle of detailed balance in nuclear reactions. Experiments were comparing the probabilities for reactions to occur in the forward and time reversed direction (see for example reference [7]). Those experiments were very difficult. They have to measure two different reactions in two separate experiments with different systematics. In addition, they have to deal with different phase space. All these and the limitations of experimental techniques available led to poor accuracy and to too low sensitivity to possibly detect time-reversal violation.

The neutral-kaon system proved to be a unique experimental environment to directly test discrete symmetries. Up to recently, it was the only system in nature where the violation of CP symmetry has been observed. In addition, experimental observation of interference effects allows direct measurements of T violation and CPT tests with a precision unachievable in other systems. Following an original proposal of Kabir [8] and Aharony [9], direct evidence for T violation can be obtained by comparing the transformation rates of $\overline{K}^0$ to K^0 and K^0 to $\overline{K}^0$ which are two T-conjugate processes. The idea was elaborated in the eighties by Tanner and

Dalitz [10] when a symmetric production of K^0 and $\overline{K}^0$ became possible at the LEAR accelerator at CERN. This test was later performed by the CPLEAR Collaboration at CERN and will be reported in more details in the present article.

3 CPLEAR experiment at CERN

3.1 Method

Since weak interactions do not conserve strangeness, a K^0 meson can transform into a $\overline{K}^0$ and vice versa. Time-reversal invariance would require all details of the second process to be deducible from the first. In particular, the probability ($\mathcal{P}$) that a K^0 at $t = 0$ is observed as a $\overline{K}^0$ at time τ should be equal to the probability that a $\overline{K}^0$ at $t = 0$ is observed as a K^0 at the same time τ [8]. Any difference between these two probabilities is a signal for T violation and can be measured through the time-reversal asymmetry

$$\frac{\mathcal{P}(\overline{K}^0 \to K^0) - \mathcal{P}(K^0 \to \overline{K}^0)}{\mathcal{P}(\overline{K}^0 \to K^0) + \mathcal{P}(K^0 \to \overline{K}^0)}. \tag{1}$$

Experimentally this requires knowledge of the strangeness of the neutral kaon at its production and its decay. The initial strangeness of the neutral kaon is tagged by the charge of the accompanying charged kaon at the production time. To tag the strangeness of the kaon at the decay time, semileptonic decays $K^0 \to e^+\pi^-\nu$ and $\bar{K}^0 \to e^-\pi^+\bar{\nu}$ are used. A measured positive lepton charge is associated to a K^0 and negative lepton charge to a $\overline{K}^0$. Details are shown in Fig. 1. There are four measurable decay rates R, labelled by the initial kaon strangeness and the final electron charge:

$$R^+(\tau) = R(K^0_{t=0} \to e^+\pi^-\nu_{t=\tau}) \qquad R^-(\tau) = R(K^0_{t=0} \to e^-\pi^+\overline{\nu}_{t=\tau})$$
$$\bar{R}^-(\tau) = R(\overline{K}^0_{t=0} \to e^-\pi^+\overline{\nu}_{t=\tau}) \qquad \bar{R}^+(\tau) = R(\overline{K}^0_{t=0} \to e^+\pi^-\nu_{t=\tau}).$$

The decay amplitudes are parametrized with quantities which have well defined symmetry properties under T, CP and CPT symmetries,

$$\langle \ell^+\pi^-\nu| \, T \, |K^0\rangle = a + b \,, \qquad \langle \ell^-\pi^+\overline{\nu}| \, T \, |\overline{K}^0\rangle = a^* - b^* \,,$$
$$\langle \ell^-\pi^+\overline{\nu}| \, T \, |K^0\rangle = c + d \,, \qquad \langle \ell^+\pi^-\nu| \, T \, |\overline{K}^0\rangle = c^* - d^* \,,$$

where ℓ denotes an electron or a muon and T the decay transition matrix. $\mathcal{R}e(a)$ is CP, T and CPT symmetric, all the imaginary parts are T violating, b and, equivalently, $y = -b/a$ violate CPT. The last two transitions (c and d) violate the $\Delta S = \Delta Q$ rule, whose validity is parametrized by $x = (c^* - d^*)/(a + b)$ and $\bar{x} = (c^* + d^*)/(a - b)$, or $x_+ = (x + \bar{x})/2$ and $x_- = (x - \bar{x})/2$, with x_- also violating CPT and $x_+ \leq 10^{-6}$ in the Standard Model.

The asymmetry

$$A_T = \frac{R(\overline{K}^0 \to K^0) - R(K^0 \to \overline{K}^0)}{R(\overline{K}^0 \to K^0) + R(K^0 \to \overline{K}^0)} \tag{2}$$

represents a direct test of T reversal through detailed balance [8].

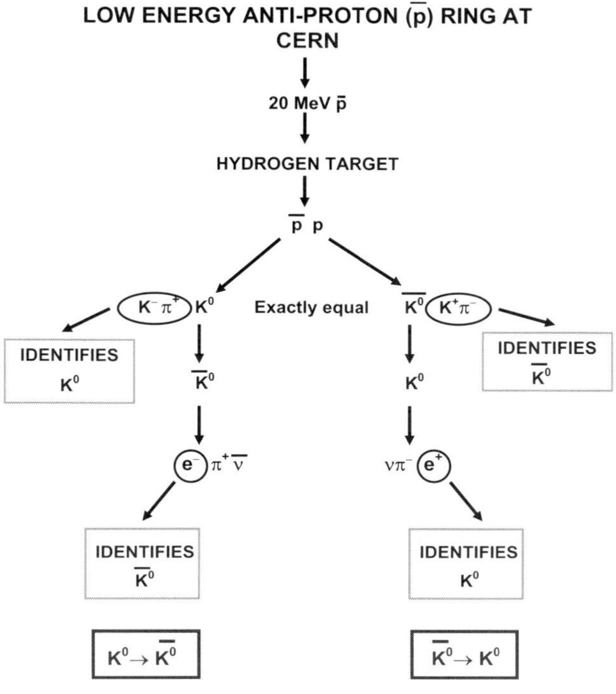

Figure 1. Schematic view of the CPLEAR method.

3.2 *Apparatus*

In the CPLEAR experiment, initial K^0 and $\overline{K}^0$ are produced concurrently via the annihilations of antiprotons at rest into $K^0 K^- \pi^+$ and $\overline{K}^0 K^+ \pi^-$. Both reactions occur at a branching ratio of $\approx 2 \times 10^{-3}$. The antiprotons are delivered by the low-energy antiproton ring LEAR at CERN and are stopped in a gaseous hydrogen target in the centre of the CPLEAR detector. The tagging of opposite strangeness states at production time maximizes the interference effects to be observed in K^0, $\overline{K}^0$ decays.

The experimental apparatus is shown in Fig. 2. Ten chamber layers (2 Proportional Chambers, 6 Drift Chambers, 2 Streamer Tubes) are used to trace charged particles resulting from annihilation and neutral-kaon decays. A 32-segment sandwich of scintillator-Cherenkov-scintillator detectors provided particle identification (kaons/pions/electrons). Photons were detected by an 18-layer fine-grain streamer tube/lead sampling calorimeter. Signals from all detectors were processed in a multilevel trigger, providing a rejection factor of over 1000 and allowing the detector to operate at a $\bar{p}$ rate of 1 MHz. The material in the decay region up to the streamer tubes was minimized by using a gas target with mylar-kevlar walls and innovative low-mass chamber construction, thus reducing regeneration effects of neutral kaons.

CPLEAR Detector

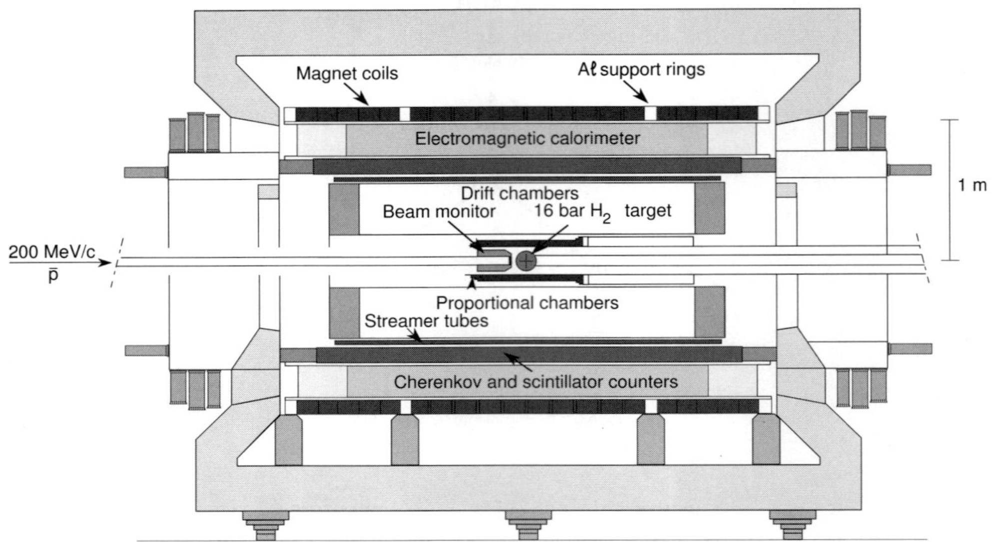

Figure 2. View of the CPLEAR detector.

In 1995 a proportional chamber was added at 1.7 cm radius to improve the trigger and tracking capabilities. A more detailed description of the detector performance can be found elsewhere [12].

In total, about 100 million K^0 and $\overline{K}^0$ decays were reconstructed. The results presented refer to the analysis of the complete data-set of 70 M $K^0, \overline{K}^0 \to \pi^+\pi^-$ decays with $\tau > 1\tau_S$, 1.3 M $K^0, \overline{K}^0 \to e\pi\nu$ decays, 0.5 M $K^0, \overline{K}^0 \to \pi^+\pi^-\pi^0$ decays, 2 M $K^0, \overline{K}^0 \to \pi^0\pi^0$ decays and 17 k $K^0, \overline{K}^0 \to \pi^0\pi^0\pi^0$ decays, extracted from about 50 000 data tapes with 100 000 events each.

3.3 Analysis

Time-reversal invariance was studied in the semileptonic decay channel of the neutral kaons. The desired $\bar{p}$ p annihilations followed by the decay of the neutral kaon into $e\pi\nu$ were first selected by requiring that the events had four charged tracks and zero total charge, and by identifying one of the decay tracks as that of an electron or a positron. The lepton identification was performed by a neural network algorithm making use of energy loss and time of flight in the scintillators and number of electrons in the Čerenkov detector. To further reduce the background, kinematically constrained fits were used. A total of 1.3×10^6 $e\pi\nu$ events, with decay times $\tau > 1\,\tau_S$ survived the above criteria. The signal to background ratios for the different background channels as a function of the decay time were obtained by a Monte Carlo simulation. There is excellent agreement between real and simulated data as can be seen in Fig. 3. A conservative estimate of 10 % uncertainty was used for

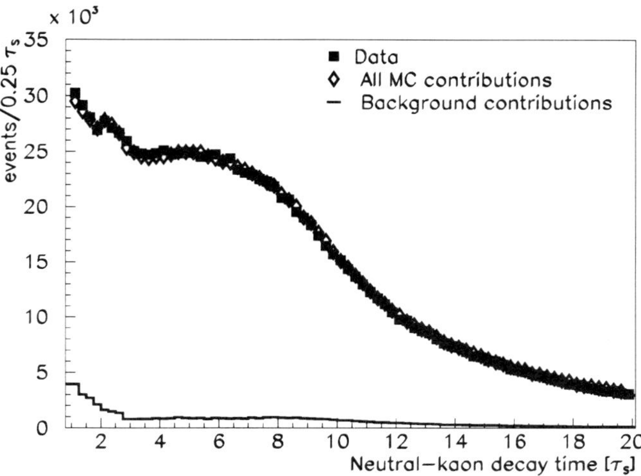

Figure 3. Decay time distribution for real and simulated data.

the systematic errors on the background levels. Regeneration was corrected using the amplitudes measured by CPLEAR [11] on an event-by-event basis.

Measurement of the asymmetry A_T from Eq. 2 requires knowledge of the strangeness of the neutral kaon at its production and its decay. The K^0 and $\overline{K}^0$ tagging efficiencies differ since the interactions of their tagging particles in the detector materials are different (the CPLEAR detector was not made of equal parts of matter and antimatter). The efficiency correction for the initial strangeness tag is parametrized by the efficiency ratio $\xi = \epsilon(K^+\pi^-)/\epsilon(K^-\pi^+)$ while the correction for the relative detection efficiency of $e^+\pi^-$ versus $e^-\pi^+$ is applied through $\eta = \epsilon(\pi^+e^-)/\epsilon(\pi^-e^+)$. With the measured number of events N corrected for η and ξ, the time-reversal asymmetry from Eq. 2 reads

$$A_T^{exp} = \frac{\eta\bar{N}^+ - \xi N^-}{\eta\bar{N}^+ + \xi N^-}.$$

(3)

Calibration data were used to obtain η in bins of (p_e, p_π). Its average value was $< \eta > = 1.014 \pm 0.002$. The parameter ξ was taken from $\pi^+\pi^-$ decays through $\alpha_{2\pi} = [1 + 4\mathcal{R}e(\epsilon_T - \delta)] \times \xi$, where $\alpha_{2\pi}$ is obtained by using the measured (un-weighted) number of K^0, $\overline{K}^0 \to \pi^+\pi^-$ decays in the decay-time interval 1-4 τ_S, as described in details in [13]. External information on $\mathcal{R}e(\epsilon_T - \delta_{CPT})$ was taken from the semileptonic charge asymmetry [14]

$$\delta_\ell = 2\mathcal{R}e(\epsilon_T - \delta_{CPT}) - 2\mathcal{R}e(x_- + y) = (3.27 \pm 0.12) \times 10^{-3},$$

which is equal to $2\mathcal{R}e(\epsilon_T - \delta_{CPT})$ in the limit of CPT symmetry in decay amplitude, and an event-by-event correction was applied according to primary pair kinematics. The average value of ξ is $< \xi > = 1.12023 \pm 0.00043$.

3.4 Results

The experimental asymmetry obtained from the complete CPLEAR data-set is shown in Fig. 4. The data points scatter around a positive and constant offset

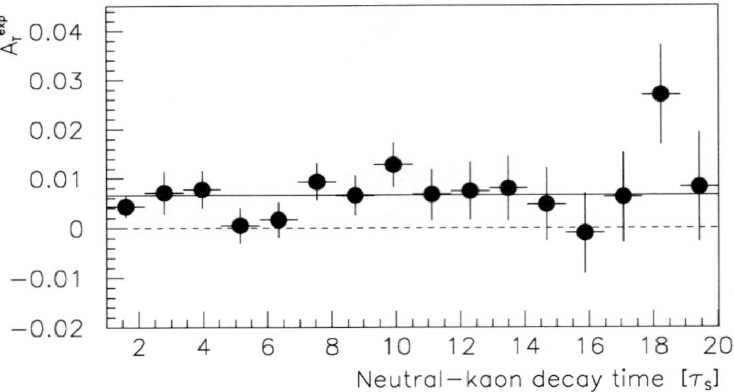

Figure 4. Time-reversal violation asymmetry A_T^{exp}. Full line represents a constant fit in the decay-time interval 1-20 τ_S.

from zero, representing a surplus of $\bar{K}^0 \rightarrow K^0$ transitions. Its average value

$$< A_T^{exp} >_{(1-20)\tau_S} = (6.6 \pm 1.3_{stat} \pm 1.0_{syst}) \times 10^{-3}$$

represents the first direct measurement of time-reversal violation. Details of the analysis and a full description of systematic errors are given elsewhere [15].

To clarify the eventual influence of $\Delta Q \neq \Delta S$ or CPT-violating transitions one has to consider the phenomenological expression for A_T^{exp}

$$A_T^{exp} = 4(\mathcal{R}e(\epsilon_T - y) - \mathcal{R}e(x_-)) +$$
$$+ 2\frac{\mathcal{R}e(x_-)(e^{-\frac{1}{2}\Delta\Gamma\tau} - \cos(\Delta m\tau)) + \mathcal{I}m(x_+)\sin(\Delta m\tau)}{\cosh(\frac{1}{2}\Delta\Gamma\tau) - \cos(\Delta m\tau)} .$$

In the long lifetime limit the expression for A_T^{exp} reads

$$A_T^{exp} \overset{\tau \gg \tau_S}{\longrightarrow} 4\mathcal{R}e(\epsilon_T) - 4\mathcal{R}e(y + x_-), \tag{4}$$

where the parameter y describes CPT violation in semileptonic decays while x_- is a CPT and $\Delta S = \Delta Q$ rule violating parameter.

With the replacement $\alpha_{2\pi} \rightarrow \xi$ using δ_ℓ, an additional term $-4\mathcal{R}e(y + x_-)$ entered into A_T^{exp} leading to the possibility that the observed asymmetry might be a consequence of CPT or both T and CPT violation. This CPT violating term can be determined by using the Bell-Steinberger relation which relates all decay channels of neutral kaons to the parameters describing T and CPT non-invariance. By using data from the CPLEAR experiment, together with the most recent world averages for some of the neutral-kaon parameters and the Bell-Steinberger relation CPLEAR has determined [16] that

$$\mathcal{Re}(y + x_-) = (-0.2 \pm 0.3) \times 10^{-3}$$

This result, based on the sole assumption of unitarity, confirms the interpretation of A_T^{exp} as the first direct measurement of time-reversal violation.

4 Discussion

The CPLEAR result on time-reversal violation triggered a vivid scientific discussion. Criticism of the CPLEAR interpretation of A_T^{exp} result (see for example [17]) was confronted with arguments in favor of their interpretation (see for example [18], [19]). Our main arguments that CPLEAR did directly measure the time-reversal non-invariance are as follows:

- CPLEAR has independently tested CPT invariance. Their free of assumption result on CPT violating parameter δ_{CPT} is $\mathcal{Re}(\delta_{CPT}) = (3.0 \pm 3.3_{stat} \pm 0.6_{syst}) \times 10^{-4}$ which is compatible with zero with an accuracy of the order of 10^{-4}. In the absence of a CPT violating theory we can expect $\mathcal{Re}(y)$ to be equal to or smaller than δ_{CPT}.

- parameter x_-, as a CPT and $\Delta S = \Delta Q$ rule violating parameter, is expected to be even much smaller than y.

Other CPLEAR experimental data, constraining possible violations of CPT and $\Delta S = \Delta Q$ rule, exclude the possibility that the semileptonic-decay asymmetry A_T could be solely due to CPT violation. Together with arguments from the paper by L. Alvarez-Gaumé et al. [19] we confirm that CPLEAR experiment constitute direct evidence for time-reversal violation without any assumption on unitarity and CPT violation.

5 Conclusion

The CPLEAR experiment has directly measured for the first time the violation of time-reversal invariance in the neutral-kaon system. Within the experimental accuracy, the amount of T violation is found to be equal to the amount of CP violation observed in the neutral-kaon system and is consistent with CPT conservation.

Our results show that transformation of anti-matter to matter is slightly more probable than the reverse process. This represents the first measurement of irreversibility of a process in the microcosm. In other words, it means that laws of physics are not symmetric with respect to time-reversal. However, this discovery does not tell us if entropy is produced in the microcosm. Increase of entropy would require CPT violation and probably evidence of new fundamental interactions in nature. Nevertheless, the discovery of the violation of time-reversal invariance might have far-reaching consequences for our understanding of the cosmos and the concept of time.

References

1. S.W. Hawking, A brief history of time: from big bang to black holes, (reprinted ed.), (Bantam Books, Toronto, 1996).
2. S.M. Barr, *Int. J. Mod. Phys.* A **8**, 209 (1993).
3. P.G. Harris *et al.*, *Phys. Rev. Lett.* **82**, 904 (1999).
4. B.C. Regan *et al.*, *Phys. Rev. Lett.* **88**, 071805 (2002).
5. K.R. Schubert *et al*, *Phys. Lett.* B **31**, 662 (1970),
   J.C. Chollet *et al*, *Phys. Lett.* B **31**, 658 (1970).
6. A. Alavi-Harati *et al.*, *Phys. Rev. Lett.* **84**, 408 (2000).
7. U. Wimmersperg *et al.*, *Phys. Lett.* B **33**, 291 (1970).
8. P. Kabir, *Phys. Rev.* D **2**, 540 (1970).
9. A. Aharony, *Lett. Nuovo Cimento* **3**, 791 (1970).
10. N.W. Tanner, R.H. Dalitz, *Ann. of Phys.* **171**, 463 (1986).
11. CPLEAR Collaboration: A. Angelopoulos *et al.*, *Phys. Lett.* B **413**, 422 (1997).
12. CPLEAR Collaboration: R. Adler *et al.*, *Nucl. Instrum. Methods* A **379**, 76 (1996).
13. CPLEAR Collaboration: A. Angelopoulos *et al.*, *Phys. Lett.* B **444**, 52 (1998).
14. K. Hagiwara *et al.*, Particle Data Group, *Phys. Rev.* D **66**, 010001 (2002).
15. CPLEAR Collaboration: A. Angelopoulos *et al.*, *Phys. Lett.* B **444**, 43 (1998).
16. CPLEAR Collaboration: A. Apostolakis *et al.*, *Phys. Lett.* B **456**, 297 (1999).
17. L. Wolfenstein, *Phys. Rev. Lett.* **83**, 911 (1999),
    P.K. Kabir, *Phys. Lett.* B **459**, 335 (1999),
    I.I. Bigi and A.I. Sanda, *Phys. Lett.* B **466**, 33 (1999).
18. J. Ellis and N.E. Mavromatos, *Phys. Rep.* **320**, 341 (1999).
19. L. Alvarez-Gaumé *et al.*, *Phys. Lett.* B **458**, 347 (1999).

'PER ASPERA AD ASTRA'* — A SHORT ESSAY ON THE LONG QUEST FOR CP VIOLATION

I.I. BIGI

Dept. of Physics, University of Notre Dame du Lac, Notre Dame, IN 46556, USA
E-mail: ibigi@nd.edu

After briefly explaining the special role played by violations of CP and T invariance and their connection with the baryon number of the Universe, I sketch the history of CP violation studies since its totally unexpected discovery in 1964. For about 30 years CP violation could be described by a single number; this has changed dramatically in the years around the turn of the millenium: (i) The existence of *direct* CP violation was unequivocally established in the decays of long lived kaons. (ii) For the first time CP violation was observed in a system other than that of neutral kaons, namely in $B \to \psi K_S$. The findings are in impressive agreement with the prediction of the CKM ansatz, which thus has been promoted to the status of a tested theory. These new insights were made possible by close feedback between theory and experiment as well as advances in detector design and a novel machine concept, namely that of an asymmetric collider. We also have direct *experimental* evidence that the observed CP violation in K_L and B decays is matched by a violation of microscopic time reversal violation, as required by CPT symmetry. More recently CP violation has been observed also in $B \to \pi^+\pi^-$ and $B \to K^\mp \pi^\pm$. A few comments are added on subtle aspects of direct CP violation. While we know that the CKM dynamics are irrelevant for generating the baryon number of the Universe – i.e. hitherto unknown forces have to be driving it – we have also learnt that such 'New Physics' is likely to contain CP violation of sufficient strength. [a]

---

[a]This essay covers material from the talks by C. Jarlskog and D. Hitlin, for which no manuscript was received. Numbers have been updated to reflect the status of 2004, as described at ICHEP 2004, http://ichep04.ihep.ac.cn/program.htm.

Prologue

The conference 'Time and Matter' has as subtitle 'An International Colloquium on the Science of Time'. The tale of the physicists' quest for CP violation fits naturally into this frame, since it presents us with several variations on the theme of *time*: the time it took to perform the experimental studies, the uneven rate of progress in our understanding, the time that had to be measured to reveal the sought-after CP asymmetry and the preference Nature shows on the microscopic level for the flow of time. It involves glorious applications of fundamental quantum mechanics, its superposition principle and of EPR correlations [1] with their effects building up over macroscopic distances of centimeters, meters and even hundreds of meters. Finally it connects the "heavens" to the "earth" in that it provides us with a scenario where the seeds for the preponderance of matter over antimatter observed today can be generated *dynamically* in the very early Universe.

---

*"THROUGH THE ROUGHS TO THE STARS"

1 On the Special Role of CP Violation

There are three *discrete* transformations of general interest, namely parity P, *microscopic* time reversal T (operationally amounting to reversal of motion $\vec{p} \to -\vec{p}$) and charge conjugation C, which replaces particles by their antiparticles. Originally it had been assumed without much reflection that all three represent symmetries of nature, since they were known to be conserved by the strong and electromagnetic forces. The first to fall from this pedestal were P and C. The 1957 discovery of P (and subsequently also of C) being violated by the weak forces did cause a paradigm shift. It was, however, realized that even *maximal* parity violation – meaning there are left-, but *no* right-handed neutrinos – does not necessarily imply that nature exhibits a genuine preference for left over right. For while the decay $\pi^+ \to \mu^+ \nu_L$ produces only left-handed neutrinos, the antiparticle decay $\pi^- \to \mu^- \bar{\nu}_R$ yields right-handed neutrinos. They are referred to as antineutrinos, but at this point what is called particle and antiparticle is *pure convention*. For CP transformations relate the two processes; as long as CP invariance holds, they exhibit identical rates, and "left" and "right" is defined in terms of what one calls "positive" or "negative". This is reminiscent of the definition "the thumb is left on your right hand" – which is as correct as it is circular and thus useless.

The observation of CP violation in 1964 by the Fitch-Cronin experiment [2] then came as another shock. [a] For

$$\frac{\Gamma(K_L \to \mu^+ \nu \pi^-)}{\Gamma(K_L \to \mu^- \bar{\nu} \pi^+)} \simeq 1.006 \neq 1 , \tag{1}$$

which is related to $K_L \to \pi\pi$ [3], allows distinguishing a positive charge from a negative one through observation rather than convention. The discovery of CP violation thus changes our picture of Nature's structure even more profoundly than that of parity violation. At the same time it is quite 'frustrating' that a CP invariant – i.e. matter-antimatter symmetric – world is such a 'near-miss' with the difference on the 10^{-3} level, Eq.(1), in contrast to 'maximal' parity violation: $\Gamma(\pi^+ \to \mu^+ \nu_R)/\Gamma(\pi^+ \to \mu^+ \nu_L) = 0$.

There are more features singling out CP violation as particularly special and more fundamental than parity violation:

(i) Almost any Lorentz invariant local quantum field theory has to possess CPT invariance. CP violation thus has to be matched by a commensurate T violation. I.e., nature distinguishing 'left' and 'right' implies her to do likewise between a 'forward' and 'backward' flow of time already on the *microscopic* level *beyond* the macroscopic statistical consideration expressed through thermodynamics' second law of entropy increase.

(ii) The leading contribution to the CP asymmetry in kaon decays involves $K^0 - \bar{K}^0$ oscillations, which represent quantum mechanical interference on a *macroscopic* scale. This allows the accurate measurement of truly tiny effects. The deviation from unity in Eq.(1) is driven by a tiny difference in the off-diagonal element of the

---

[a]To my knowledge only Okun had stated explicitly *before* the Fitch-Cronin experiment that the question of CP invariance is one to be decided by experiment rather than theory.

$K^0 - \bar{K}^0$ mass difference:

$$\mathrm{Im}M_{12} \simeq 1.1 \cdot 10^{-8} \text{ eV} \quad \Leftrightarrow \quad \frac{\mathrm{Im}M_{12}}{M_K} \simeq 2.2 \cdot 10^{-17} \qquad (2)$$

CP violation can thus be seen as the smallest observed (rather than hypothesized) violation of a symmetry. In turn this means that searches for CP violation in different systems serve as very high sensitivity probes for hitherto unknown physics. (iii) In a 1967 paper [4] the famous Russian physicist (and dissident) Andrei Sakharov noted that the existence of CP violation opened the path towards a new paradigm with a cosmic connection, namely to understand the baryon number of the Universe, for which data yield $\sim 10^{-9}$, as a *dynamically generated* number rather than an arbitrary initial value [b]; this number can be qualitatively characterized through the following dual phrase: while the Universe is not empty, it is almost empty.

To achieve this goal, three requirements have to be met:

- There are forces changing the baryon number;

- CP invariance is broken;

- the Universe is out of thermal equilibrium.

The probably most ambitious goal in CP studies is to identify this cosmic connection, i.e. to find other manifestations of the CP violating forces that drive the baryon number of the Universe, which could be probed in reproducible laboratory experiments.

(iv) A particularly subtle feature is the following [3]. Since the time reversal operator T is *anti*unitary, we have $T^2 = \pm 1$. Thus the Hilbert space for systems invariant under time reversal consists of two disjoint sectors, one with $T^2 = +1$ and the other with $T^2 = -1$. Consider an energy eigenstate $|E\rangle$ belonging to the latter; it can be shown that the degenerate state $T|E\rangle$ is *orthogonal* to $|E\rangle$, which is referred to as 'Kramers' degeneracy' [5]. This implies that $|E\rangle$ carries an internal degree of freedom that is changed by T, *if $T^2 = -1$*. One should note that this property was derived *without* any reference to spin, half-integer or otherwise! It is of course completely consistent with the transformation behaviour of spin degrees of freedom. This conceptual consequence of Kramers' degeneracy can be rephrased as follows: a world where all systems have $T^2 = +1$ would be conceivable – as is one with only integer-spin or only with states obeying Bose-Einstein statistics; yet once again nature reveals its clear tendency to realize dynamical structures that are mathematically admissible – and to do it in a very efficient way: odd-integer states double as fermions and as systems with $T^2 = -1$. There is a more practical consequence of Kramers' degeneracy as well: an odd-number electron system placed inside an external electrostatic field will always exhibit (at least) a *two*-fold degeneracy, no matter how complicated that field is; this property does *not* hold for an even-number electron system.

---

[b]By baryon number of the Universe we mean effectively the ratio of the number of baryons in the Universe relative to that of photons in the microwave background radiation; we know of no significant source of *primary* antibaryons in the Universe.

While the Fitch-Cronin result was initially greeted with surprise, dismay and even shock, attempts at 'denial' [c] were soon abandoned. It was realized that novel insights and perspectives onto Nature's Grand Design could be gained by continuing a dedicated and comprehensive study of CP violation in a wide array of different reactions. The ensuing story is a fascinating one. It can be seen as describing the high energy physics paradigm in a nutshell: a fundamental question is at stake; long periods of *seeming* stagnation are often followed by intervals of unexpected twists and turns, even breakthroughs; the conclusion of one chapter often comes with the first message from the next chapter; progress is achieved through the interplay and constructive interference of theory, experiment and new technologies in turn taking the lead. The outcome is one where participants can take a great deal of pride and others feel a great deal of envy.

2   1964 - 1998: The Long Wait

On the surface the long period between just after the discovery of CP violation in 1964 and 1998 appeared one of stagnation; in retrospect it can be seen rather as one of fermentation, with many things of future importance developing just below the surface.

2.1   Experimental Searches

After a long period of dedicated and ingenious experimentation described in more detail in the talks by Blucher [7] and Zavrtanik [8] CP violation had been found in three classes of processes all starting from neutral kaons: (i) $\Gamma(K_L \to \pi\pi) \neq 0$; (ii) $\Gamma(K_L \to l^+\nu\pi^-) \neq \Gamma(K_L \to l^-\nu\pi^+)$; (iii) rate$(K^0(t) \to \pi\pi) \neq$ rate$(\bar{K}^0(t) \to \pi\pi)$. Assuming CPT invariance one finds that these transitions can be described by two ratios of transition amplitudes:

$$\eta_{+-} \equiv \frac{T(K_L \to \pi^+\pi^-)}{T(K_S \to \pi^+\pi^-)} \, , \; \eta_{00} \equiv \frac{T(K_L \to \pi^0\pi^0)}{T(K_S \to \pi^0\pi^0)} \tag{3}$$

Using the notation

$$\eta_{+-} = \epsilon + \epsilon' \, , \; \eta_{00} = \epsilon - 2\epsilon' \tag{4}$$

one finds that $\epsilon \neq 0$ describes 'indirect' CP violation, i.e. CP violation in the $\Delta S = 2$ dynamics driving $K^0 - \bar{K}^0$ mixing, which prepares the initial K_L state and $\epsilon' \neq 0$ 'direct' CP violation in the $\Delta S = 1$ decay sector. Up to 1998 all data could be described by a single non-vanishing real number, namely $|\epsilon|$ (or arg(M_{12}/Γ_{12})). The situation was quite unsettled concerning *direct* CP violation:

$$\mathrm{Re}\frac{\epsilon'}{\epsilon_K} = \begin{cases} (2.3 \pm 0.65) \cdot 10^{-3} & NA \, 31 \\ (1.5 \pm 0.8) \cdot 10^{-3} & PDG \, '96 \text{ average} \\ (0.74 \pm 0.52 \pm 0.29) \cdot 10^{-3} & E \, 731 \end{cases} \tag{5}$$

---

[c] At first there were suggestions to make the observation of $K_L \to \pi\pi$ compatible with CP symmetry by either introducing *non*linear terms into the Schrödinger equation thus abandoning the superposition principle of quantum mechanics [6] or by postulating that actually $K_L \to \pi\pi U$ had occurred with a hypothetical neutral pseudoscalar particle U having escaped detection. Additional data sealed off these escape routes.

The CPLEAR experiment had performed the socalled Kabir Test of T invariance and under very general assumptions found an asymmetry commensurate with the observed CP violation as predicted by CPT invariance [8]:

$$\frac{\Gamma(K^0 \Rightarrow \bar{K}^0) - \Gamma(\bar{K}^0 \Rightarrow K^0)}{\Gamma(K^0 \Rightarrow \bar{K}^0) + \Gamma(\bar{K}^0 \Rightarrow K^0)} = (6.3 \pm 2.1 \pm 1.8) \cdot 10^{-3} \quad CPLEAR \qquad (6)$$

Truly impressive sensitivities had been achieved in searches for T violation through electric dipole moments for neutrons and electrons, respectively [3]:

$$d_N < 6.3 \cdot 10^{-26} \; ecm \quad \text{from ultracold neutrons} \qquad (7)$$
$$d_e = (-0.3 \pm 0.8) \cdot 10^{-26} \; ecm \quad \text{from atomic EDM} \qquad (8)$$

I.e., the upper bound on d_N amounts to a shift 12 orders of magnitude smaller than the radius of the neutron. This corresponds to searching for a shift of less than the width of human hair in an object the size of the earth !

2.2 *Theoretical Models*

The 'superweak' ansatz positing that CP violation resides only in $\Delta S = 2$ transitions – and thus $\epsilon/\epsilon' = 0$ – was put forward by Wolfenstein already in 1964 [9]. Yet it has to be kept in mind that it constitutes a *classification* scheme rather than a *dynamical* model, let alone a theory. The community might be forgiven for not worrying unduly over a tiny effect – characterized by BR$(K_L \to \pi^+\pi^-) \simeq 2.3 \cdot 10^{-3}$ – at a time where there was no renormalizable theory for the weak forces, and one had to deal with *infinities* in decay widths. However it is quite remarkable that after the emergence of a renormalizable theory in the late 1960's – namely the Standard Model (SM) based on $SU(2)_L \times U(1)$ gauge interactions – it was not realized for several years that New Physics, i.e. physics beyond the standard model *of that time* had to exist. This lack of a theory was stated unequivocally in the 1973 paper by Kobayashi and Maskawa [10]. They also listed the various scenarios that can support CP breaking (while maintaining CPT invariance): right-handed charged currents [d], extra Higgs fields or – the existence of (at least) a third quark family. This last of their options is now referred to as the KM description[e].

They made their suggestions when neither right-handed currents nor any Higgs states had been found – a fact that still holds today – and when only three quark 'flavours' were known, namely u, d and s; i.e. even the second family was not complete[f]!

Kobayashi and Maskawa extended an earlier observation by Cabibbo that *mass* eigenstates are not necessarily *interaction* or *'flavour'* eigenstates as well. Then

---

[d]Mohapatra had suggested this option already in 1972 [11].

[e]The only option they missed is the possibility that a nonabelian gauge theory like QCD can break CP invariance through topological effects in its ground state [3].

[f]Kobayashi and Maskawa had benefitted in two ways from the 'genius loci' of Nagoya University where they both worked at that time: (i) While the notion of quarks as truly dynamical objects rather than convenient mathematical entities had not been universally accepted, this was not doubted at Nagoya, which was the birth place of the Sakata model. (ii) Niu, a prominent professor at Nagoya, had found evidence for a charm hadron in his cosmic ray data in 1971 [12]. At Nagoya it was thus 'known' that at least two full families of quarks existed in nature.

there are nontrivial transformations T_L^U and T_L^D relating the left-handed flavour eigenstates of $U = u, c, ...$ and $D = d, s, ...$ quarks, respectively, to mass eigenstates; the charged current couplings of pairs of U and D quarks are described by a matrix, the CKM matrix

$$V^{CKM} = T_L^U (T_L^D)^\dagger \tag{9}$$

This matrix has to be unitary as long as the weak interactions are described by a *single* $SU(2)$ gauge theory. Kobayashi and Maskawa pointed out that with two families only – (u, d) and (c, s) – V_{CKM} cannot contain a *physical* complex phase, which is required to implement CP violation. Yet for three families, where one has a 3×3 matrix, it contains three mixing angles – in analogy to the Euler angles of rotation matrices – plus one complex phase, the KM phase ϕ_{KM}.

This observation has lead to enhanced visibility of basic trigonometry in fundamental physics. The unitarity of the 3×3 CKM matrix implies two classes of constraints:

$$\sum_{l=1}^{3} (V_{il}^{CKM})^* V_{lj}^{CKM} = \delta_{ij} \tag{10}$$

For $i = j$ this represents 'weak universality' (for example: $|V_{ud}|^2 + |V_{us}|^2 + |V_{ub}|^2 = 1$) , for $i \neq j$ a *triangle* relation in the complex plane. There are actually six triangles: they are of very different shapes, yet share one important feature: they all have the same area, which can be expressed through the so-called Jarlskog variable [13]:

$$\text{area} = \frac{1}{2} J , \quad J = |\text{Im} V_{km}^* V_{lm} V_{kn} V_{ln}^*| \tag{11}$$

irrespective of the indices k, l, m, n. This feature of equal areas reflects the fact that there is a single irreducible complex phase for three families. The angles in these triangles control CP asymmetries in the different charged current transitions like the decays of kaons and B mesons, and they are all driven by the KM phase ϕ_{KM}. The main point here is that the CKM description involves a *high degree of overconstraints*:

- The angles which control the CP asymmetries can be determined by the sides of the triangles, which in turn can be inferred from CP *in*sensitive rate measurements.

- The different angles are related to each other, and all depend on ϕ_{KM}.

The KM ansatz naturally predicts direct CP violation, i.e. it does *not* represent a superweak description.

2.3 The Distinct Attraction of Beauty

According to the KM ansatz CP violation is due to the interplay of (at least) three quark families. Kaons made up from quarks of the first two families (d, u and s) are sensitive to the third family due to quantum corrections, yet those are suppressed, since the quarks of the third family are so much heavier; this provides a natural qualitative explanation, why CP violation is so feeble in kaons, i.e. why CP invariance is such a 'near miss' in strange decays.

The situation changes very significantly for the decay of B mesons, since the b quark already belongs to the third family. Right after the first evidence for b quarks was found, it was recognized that B decays had the potential to exhibit sizable CP asymmetries. In 1980 it was realized how this potential could be tapped [14]. More specifically it was pointed out that some CP asymmetries in B decays could be two orders of magnitude larger than what had been found in $K_L \to \pi\pi$ – even approaching 100 % – and that they could be related reliably to basic parameters of the CKM description. This applies in particular to the mode $B_d \to \psi K_S$. The CP asymmetry was predicted to exhibit another striking signature beyond its size, namely a peculiar dependence on the time of decay t

$$\mathrm{rate}(B_d(t)[\bar{B}_d(t)] \to \psi K_S) \propto e^{-t/\tau_B}(1 - [+]A\sin\Delta m_B t) , \qquad (12)$$

since it involves $B_d - \bar{B}_d$ oscillations in an essential way. The asymmetry parameter A for this transition can reliably be expressed through one angle of the unitarity triangle, see Fig.1:

$$A = \sin2\phi_1 \qquad (13)$$

The final state ψK_S does not reveal whether it came from a B_d or $\bar{B}_d$ decay; that principal ambiguity is actually essential for the asymmetry in Eq.(12) to arise, since the asymmetry is due to the interference of two *coherent* amplitudes. Thus one needs *independent* information on the 'flavour' identity of the decaying B meson; this can be achieved by 'associated' production, as sketched later.

At first not much attention was paid to these suggestions, partly because the lifetime τ_B of B mesons was not known, let alone Δm_B, the rate of $B^0 - \bar{B}^0$ oscillations. A highly significant change in perception occurred when first the B lifetime was measured and found to be surprisingly "long" at $\mathcal{O}(psec)$ followed by the observation of $B^0 - \bar{B}^0$ oscillations with $\Delta m_B \sim 0.7/\tau_B$. These observations established a new paradigm:

- It revealed a peculiar hierarchical pattern in the CKM matrix most transparently expressed through the Wolfenstein representation:

$$V^{CKM} \equiv \begin{pmatrix} V_{ud} & V_{us} & V_{ub} \\ V_{cd} & V_{cs} & V_{cb} \\ V_{td} & V_{ts} & V_{tb} \end{pmatrix}$$

$$\simeq \begin{pmatrix} 1 - \frac{1}{2}\lambda^2 & \lambda & A\lambda^3(\rho - i\eta + \frac{i}{2}\eta\lambda^2) \\ -\lambda & 1 - \frac{1}{2}\lambda^2 - i\eta A^2\lambda^4 & A\lambda^2(1 + i\eta\lambda^2) \\ A\lambda^3(1 - \rho - i\eta) & -A\lambda^2 & 1 \end{pmatrix} \qquad (14)$$

- This pattern lead to the realization there is a unitarity triangle where all three sides are of order λ^3 in length; accordingly all three angles are naturally large. This special triangle, shown in Fig. 1, describes transitions of B mesons. Thus one predicts CP asymmetries of several$\times$ 10% [15].

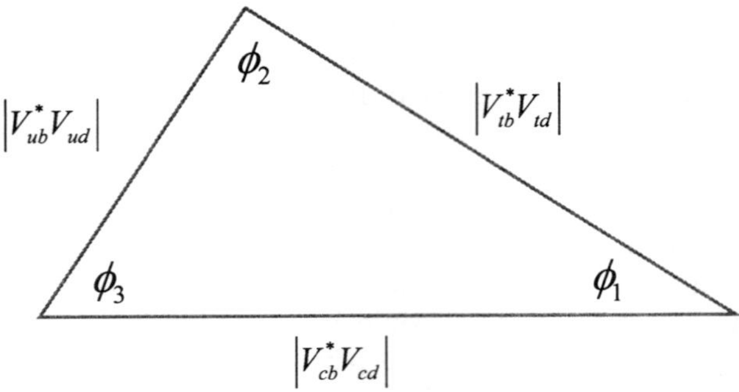

Figure 1. The CKM Unitarity Triangle for B decays.

- It just happened that the detector technology for tracking decay times ~ 1 psec was available 'off the shelf', since it had been developed for prior studies of charm hadrons.

- Events with times of decay much shorter than 1 psec, which largely escape tracking, contribute little to the asymmetry in Eq.(12).

There are various experimental setups for undertaking such measurements that have been and will be pursued in the future [16,17], in particular at hadronic colliders. Yet the cleanest experimental stage is provided by colliding electron and positron beams:

$$e^+e^- \to \Upsilon(4S) \to B_d\bar{B}_d \tag{15}$$

For the $\Upsilon(4S)$ resonance yields an enhanced production rate for B mesons, and all final state particles are decay products of one or the other B meson.

There is another equally important point, which at first represents a serious challenge. To *experimentally define* a CP asymmetry, one can *flavour tag* an event by observing a flavour specific transition like $B_d \to l^+\nu X$ or $\bar{B}_d \to l^-\nu X$ *in conjunction* with the *non*-specific mode $B_d/\bar{B}_d \to \psi K_S$ to compare

$$B_d\bar{B}_d \to (l^+X)_B(\psi K_S)_{B/\bar{B}} \; vs. \; B_d\bar{B}_d \to (l^-X)_{\bar{B}}(\psi K_S)_{B/\bar{B}} \tag{16}$$

Denoting the times of the two decays by t_1 and t_2, respectively, one finds

$$\mathrm{rate}(B_d(t_1) \to l^+\nu X), \bar{B}_d(t_2) \to \psi K_S) \propto e^{-(t_1+t_2)/\tau_B}[1 - A\sin\Delta m_B(t_1-t_2)] \; ; \tag{17}$$

likewise for rate$(\bar{B}_d(t_1) \to l^-\nu X), B_d(t_2) \to \psi K_S)$. The relative minus sign between t_1 and t_2 is due to the fact that the $B\bar{B}$ pair in $\Upsilon(4S) \to B_d\bar{B}_d$ forms a C *odd* configuration. There was a significant fly-in-the-ointment, though: One cannot measure the time of decay *directly*. Silicon microvertex detectors instead allow to identify the *location* of the *decay* vertex. Knowing the *production* vertex as well one can infer the distance the B meson traveled; from it and the B momentum one obtains the lifetime. Space distances are thus translated into time intervals. However the $\Upsilon(4S)$ is barely above the $B_d\bar{B}_d$ production threshold; therefore the B mesons move slowly in the $\Upsilon(4S)$ rest frame. The distance they cover before decaying is too short to be resolved by existing detectors. Yet integrating Eq.(17)

over all times $t_{1,2}$ removes the asymmetry term. I.e., a search for such a CP asymmetry would amount merely to a test for .

An intriguing answer to this challenge would utilize quantum mechanical coherence as follows: in the reaction

$$e^+e^- \to B_d^* \bar{B}_d + h.c. \to B_d \bar{B}_d \gamma \ . \tag{18}$$

The $B\bar{B}$ pair now forms a C *even* state; accordingly the time difference $t_1 - t_2$ in the $\sin\Delta m_B(t_1 - t_2)$ term of Eq.(17) is replaced by the sum $t_1 + t_2$; integrating over all time t_1, t_2 then yields a non-vanishing contribution. I.e., one then could search for a CP asymmetry in $B_d \to \psi K_S$ *without* resolving decay vertices. Yet such an undertaking suffers from a lack of the necessary statistics.

Oddone [18] suggested to cut the Gordian knot for $e^+e^- \to \Upsilon(4S) \to B_d\bar{B}_d$ by using *asymmetric* colliding beams, which makes the $\Upsilon(4S)$ move in the lab frame. With a sufficiently large beam energy asymmetry the B mesons receive a boost in the lab frame making them travel a distance long enough to be resolved.

This was quite an innovative and daring proposal, since there was no experience with asymmetric colliders; furthermore integrating two separate beam lines of greatly different energies – and high intensity on top of that – into a detector posed novel problems. Yet it was a challenge that was thought, at least by some adventurous souls, worth the effort.

This is best demonstrated by the fact that a race ensued to build an asymmetric high intensity $e^+e^- \to \Upsilon(4S)$ collider together with a detector capable of operating in a high radiation environment. It was joined by two teams in the late 1980's: the BABAR collaboration based at SLAC, the Stanford Linear Accelerator Center, in the USA, and the BELLE collaboration based at KEK, the Japanese High Energy Accelerator Research Organization. While the designs of the two teams for the detectors and accelerators differed in many details – like energy asymmetries, injection systems for the beams, luminosities, particle id for the detectors etc.– both were truly ambitious: the SLAC project aimed for a luminosity of $3 \cdot 10^{33} \ cm^{-2}s^{-1}$ and KEK even for $10^{34} \ cm^{-2}s^{-1}$, when the 10^{33} level had never been achieved before even with conventional colliders! These design intensities amount to the production of about three and ten $B\bar{B}$ pairs per second, respectively.

2.4 On the Eve of a 'Phase Transition'

In 1998 the theoretical status of CP violation could be summarized as follows::

- The strength of indirect CP violation observed in K_L decays could be reproduced within the CKM description without forcing any parameter outside the range inferred for it from other measurements.

- There appeared to be more consensus about the strength of *direct* CP violation among the theory predictions than between the two sets of data from NA31 and E731, Eq.(5): most authors predicted $\text{Re}(\epsilon'/\epsilon)$ not to exceed 10^{-3}, while some heretics –early ones [19] and 'just in time' ones predicted larger values: $(1.7^{+1.4}_{-1.0}) \cdot 10^{-3}$ [20].

- The KM ansatz predicted unobservably small values for electric dipole moments of neutrons, electrons and atoms, i.e. below 10^{-30} *ecm*.

- There is one caveat: it had been realized that QCD, the nonabelian gauge theory of the strong interactions, does *not* conserve CP invariance *naturally*. The nontrivial topological structure of its ground state induces a P and T *odd* term in QCD's effective Lagrangian with an a priori unknown coefficient θ_{QCD}, for which the 'natural' expectation is $\theta_{QCD} \sim \mathcal{O}(1)$. This term would generate an EDM for the neutron; the *non*observation of the later imposes $\theta_{QCD} \leq \mathcal{O}(10^{-9})$ – i.e. smaller by many orders of magnitude than the 'natural' expectation! This means that either the neutron's EDM could be 'just around the corner' – i.e. any improvement in experimental sensitivity might reveal an effect – or again unobservably small, since 'natural' explanations of $\theta_{QCD} \leq \mathcal{O}(10^{-9})$ based on a Peccei-Quinn symmetry drive θ_{QCD} down by several additional orders of magnitude [3].

- The CKM description actually *post*dicted ϵ (as it did for the K_L-K_S mass difference Δm_K), and its predictions for other observables like Δm_B suffer from considerable uncertainties. Nevertheless it was argued [3] that the ability of the CKM scheme to accommodate a body of observables spanning six or seven orders of magnitude in energy has to be seen as highly nontrivial, in particular since it was achieved with fundamental quantities like quark masses and the CKM parameters shown in Eq.(14) that would have seen frivolous – if not forced upon us by the data. Thus some of us had considerable confidence that the CKM prediction for a large CP asymmetry in $B_d(t) \to \psi K_S$ would be confirmed.

This confidence was expressed by saying that there was no 'plausible deniability' for the KM ansatz, if no large CP asymmetries were found in B decays; in 1991/92, i.e. before the top quark was discovered and its mass measured directly, the expectation [21]

$$\sin 2\phi_1 \sim 0.6 \div 0.7 \tag{19}$$

was formulated, and in 1998 an even more specific prediction was given [22]:

$$\sin 2\phi_1 \sim 0.72 \pm 0.07 \tag{20}$$

The point of listing this last number is to emphasize that it *was made* as a true *pre*diction rather than to endorse the error quoted there. It should be added that us Bavarians always admire courage, in particular of the somewhat reckless kind.

- The first suggestion of sizable CP asymmetries in B decays like $B \to K\pi$ was actually made by the authors of Ref.[23]. A re-analysis by the authors of Ref.[24] lead to

$$A_{CP}^{K\pi} \equiv \frac{\Gamma(\bar{B}^0 \to K^- \pi^+) - \Gamma(B^0 \to K^+ \pi^-)}{\Gamma(\bar{B}^0 \to K^- \pi^+) + \Gamma(B^0 \to K^+ \pi^-)} = -0.10 \tag{21}$$

as a reasonable, though not firm prediction.

3 The 'Phase Transition' at the Turn of the Millenium

During a relatively short time interval around the year 2000 data provided us with several seminal insights answering old questions – and raising new ones.

3.1 The Conclusion of an Epoch

In 1999, after more than thirty years of dedicated and ingenious experimentation [7], KTEV [25] and NA48 [26] *conclusively* confirmed earlier evidence from NA31 that indeed there is *direct* CP violation in K_L decays. The 2003 world average reads

$$\mathrm{Re}\frac{\epsilon'}{\epsilon} = (1.66\pm0.16)\cdot10^{-3} \; \hat{=} \; \frac{\Gamma(K^0 \to \pi^+\pi^-) - \Gamma(\bar{K}^0 \to \pi^+\pi^-)}{\Gamma(K^0 \to \pi^+\pi^-) + \Gamma(\bar{K}^0 \to \pi^+\pi^-)} = (5.5\pm0.6)\cdot10^{-6} \tag{22}$$

The second number even more than the first one indicates what kind of achievement lies behind these data. The physicists involved in these experiments have earned our respect, and they certainly have my admiration. Establishing direct CP violation for the first time is a discovery of the first rank irrespective of what theory does or does not say. At the present status of our knowledge (or the lack thereof) it is not inconsistent with the CKM description. Nature has exhibited its slightly more malicious side here, since ϵ' receives several contributions with the two largest ones coming in with the opposite sign. Thus we cannot count on theory yielding a *definitive* answer soon. Yet again, I find it highly nontrivial that theory yields the correct number to within a factor of two or so.

3.2 The Beginning of a New Era

As already mentioned, the two B factories had very ambitious goals concerning their luminosities and reliabilities. They have actually met them – and surpassed them. BELLE and BABAR with design luminosities of $1\cdot10^{34}$ and $3\cdot10^{33}cm^{-2}s^{-1}$, respectively, in 2004 have achieved running at $1.2\cdot10^{34}cm^{-2}s^{-1}$ and $8\cdot10^{33}cm^{-2}s^{-1}$, respectively. They presented their first still inconclusive data on $B_d \to \psi K_S$ in 2000:

$$\sin2\phi_1 = 0.45 \pm 0.44 \pm 0.09 \quad \mathrm{BELLE} \, '00 \tag{23}$$

$$\sin2\phi_1 = 0.12 \pm 0.37 \pm 0.09 \quad \mathrm{BABAR} \, '00 \tag{24}$$

In the summer of 2001 they established the existence of an asymmetry, the first one outside the decays of neutral kaons and a truly large one [9]:

$$\sin2\phi_1 = 0.99 \pm 0.14 \pm 0.06 \quad \mathrm{BELLE} \, '01 \tag{25}$$

$$\sin2\phi_1 = 0.59 \pm 0.14 \pm 0.05 \quad \mathrm{BABAR} \, '01 \tag{26}$$

---

[9] To obtain these small errors various other flavour tagging modes beyond semileptonic decays had to be used and other final states like in particular $B_d/\bar{B}_d(t) \to \psi K_L$; for the latter one predicts an asymmetry of equal size, yet opposite sign to that in $B_d/\bar{B}_d(t) \to \psi K_S$.

Two years later the data had converged to an amazing degree:

$$\sin 2\phi_1 = \begin{cases} 0.733 \pm 0.057 \pm 0.028 & \text{BELLE '03} & \text{with} \sim 1.2 \cdot 10^8 \ B\bar{B} \\ 0.741 \pm 0.067 \pm 0.030 & \text{BABAR '03} & \text{with} \sim 0.8 \cdot 10^8 \ B\bar{B} \\ 0.736 \pm 0.049 & \text{world average '03} \\ 0.726 \pm 0.037 & \text{world average '04} \end{cases} \quad (27)$$

leading to the following general statements: the CP asymmetry in $B_d \to \psi K_S$ is there, and it is huge, fully as expected, see Eq.(20) [h] !

Hence I conclude:

- The CKM paradigm has been promoted from an *ansatz* to a *tested theory*.

- CP violation has actually been 'demystified': *if* the dynamics are sufficiently multilayered such that they can support CP violation (like the existence of at least three quark families), the latter can be truly large; i.e. there is no intrinsic reason why the complex phases should be small.

3.3   The Unsung Hero

Hadronization – the formation of hadrons out of quarks – is usually listed as an unwelcome complication greatly impeding our description of CP violation, since the strong forces have not been brought under full theoretical control. The latter is certainly true – yet so is the fact that hadronization greatly enhances the features of CP breaking and thus facilitates its observability through three effects:

- The existence of pions and kaons with the latter only moderately above the three pion threshold reduces the rate for the CP conserving $K_L \to 3\pi$ process relative to the CP violating $K_L \to 2\pi$ one by a factor close to $\Gamma(K_S)/\Gamma(K_L) \sim 500$.

- It awards 'patience'; i.e., an initial beam of K^0 and $\bar{K}^0$ turns into a pure K_L beam, since the K_S component decays away quickly.

- CP violation can be established through the *existence* of a transition – here $K_L \to \pi\pi$ – rather than an asymmetry between two allowed processes.

Hadronization should thus be recognized as the hero of the tale of CP violation rather than the villain it is usually depicted.

3.4   EPR Correlations – a Precision Tool Rather than a Paradox

The BABAR and BELLE analyses are based on a glorious application of quantum mechanics and in particular EPR correlations[1]. At first it would seem that an asymmetry of the form given in Eq.(12) could not be measured for practical reasons. For in the reaction

$$e^+ e^- \to \Upsilon(4S) \to B_d \bar{B}_d \quad (28)$$

---

[h] The procession of these numbers reflects a better understanding of the detectors in addition to increasing statistics. It should also remind theorists to consider experimental uncertainties when interpreting data.

the point where the B meson pair is produced is ill determined due to the finite size of the electron and positron beam spots: the latter amounts to about 1 mm in the longitudinal direction, while a B meson typically travels only about a quarter of that distance before it decays. It would then seem that the length of the flight path of the B mesons is poorly known and that averaging over this ignorance would greatly dilute or even eliminate the signal.

It is here where the existence of a EPR correlation comes to the rescue. While the two B mesons in the reaction of Eq.(28) oscillate back and forth between a B_d and $\bar{B}_d$, they change their flavour identity in a *completely correlated* way. For the $B\bar{B}$ pair forms a C odd state; Bose statistics then tells us that there cannot be two identical flavour hadrons in the final state:

$$e^+e^- \to \Upsilon(4S) \to B_d\bar{B}_d \not\to B_dB_d,\ \bar{B}_d\bar{B}_d \tag{29}$$

Once one of the B mesons decays through a flavour specific mode, say $B_d \to l^+\nu X$ $[\bar{B}_d \to l^-\bar{\nu}X]$, then we know unequivocally that the other B meson was a $\bar{B}_d$ $[B_d]$ at *that* time. The time evolution of $\bar{B}_d(t)[B_d(t)] \to \psi K_S$ as described by Eq.(12) starts at *that* time as well; i.e., the relevant time parameter is the *interval between* the two times of decay, not those times themselves. That time interval is related to – and thus can be inferred from – the distance between the two decay vertices, which is well defined and can be measured.

The great *practical* value of the EPR correlation is instrumental for another consideration as well, namely how to see directly from the data that CP violation is matched by T violation. Fig.2 shows two distributions, one for the interval Δt between the times of decays $B_d \to l^+X$ and $\bar{B}_d \to \psi K_S$ and the other one for the CP conjugate process $\bar{B}_d \to l^-X$ and $B_d \to \psi K_S$. They are clearly different proving that CP is broken. Yet they show more: the shape of the two distributions is actually the same the only difference being that the average of Δt is *positive* for $(l^-X)_{\bar{B}}(\psi K_S)$ and *negative* for $(l^+X)_B(\psi K_S)$ events. I.e., there is a (slight) preference for $B_d \to \psi K_S$ $[\bar{B}_d \to \psi K_S]$ to occur *after* [*before*] and thus more [less] slowly (rather than just more rarely) than $\bar{B} \to l^-X$ $[B \to l^+X]$. Invoking CPT invariance merely for semileptonic B decays – yet not for nonleptonic transitions – synchronizes the B and $\bar{B}$ decay 'clocks'. We thus see that CP and T violation are 'just' different sides of the same coin. As explained above, EPR correlations are essential for this argument!

The reader can be forgiven for feeling that this argument is of academic interest only, since CPT invariance of all processes is based on very general arguments. Yet the main point to be noted is that EPR correlations, which represent some of quantum mechanics' most puzzling features, serve as an essential precision tool, which is routinely used in these measurements. I feel it is thus inappropriate and misleading to refer to EPR correlations as a paradox.

3.5 Direct CP Violation and "Yesterday's Sensation, Today's Calibration ..."

After the discovery of $K_L \to \pi\pi$ it took 35 years to observe and confirm the existence also of *direct* CP violation in the kaon sector. The analogous development took much less time in the beauty sector. Direct CP violation has been established in

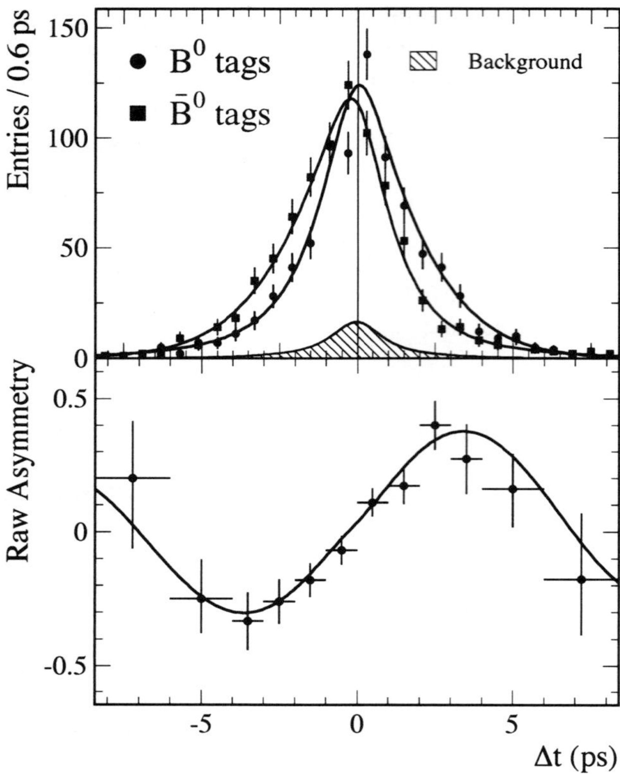

Figure 2. The observed decay time distributions for B^0 (red) and $\bar{B}^0$ (blue) decays.

2004 by both BABAR and BELLE; averaging their results yields:

$$A_{CP}^{K\pi} \equiv \frac{\Gamma(\bar{B}^0 \to K^-\pi^+) - \Gamma(B^0 \to K^+\pi^-)}{\Gamma(\bar{B}^0 \to K^-\pi^+) + \Gamma(B^0 \to K^+\pi^-)} = -0.101 \pm 0.025 \pm 0.005 \qquad (30)$$

in amusing agreement with the expectation given in Ref.[24] based on the ansatz of Ref.[23].

It is not widely appreciated that the first strong experimental evidence for direct CP violation had actually emerged in $B_d(t) \to \pi^+\pi^-$. It provides an example of high energy physics' adage "Yesterday's sensation is today's calibration and tomorrow's background". To analyze $B_d(t)/\bar{B}_d(t) \to \pi^+\pi^-$ one again exploits EPR correlations and flavour tagging by a flavour specific decay of the other B meson. With the notation

$$R^{+[-]}(\Delta t) \equiv \mathrm{rate}((l^{+[-]}X)_B \text{ at } t; (\pi^+\pi^-)_{B/\bar{B}} \text{ at } t + \Delta t) \qquad (31)$$

one can write the asymmetry between $B_d \to \pi^+\pi^-$ and $\bar{B}_d \to \pi^+\pi^-$ in terms of two contributions distinguishable through their dependance on Δt:

$$\frac{R^+(\Delta t) - R^-(\Delta t)}{R^+(\Delta t) + R^-(\Delta t)} = S\sin(\Delta m_B \Delta t) + C\cos(\Delta m_B \Delta t) \tag{32}$$

with CP invariance requiring $S = 0 = C$. These coefficients depend on the angles of the CKM triangle, its sides and other hadronic quantities, over which theoretical control has not been established yet beyond the general constraint $S^2 + C^2 \leq 1$. The status in the summer of 2004 is as follows:

$$C = \begin{cases} +0.58 \pm 0.15 \pm 0.07 & \text{BELLE} \\ -0.09 \pm 0.15 \pm 0.04 & \text{BABAR} \end{cases} \tag{33}$$

$$S = \begin{cases} -1.00 \pm 0.21 \pm 0.07 & \text{BELLE} \\ -0.30 \pm 0.17 \pm 0.03 & \text{BABAR} \end{cases} ; \tag{34}$$

BELLE observes a CP asymmetry with a significance of 5.2 σ; historically it was the second case of CP violation found in B decays. As explained in the Appendix, in a superweak scenario $B_d(t) \to \psi K_S$ provides the calibration; i.e. one would have

$$S = -\sin 2\phi_1 = -0.736 \pm 0.049 \, , \, C = 0 \, . \tag{35}$$

This is ruled out by BELLE's numbers, and *direct* CP violation thus established with at least 3.2 σ significance, which is significant, but not conclusive yet. No firm conclusions can be derived from the BABAR data at present.

3.6   The Fly in the CKM Ointment

The observed asymmetry in $B_d \to \psi K_S$ constitutes a striking success for the CKM description and the data on $B_d \to \pi\pi$ are quite compatible with it. Yet a potential discrepancy has arisen in $B \to \phi K_S$, a channel that had actually been recognized before as having a good potential to reveal physics beyond the SM in general and SUSY in particular. In the SM one predicts very confidently:

$$C_{\phi K_S} \simeq 0 \, , \, S_{\phi K_S} \simeq S_{\psi K_S} = 0.736 \pm 0.049 \tag{36}$$

In the summer of 2004 the data read as follows:

$$C_{\phi K_S} = \begin{cases} -0.08 \pm 0.22 \pm 0.09 & \text{BELLE} \\ +0.00 \pm 0.23 \pm 0.05 & \text{BABAR} \end{cases} \tag{37}$$

$$S_{\phi K_S} = \begin{cases} +0.06 \pm 0.33 \pm 0.09 & \text{BELLE} \\ +0.50 \pm 0.25^{+0.07}_{-0.04} \pm 0.03 & \text{BABAR} \end{cases} . \tag{38}$$

While BABAR's findings are consistent with the predictions, BELLE's results point to a 2.2 σ discrepancy. This has attracted considerable attention since the CP asymmetries in a whole class of related modes all seem to show a shift relative to the CKM predictions, and this class is expected to exhibit a high sensitivity to the intervention of New Physics. It will take some time, though, to clarify the experimental situation.

3.7 "... and Tomorrow's Background" – the Cosmic Connection

As mentioned before the most ambitious goal in CP studies is to understand the observed baryon number of the Universe as a dynamically generated quantity, for which CP violation is one of the three central ingredients. We know now that the standard CKM dynamics while successful in describing CP breaking observed in particle decays is quite incapable to provide this cosmic connection: for its effective strength is too feeble, and it cannot induce the required *first* order phase transition at the electroweak symmetry breaking. Thus this program requires the intervention of New Physics, for which many interesting scenarios have been put forward.

Of course one wants to identify manifestations of such hypothetical New Physics in processes that can be probed in reproducible laboratory experiments. In general such new dynamics will affect CP asymmetries in B decays; yet those 'suffer' from the background of large asymmetries due to CKM dynamics that are present. This poses considerable – though hopefully not insurmountable – challenges on the experimental as well as theoretical side.

The encouraging news is the aforementioned demystification of CP violation: as the example of CKM dynamics shows in general there is no impediment to the CP violating complex phase being large; this will presumably be needed to generate the baryon number of the Universe.

3.8 Gateway to a New World: Neutrino Oscillations

A particular intriguing class of models interprets the baryon number as a secondary effect derived from the primary phenomenon of leptogenesis; i.e., first a non-vanishing lepton number is generated for the Universe, which is then transmogrified into a baryon number. This provides new impetus – actually makes it mandatory – to search for manifestations of CP violation in the dynamics of leptons. Relevant processes are

- neutrino oscillations,

- the decays of τ (and μ) leptons [i] and

- atomic electric dipole moments.

The observation of neutrino oscillations through solar and 'atmospheric' neutrinos – which constitutes the third column of the 'phase transition' at the turn of the millenium referred to before – opens up a new world to probe fundamental physics in general and CP violation in particular, which is of direct relevance for leptogenesis.

Neutrino oscillations can occur only when the different neutrino types are *not* mass degenerate. The leptonic charged current couplings are then described by the so-called PMNS matrix [28] in analogy to the CKM matrix for quarks. The 3×3 PMNS matrix is far from exhibiting the strictly hierarchical form of Eq.(14) – yet that is not surprising since there is a fundamental distinction between charged and

---

[i] The transition $\tau^- \to \mu^- \mu^+ \mu^-$ forbidden in the SM is the leptonic analogue of the quark transition $b \to s\bar{s}s$ driving $B \to \phi K_S$.

neutral fermion fields: for the latter can acquire a Majorana mass term in addition to a Dirac mass; through the so-called see-saw mechanism [29,3] this can provide a natural explanation why the neutrino masses are so tiny compared to all other fermion masses. The fact that the form of the CKM and PMNS matrices does not exhibit a unified pattern should therefore not be seen as a drawback – on the contrary! This can be illustrated by the following true anecdote: a long time ago a French politician was asked whether his opposition to German unification does not reveal his basic dislike of Germany. He rejected this assertion by saying that he truly loves Germany and he is therefore overjoyed that there are two Germanies he can love [j].

Irrespective of that connection finding CP violation in leptodynamics would complete the 'de-mystification' of CP violation that, as repeatedly mentioned before, has occurred in quark dynamics.

4 Reflections about the Past and the Future

After a long gestation several important developments have come to fruition starting in 1999:

- A second qualitatively different source of CP violation, namely direct CP violation, has unequivocally been established experimentally in $K_L \to \pi^+\pi^-$ vs. $K_L \to \pi^0\pi^0$ decays.

- Data together with truly *minimal* theoretical assumptions confirm that the CP asymmetry observed in neutral kaon decays is fully matched by violation of time reversal invariance T as required by CPT invariance.

- The large CP asymmetry predicted for $B_d \to \psi K_S$ by the CKM description in the old-fashioned sense – i.e., the prediction was made well before the experimental findings were known – has been confirmed by the data to an amazing degree.

- The measurements show – again with minimal assumptions – that the observed asymmetry is matched by a commensurate violation of T invariance.

- A CP asymmetry has been observed in a second class of channels, namely $B_d \to \pi^+\pi^-$, yet its detailed interpretation is still open to debate.

- EPR correlations provide an indispensable *precision* tool for the experimental analyses; these are actually a novel type of EPR correlations, where the two correlated states change their identity on a time scale of picoseconds! The data are fully consistent with quantum mechanics' specific predictions.

- With all these observations (and others before) consistent with predictions and expectations based on the CKM description, the latter has been promoted from an ansatz to a tested theory.

---

[j] There is another lesson to be learnt from this analogy: even when unification seems impossible, it can happen in due course.

- This progress was achieved through an intimate interplay between theoretical suggestions, experimental results and novel concepts in detector and accelerator design.

The experimenters will not rest on their laurels, however well-deserved that would be. The B factories at KEK and SLAC will run with ever increasing statistics and refined experimental techniques to probe B, charm and τ decays with higher and higher sensitivity; hopefully one of them will be upgraded to a 'Super-B' factory [30]. In a few years they will be joined and pushed by experiments performed at the hadronic colliders of Fermilab in the US and CERN in Europe. Such further efforts are actually mandatory. For the presently achieved successes do not resolve central mysteries of the SM: Why are there families of quarks and leptons? Why three – or are there more? Why has the CKM matrix its highly unusual hierarchical structure? Why does QCD conserve CP invariance in flavour diagonal transitions to such a high degree of precision?

In addition to these indirect arguments for the incompleteness of the SM, there are more direct ones as well:

- At present there are intriguing indications from BELLE's measurements that the CP asymmetry in $B_d \to \phi K_S$ exhibits a stunning deviation from the CKM prediction.

- What is the nature of the new dynamics needed to generate the observed matter-antimatter asymmetry in the Universe?

- What are the forces driving the observed neutrino oscillations? Will they exhibit CP violation as well?

Continuing dedicated and comprehensive studies should provide us with information that will facilitate our searches for answers to these fundamental questions. I consider further experimental information as crucial in this endeavour, since it will point us in the right direction – yet it will not be sufficient: it will be nontrivial to digest the experimental information theoretically.

Epilogue

The physicists' tale of CP violation is a profound one that is teaching us important lessons that go beyond identifying Nature's fundamental forces. It has lead us to formulate questions about 'Nature's Grand Design' that we did not think about at the beginning of our journey: Is it really possible to create a Universe with only matter, but no domains of antimatter? What about other Universes that might exist besides ours? Maybe CP violation and the whole family structure embedded into the SM carries a coded message about extra dimensions beyond the well-known 1+3 time-space dimensions. What about the very structure of time? What made it one-dimensional – or are there additional though hidden dimensions of time. What about the arrow of time in such exotic scenarios?

These are questions, where we still have not even clues about the answers. Yet this is probably quite appropriate for the subjects discussed at this 'International Colloquium on the Science of Time'.

A final observation: for me it reflects one of the noblest and thus most encouraging features of the human race that there is a continuous stream of young people eager to commit themselves to exploring Nature for the gain of knowledge for knowledge's sake and that they bring to the table unusual amounts of talent, dedication, persistence and creativity.

Appendices

A.1 *On* direct *CP Violation*

It is often said – or at least implied – that $S \neq 0$ and $C \neq 0$ reflect two distinct sources of CP violation. Indeed $C \neq 0$ reveals unequivocally *direct* CP violation. Yet the situation with $S \neq 0$ is more complex, as can be read off from the explicit expression for S:

$$S = \mathrm{Im}\frac{q}{p}\frac{T(\bar{B}_d \to \pi^+\pi^-)}{T(B_d \to \pi^+\pi^-)} \tag{A.1}$$

$\frac{q}{p}$ reflects $\Delta B = 2$ dynamics driving $B^0 - \bar{B}^0$ oscillations, and its phase provides a measure for *indirect* CP violation; yet the ratio of the instantaneous transition amplitudes $T(\bar{B}_d[B_d] \to \pi^+\pi^-)$ represents $\Delta B = 1$ dynamics, including their CP features. As a further complication the phases of $\frac{q}{p}$ and $T(\bar{B}_d \to \pi^+\pi^-)/T(B_d \to \pi^+\pi^-)$ depend on the phase *convention* adopted for the definition of $\bar{B}_d$ – only their product does not. Therefore as long as CP violation is studied in a *single* channel, it is a matter of convention whether $S \neq 0$ is called an indirect or a direct CP violation. However once one can compare it in two final states common to B_d and $\bar{B}_d$ decays – like in $B_d(t) \to \psi K_S$ vs. $B_d(t) \to \pi^+\pi^-$ – the two cases can be distinguished. For if there is *no direct* CP violation – i.e. for a superweak scenario – one has

$$S^{\pi\pi} = -S^{\psi K_S} \, , \tag{A.2}$$

where the minus sign is due to the final states $\pi\pi$ and ψK_S having opposite CP parity. Finding instead

$$S^{\pi\pi} \neq -S^{\psi K_S} \tag{A.3}$$

establishes unequivocally the intervention of direct CP violation, since it shows there is a relative phase between $qT(\bar{B}_d \to \pi^+\pi^-)/pT(B_d \to \pi^+\pi^-)$ and $qT(\bar{B}_d \to \psi K_S)/pT(B_d \to \psi K_S)$ and thus also between $T(\bar{B}_d \to \pi^+\pi^-)/T(B_d \to \pi^+\pi^-)$ and $T(\bar{B}_d \to \psi K_S)/T(B_d \to \psi K_S)$, i.e. in pure $\Delta B = 1$ amplitudes. Likewise for

$$S^{\phi K_S} \neq S^{\psi K_S} \text{ or } S^{\eta K_S} \neq S^{\psi K_S} \text{ or } S^{K_S \pi^0} \neq S^{\psi K_S} \text{ etc.} \tag{A.4}$$

One should also note that such direct CP violation might *not* generate $C \neq 0$, since it does not require the presence of two different amplitudes with a nontrivial phase shift between them.

A.2 *A New Opening for 'Patience'*

The CDF experiment at FNAL has obtained the intriguing, though preliminary result that the two B_s mass eigenstates might possess significantly different widths

27:

$$\frac{\Delta\Gamma(B_s)}{\Gamma(B_s)} = 0.65^{+0.25}_{-0.33} \pm 0.01 \ . \tag{A.5}$$

If the true number is close to this central value, which is about four times larger than predicted, then history could repeat itself. For in qualitative analogy to the K_L - K_S case 'patience' would be awarded; i.e., an initial beam of B_s and $\bar{B}_s$ mesons would turn itself into an increasingly pure beam of the long lived meson, since the short lived one would decay away faster. This would open the door for novel searches for CP violation in B_s decays, where an inability to resolve the fast $B_s - \bar{B}_s$ oscillations driven by $\Delta M(B_s)$ might turn into a virtue rather than a vice. The fact that this could be achieved when the short and long lived components have lifetimes of close to one and two picoseconds rather than ~ 0.1 and ~ 50 nanoseconds, as it was the case for K_S and K_L, exemplifies the impressive progress in detector technology.

References

1. A. Einstein, B. Podolsky, N. Rosen, *Phys.Rev.* **47** (1935) 777.
2. J.H. Christensen *et al.*, *Phys.Rev.Lett.* **13** (1964) 138.
3. A very detailed discussion of all aspects of CP violation can be found in: I.I. Bigi, A.I. Sanda, 'CP Violation', Cambridge Monographs on Particle Physics, Nuclear Physics and Cosmology, Cambridge University Press, 2000.
4. A.D. Sakharov, *JETP Lett.* **5** (1967) 24; for an updated review see: A.D. Dolgov, hep-ph/9707419.
5. H. A. Kramers, *Proc. Acad. Sci. Amsterdam* **33** (1930) 959; see also: F.J. Dyson, *J. Math. Phys.* **3** (1962) 140.
6. B. Laurent, M. Roos, *Phys.Lett.* **13** (1964) 269; *ibid.* **15** 104.
7. E. Blucher, these Proceed.
8. D. Zavrtanik, these Proceed.
9. L. Wolfenstein, *Phys. Rev. Lett.* **13** (1964) 562.
10. M. Kobayashi, T. Maskawa, *Prog. Theor. Phys.* **49** (1973) 652; N. Cabibbo, *Phys. Rev. Lett.* **10** (1963) 531.
11. R. Mohapatra, *Phys. Rev.* **D6** (1972) 2023.
12. K. Niu, E. Mikumo, Y. Maeda, *Prog. Theor. Phys.* **46** (1971) 1644.
13. C. Jarlskog in: *CP Violation*, ed. C Jarlskog (World Scientific, Singapore, 1988).
14. A.B. Carter, A.I. Sanda, *Phys. Rev.* **D23** (1981) 1567; I.I. Bigi, A.I. Sanda, *Nucl. Phys.* **B193** (1981) 85.
15. I.I. Bigi, A.I. Sanda, *Nucl. Phys.* **B281** (1987) 41.
16. OPAL Collab., K. Ackerstaff *et al.*, *Eur. Phys. J.* **C5** (1998) 379.
17. CDF Collab., *Phys. Rev.* **D61** (2000) 072005.
18. P. Oddone, in: Proceed. of the 1987 UCLA Workshop on the Linear Collider $B\bar{B}$ Factory Conceptual Design, Los Angeles, 1987.
19. T. Morozumi, C.S. Lim, A.I. Sanda, *Phys. Rev. Lett.* **65** (1990) 404.
20. S. Bertolini, J. Eeg, M. Fabbrichesi, *Rev. Mod. Phys.* **72** (2000) 65.

21. I.I. Bigi, in: Proceed. of 'Les Rencontres de Physique de la Vallee d'Aoste, La Thuile, Italy, 1991; in: Proceed. of 'Les Rencontres de Moriond, Les Arcs, France, 1992.

22. F. Parodi, P. Roudeau, A. Stocchi, *Nuovo Cim.* **A112** (1999) 833.

23. M. Bander, D. Silverman, A. Soni, *Phys.Rev.Lett.* **43** (1979) 242.

24. I.I. Bigi, V.A. Khoze, N.G. Uraltsev, A.I. Sanda, in: *CP Violation*, ed. C Jarlskog (World Scientific, Singapore, 1988), p. 218.

25. KTeV Collab., A. Alavi-Harati *et al.*, *Phys. Rev.* **D67** (2003) 012005.

26. NA48 Collab., A. Lai *et al.*, *Eur. Phys. J.* **C22** (2001) 231.

27. http://www-cdf.fnal.gov/physics/new/bottom/040708.blessed-dgog-bsjpsiphi/

28. Z. Maki, M. Nakagawa, S. Sakata, *Prog. Theor. Phys.* **30** (1963) 727; B. Pontecorvo, *J. Exp. Theor. Phys.* **33** (1957) 549.

29. M. Gell-Mann, R. Slansky, P. Ramond, in: *Supergravity*, North Holland, 1979, p. 315; T. Yanagida, in: *Proc. Workshop on Unified Theory and Baryon Number in the Universe*, KEK, Japan, 1979.

30. I.I. Bigi, A.I. Sanda, hep-ph/0401003.

SECTION V
MACROSCOPIC TIME REVERSAL AND THE ARROW OF TIME

The Thermodynamic Arrow: Puzzles & Pseudo-Puzzles
H. Price

Arrow of Time from Timeless Quantum Gravity
C. Kiefer

The Evolution of the Universe
J. Garcia-Bellido

Did Time and Its Arrow have a Beginning?
G. Veneziano

The Wormhole Hazard
S. Krasnikov

Elementary Particles as Black Holes
G. 't Hooft

Contributed Paper

Counter-Example where Cosmic Time Keeps Its Original Role in Quantum Cosmology
E. I. Guendelman & A. B. Kaganovich

THE THERMODYNAMIC ARROW: PUZZLES AND PSEUDO-PUZZLES

HUW PRICE

Centre for Time, Main Quad A14, University of Sydney, NSW 2006, Australia.
Department of Philosophy, University of Edinburgh, David Hume Tower,
George Square, Edinburgh EH8 9JX, Scotland.
E-mail: huw@mail.usyd.edu.au

For more than a century, physics has known of a puzzling conflict between the T-asymmetry of thermodynamic phenomena and the T-symmetry of the underlying microphysics on which these phenomena depend. This paper provides a guide to the current status of this puzzle, distinguishing the central issue from various issues with which it may be confused. It is shown that there are two competing conceptions of what is needed to resolve the puzzle of the thermodynamic asymmetry, which differ with respect to the number of distinct T-asymmetries they take to be manifest in the physical world. On the preferable one-asymmetry conception, the remaining puzzle concerns the ordered distribution of matter in the early universe. The puzzle of the thermodynamic arrow thus becomes a puzzle for cosmology.

1 The puzzle of temporal bias

Late in the nineteenth century, on the shoulders of Maxwell, Boltzmann and many lesser giants, physicists saw that there is a deep puzzle behind the familiar phenomena described by the new science of thermodynamics. On the one hand, many such phenomena show a striking temporal bias. They are common in one temporal orientation, but rare or non-existent in reverse. On the other hand, the underlying laws of mechanics show no such temporal preference. If they allow a process in one direction, they also allow its temporal mirror image. Hence the puzzle: if the laws are so even-handed, why are the phenomema themselves so one-sided?

What has happened to this puzzle since the 1890s? I suspect that many contemporary physicists regard it as a dead issue, long since laid to rest. Didn't it turn out to be just a matter of statistics, after all? However, while there are certainly would-be solutions on offer—if anything, as we'll see, too many of them—it is far from clear that the puzzle has actually been solved. Late in the twentieth century, in fact, one of the most authoritative writers on the conceptual foundations of statistical mechanics could still refer to an understanding of the time-asymmetry of thermodynamics as 'that obscure object of desire'.[1]

One of the obstacles to declaring the problem solved is that there are several distinct approaches, not obviously compatible with one another. Which of these, if any, is supposed to be *the* solution, now in our grasp? Even more interestingly, it turns out that not all these would-be solutions are answers to the same question. There are different and incompatible conceptions in the literature of what the puzzle of the thermodynamic asymmetry actually is—about what *exactly* we should be trying explain, when we try to explain the thermodynamic arrow of time.

What the problem needs is therefore what philosophers do for a living: drawing fine distinctions, sorting out ambiguities, and clarifying the logical structure of difficult and subtle issues. My aim here is to bring these methods to bear on the puzzle of the time-asymmetry of thermodynamics. I want to distinguish the true puzzle from some of the appealing false trails, and hence to make it clear where physics stands in its attempt to solve it.

Little here is new, but it is surprisingly difficult to find a clear guide to these matters in the literature, either in philosophy or in physics. Accordingly, I think the paper will serve a useful purpose, in helping non-specialists to understand the true character of the puzzle discovered by those nineteenth century giants, the extent to which it has been solved, and the nature of the remaining issues.[a]

2 The true puzzle—a first approximation and a popular challenge

Everyone agrees, I think, that the puzzle of the thermodynamic arrow stems from the conjunction of two facts (or apparent facts—one way to dissolve the puzzle would be to show that one or other of the following claims isn't actually true):

1. There are many common and familiar physical processes, collectively describable as cases in which entropy is increasing, whose corresponding time-reversed processes are unknown or at least very rare.

2. The dynamical laws governing such processes show no such T-asymmetry—if they permit a process to occur with one temporal orientation, they permit it to occur with the reverse orientation.

As noted, some people will be inclined to object at this point that the conjunction is *merely* apparent. In particular, it may be objected that we now know that the dynamical laws are not time-symmetric. Famously, T-symmetry is violated in weak interactions, by the neutral K meson. Doesn't this eliminate the puzzle?

No. If the time-asymmetry of thermodynamics were associated with the T-symmetry violation displayed by the neutral K meson, then anti-matter would show the reverse of the normal thermodynamic asymmetry. Why? Because PCT-symmetry guarantees that if we replace matter by anti-matter (i.e., reverse P and C) and then view the result in reverse time (i.e., reverse T), physics remains the same. So if we replaced matter by anti-matter but didn't reverse time, any intrinsic temporal arrow or T-symmetry violation would reverse its apparent direction. In other words, physicists in anti-matter galaxies find the opposite violations of T-symmetry in weak interactions to those found in our galaxy. So if the thermodynamic arrow were tied to the T-symmetry violation, it too would have to reverse under such a transformation.

But now we have both an apparent falsehood, and a paradox. There's an apparent falsehood because (of course) we don't think that anti-matter behaves anti-thermodynamically. We expect stars in anti-matter galaxies to radiate just like our own sun (as the very idea of an anti-matter galaxy requires, in fact). And there's a paradox, because if this were the right story, what would happen to particles which are their own anti-particles, such as photons? They would have to behave both thermodynamically and anti-thermodynamically!

---

[a]For those interested in more details, I discuss these topics at greater length elsewhere.[2,3,4]

Here's another way to put the point. The thermodynamic arrow isn't just a T-asymmetry, it is a PCT-asymmetry as well. There are many familiar process whose PCT-reversed processes are equally compatible with the underlying laws, but which never happen, in our experience. We might be tempted to explain this asymmetry as due to the imbalance between matter and anti-matter, but the above reflections show that this is not so. So instead of the puzzle of the T-asymmetry of thermodynamics, we could speak of the puzzle of the PCT-asymmetry of thermodynamics. Then it would be clear to all that the strange behaviour of the neutral K meson isn't relevant. Knowing that we could if necessary rephrase the problem in this way, we can safely rely on the simpler formulation, and return to our original version of the puzzle.

3 Four things the puzzle is not

Some of the confusions common in debates about the origins of the thermodynamic asymmetry can be avoided distinguishing the genuine puzzle from various pseudo-puzzles with which it is liable to be confused. In this section I'll draw four distinctions of this kind.

3.1 The meaning of irreversibility

The thermodynamic arrow is often described in terms of the 'irreversibility' of many common processes—e.g., of what happens when a gas disperses from a pressurised bottle. This makes it sound as if the problem is that we can't make the gas behave in the opposite way—we can't make it put itself back into the bottle. Famously, Loschmidt's reversibility objection rested on pointing out that the reverse motion is equally compatible with the laws of mechanics. Some responses to this problem concentrate on the issue as to why we can't actually reverse the motions (at least in most cases).[5]

This response misses the interesting point, however. The interesting issue turns on a numerical imbalance in nature between 'forward' and 'reverse' processes, not case-by-case irreversibility of individual processes. Consider a parity analogy. Imagine a world containing many left hands but few right hands. Such a world shows an interesting parity asymmetry, even if any individual left hand can easily be transformed into a right hand. Conversely, a world with equal numbers of left and right hands is not interestingly P-asymmetric, even if any individual left or right hand cannot be reversed. Thus the interesting issue concerns the numerical asymmetry between the two kinds of structures—here, left hands and right hands—not the question whether one can be transformed into the other.

Similarly in the thermodynamic case, in my view. The important thing to explain is the numerical imbalance in nature between entropy-increasing processes and their T-reversed counterparts, not the practical irreversibility of individual processes.

3.2 Asymmetry in time versus asymmetry of time

Writers on the thermodynamic asymmetry often write as if the problem of explaining this asymmetry is the problem of explaining 'the direction of time'. This may be a harmless way of speaking, but we should keep in mind that the real puzzle concerns the asymmetry of physical processes *in* time, not an asymmetry *of time itself*. By analogy, imagine a long narrow room, architecturally symmetrical end-to-end. Now suppose all the chairs in the room are facing the same end. Then there's a puzzle about the asymmetry in the arrangement of the chairs, but not a puzzle about the asymmetry of the room. Similarly, the thermodynamic asymmetry is an asymmetry of the 'contents' of time, not an asymmetry of the container itself.

It may be helpful to make a few remarks about the phrase 'direction of time'. Although this expression is in common use, it isn't at all clear what it could actually mean, if we try to take it literally. Often the thought seems to be that there is an objective sense in which one time direction is future (or 'positive'), and the other past (or 'negative'). But what could this distinction amount to? It's easy enough to make sense of idea that time is *anisotropic*—i.e., different in one direction than in the other. For example, time might be finite in one direction but infinite in the other. But this isn't enough to give a *direction* to time, in above sense. After all, if one direction were objectively the future or positive direction, then in the case of a universe finite at one end, there would be two possibilities. Time might be finite in the past, and or finite in the future. So anisotropy alone doesn't give us *direction*.

Similarly, it seems, for any other physical time-asymmetry to which we might appeal. If time did have a direction—an objective basis for a privileged notion of positive or future time—then for any physical arrow or asymmetry in time, there would always be a question as to whether that arrow pointed forwards or backwards. And so no physical fact could answer this question, because for any candidate, the same issue arises all over again. Thus the idea that time has a real direction seems without any physical meaning. (Of course, we can use any asymmetry we like as a basis for a conventional labelling—saying, for example, that we'll regard the direction in which entropy is increasing as the positive direction of time. But this is different from discovering some intrinsic directionality to time itself.)

For present purposes, then, I'll assume that it is a conventional matter which direction we treat as positive or future time. Moreover, although it makes sense to ask whether time is anisotropic, it seems clear that this is a different issue from that of the thermodynamic asymmetry. As noted, the thermodynamic asymmetry is an asymmetry of physical processes *in* time, not an asymmetry of time itself.

3.3 Entropy gradient not entropy increase

If it is conventional which direction counts as positive time, then it is also conventional whether entropy increases or decreases. It increases by the lights of the usual convention, but decreases if we reverse the labelling. But this may seem ridiculous. Doesn't it imply, absurdly, that the thermodynamic asymmetry is merely conventional?

No. The crucial point is that while it's a conventional matter whether the entropy gradient slopes up or down, the gradient itself is objective. The puzzling asymmetry is that the gradient is monotonic—it slopes in the same direction everywhere (so far as we know).

It is worth noting that in principle there are two possible ways of contrasting this monotonic gradient with a symmetric world. One contrast would be with a world in which there are entropy gradients, but sometimes in one direction and sometimes in the other—i.e., worlds in which entropy sometimes goes up and sometimes goes down. The other contrast would be with worlds in which there are no significant gradients, because entropy is always high. If we manage to explain the asymmetric gradient we find in our world, we'll be explaining why the world isn't symmetric in one of these ways—but which one? The answer isn't obvious in advance, but hopefully will fall out of a deeper understanding of the nature of the problem.

3.4 The term 'entropy' is inessential

A lot of time and ink has been devoted to the question how entropy should be defined, or whether it can be defined at all in certain cases (e.g., for the universe as a whole). It would be easy to get the impression that the puzzle of the thermodynamic asymmetry depends on all this discussion—that whether there's really a puzzle depends on how, and whether, entropy can be defined, perhaps.

But in one important sense, these issues are beside the point. We can see that there's a puzzle, and go a long way towards saying what it is, without ever mentioning entropy. We simply need to describe in other terms some of the many processes which show the asymmetry—which occur with one temporal orientation but not the other. For example, we can point out that there are lots of cases of big difference in temperatures spontaneously equalising, but none of big differences in temperature spontaneously arising. Or we can point out that there are lots of cases of pressurised gas spontaneously leaving a bottle, but none of gas spontaneously pressurising by entering a bottle. And so on.

In the end, we may need the notion of entropy to generalise properly over these cases. However, we don't need it to see that there's a puzzle—to see that there's a striking imbalance in nature between systems with one orientation and systems with the reverse orientation. For present purposes, then, we can ignore objections based on problems in defining entropy. (Having said that, of course, we can go on using the term entropy with a clear conscience, without worrying about how it's defined. In what follows, talk of entropy increase is just a placeholder for a list of the actual phenomena which display the asymmetry we're interested in.)

3.5 Summary

For the remainder of the paper, then, I take it (i) that the asymmetry in nature is a matter of numerical imbalance between temporal mirror images, not of literal reversibility; (ii) that we are concerned with an asymmetry of physical processes in time, not with an asymmetry in time itself; (iii) that the objective asymmetry concerned is a monotonic *gradient*, rather than an increase or a decrease; and (iv) that if need be the term 'entropy' is to be thought of as a placeholder for the relevant properties of a list of actual physical asymmetries.

4 What would a solution look like? Two models

With our target more clearly in view, I now want to call attention to what may be the most useful distinction of all, in making sense of the many things that physicists and philosophers say about the thermodynamic asymmetry. This is a distinction between two very different conceptions of *what it would take* to explain the asymmetry—so different, in fact, that they disagree on *how many* distinct violations of T-symmetry it takes to explain the observed asymmetry. On one conception, an explanation needs two T-asymmetries. On the other conception, it needs only one.

Despite this deep difference of opinion about what a solution would look like, the distinction between these two approaches is hardly ever noted in the literature—even by philosophers, who are supposed to have a nose for these things. So it is easy for advocates of the different approaches to fail to see that they are talking at cross-purposes—that in one important sense, they disagree about what the problem is.

4.1 The two-asymmetry approach

Many approaches to the thermodynamic asymmetry look for a dynamical explanation of the second law—a dynamical cause or factor, responsible for entropy increase. Here are some examples, old and new:

1. *The H-theorem.* Oldest and most famous of all, this is Boltzmann's development of Maxwell's idea that intermolecular collisions drive gases towards equilibrium.

2. *Interventionism.* This alternative to the H-theorem, apparently first proposed by S. H. Burbury in the 1890s,[6,7] attributes entropy increase to the effects of random and uncontrollable influences from a system's external environment.

3. *Indeterministic dynamics.* There are various attempts to show how an indeterministic dynamics might account for the second law. A recent example is a proposal that the stochastic collapse mechanism of the GRW approach to quantum theory might also explain entropy increase.[8,9]

I stress two points about these approaches. First, if there is something dynamical which makes entropy increase, then it needs to be time-asymmetric. Why? Because otherwise it would force entropy to increase (or at least not to decrease) in both directions—in other words, entropy would be constant. In the H-theorem, for example, this asymmetry resides in the assumption of molecular chaos. In interventionism, it is provided by the assumption that incoming influences from the environment are 'random', or uncorrelated with the system's internal dynamical variables.

The second point to be stressed is that this asymmetry alone isn't sufficient to produce the observed thermodynamic phenomena. Something which forces entropy to be non-decreasing won't produce an entropy gradient unless entropy starts low. To give us the observed gradient, in other words, this approach also needs a low entropy boundary condition—entropy has to be low in the past. This condition, too, is time-asymmetric, and it's a separate condition from the dynamical asymmetry. (It is not guaranteed by the assumption of molecular chaos, for example.)

So this approach is committed to the claim that it takes *two* T-asymmetries— one in the dynamics, and one in the boundary conditions—to explain the observed asymmetry of thermodynamic phenomena. If this model is correct, explanation of the observed asymmetry needs an explanation of both contributing asymmetries, and the puzzle of the thermodynamic arrow has become a double puzzle.

4.2 The one-asymmetry model

The two-asymmetry model isn't the only model on offer, however. The main alternative was first proposed by Boltzmann in the 1870s,[10] in response to Loschmidt's famous criticism of the H-theorem. To illustrate the new approach, think of a large collection of gas molecules, isolated in a box with elastic walls. If the motion of the molecules is governed by deterministic laws, such as Newtonian mechanics, a specification of the microstate of the system at any one time uniquely determines its entire trajectory. The key idea of Boltzmann's new approach is that in the overwhelming majority of possible trajectories, the system spends the overwhelming majority of the time in a high entropy macrostate—among other things, a state in which the gas is dispersed throughout the container. (Part of Boltzmann's achievement was to find the appropriate way of counting possibilities, which we can call the *Boltzmann measure*.)

Importantly, there is no temporal bias in this set of possible trajectories. Each possible trajectory is matched by its time-reversed twin, just as Loschmidt had pointed out, and the Boltzmann measure respects this symmetry. Asymmetry arises only when we apply a low entropy condition at one end. For example, suppose we stipulate that the gas is confined to some small region at the initial time t_0. Restricted to the remaining trajectories, the Boltzmann measure now provides a measure of the likelihood of the various possibilities consistent with this boundary condition. Almost all trajectories in this remaining set will be such that the gas disperses after t_0. The observed behaviour is thus predicted by the time-symmetric measure, once we conditionalise on the low entropy condition at t_0.

On this view, then, there's no time-asymmetric factor which causes entropy to increase. This is simply the most likely thing to happen, given the combination of the time-symmetric Boltzmann probabilities and the single low entropy restriction in the past. More below on the nature and origins of this low entropy boundary condition. For the moment, the important thing is that although it is is time-asymmetric, so far as we know, this is the only time-asymmetry in play, according to Boltzmann's statistical approach. There's no need for a second asymmetry in the dynamics.

5 Which is the right model?

It is important to distinguish these two models, but it would be even more useful to know which of them is right. How many time-asymmetries should we be looking for, in trying to account for the thermodynamic asymmetry? This is a big topic, but I'll mention two factors, both of which seem to me to count in favour of the one-asymmetry model.

The first factor is simplicity, or theoretical economy. If the one-asymmetry approach works, it simply does more with less. In particular, it leaves us with only one time-asymmetry to explain. True, this would not be persuasive if the two-asymmetry approach actually achieved more than the one-asymmetry approach—if the former had some big thoeretical advantage that the latter lacked. But the second argument I want to mention suggests that this can't be the case. On the contrary, the second asymmetry seems redundant.

Redundancy is a strong charge, but consider the facts. The two-asymmetry approach tries to identify some dynamical factor (collisions, or external influences, or whatever) that causes entropy to increase—that makes a pressurised gas leave a bottle, for example. However, to claim that one of these factors *causes* the gas to disperse is to make the following 'counterfactual' claim: *If the factor were absent, the gas would not disperse* (or would do so at a different rate, perhaps). But how could the absence of collisions or external influences *prevent* the gas molecules from leaving the bottle?

Here's a way to make this more precise. In the terminology of Boltzmann's statistical approach, we can distinguish between *normal* initial microstates (for a system, or for the universe as a whole), which lead to entropy increases much as we observe, and *abnormal* microstates, which are such that something else happens. The statistical approach rests on the fact that normal microstates are vastly more likely than abnormal microstates, according to the Boltzmann measure.

In these terms, the above point goes as follows. The two-asymmetry approach is committed to the claim that the universe begins in an abnormal microstate. Why? Because in the case of normal initial microstates, entropy increases anyway, without the mechanism in question—so the required counterfactual claim isn't true.

It is hard to see what could justify this claim about the initial microstate. At a more local level, why should we think that the initial microstate of a gas sample in an open bottle is normally such that if it weren't for collisions (or external influences, or whatever), the molecules simply wouldn't encounter the open top of the bottle, and hence disperse?

Thus it is doubtful whether there is really any need for a dynamical asymmetry, and the one-asymmetry model seems to offer the better conception of what it would take to solve the puzzle of the thermodynamic asymmetry. But if so, then the various two-asymmetry approaches—including Boltzmann's own *H*-theorem, which he himself defended in the 1890s, long after he first proposed the statistical approach—are looking for a solution to the puzzle in the wrong place, at least in part.

For present purposes, the main conclusion I want to emphasise is that we need to make a choice. The one-asymmetry model and the two-asymmetry model represent are two very different views of *what it would take* to explain the thermodynamic arrow—of what the problem is, in effect. Unless we notice that they are different approaches, and proceed to agree on which of them we ought to adopt, we can't possibly agree on whether the old puzzle has been laid to rest.

6   The Boltzmann-Schuetz hypothesis—a no-asymmetry solution?

If the one-asymmetry view is correct, the puzzle of the thermodynamic arrow is really the puzzle of the low entropy boundary condition. Why is entropy so low in the past? After all, in making it unmysterious why entropy doesn't decrease in one direction, the Boltzmann measure equally makes it mysterious why it does decrease in the other—for the statistics themselves are time-symmetric.

Boltzmann himself was one of the first to see the importance of this issue. In a letter to *Nature* in 1895, he suggests an explanation, based on an idea he attributes to 'my old assistant, Dr Schuetz'.[11] He notes that although low entropy states are very unlikely, they are very likely to occur eventually, given enough time. If the universe is very old, it will have had time to produce the kind of low entropy region we find ourselves inhabiting simply by accident. 'Assuming the universe great enough, the probability that such a small part of it as our world should be in its present state, is no longer small,' as Boltzmann puts it.

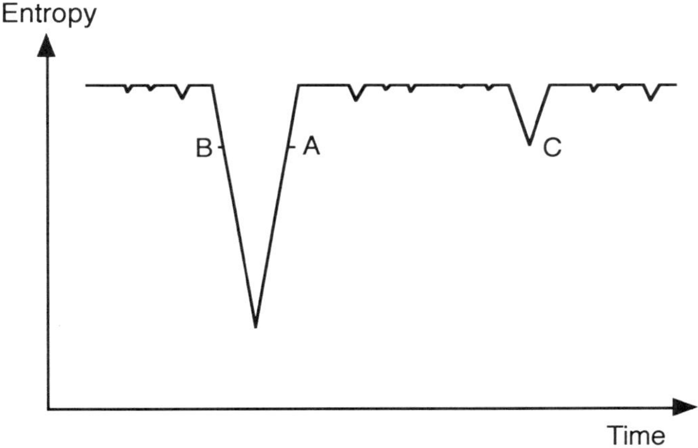

Figure 1. Boltzmann's entropy curve.

It is one thing to explain why the universe contains regions like ours, another to explain why we find ourselves in such a region. If they are so rare, isn't it more likely that we'd find ourselves somewhere else? But Boltzmann suggests an answer to this, too. Suppose, as seems plausible, that creatures like us couldn't exist in the vast regions of near-equilibrium between such regions of low entropy. Then it's no surprise that we find ourselves in such an unlikely place. As Boltzmann himself puts it, 'the ... *H* curve would form a representation of what takes place in the universe. The summits of the curve would represent the worlds where visible motion and life exist.'

Figure 1 shows what Boltzmann calls the *H* curve, except that this diagram plots entropy rather than Boltzmann's quantity *H*. Entropy is low when *H* is high, so the summits of Boltzmann's *H* curve are the troughs of the entropy curve. The universe spends most of its time very close to equilibrium. But occasionally—much more rarely than this diagram actually suggests—a random re-arrangement of matter produces a state of low entropy. As the resulting state returns to equilibrium, there's an entropy slope, such as the one on which we (apparently) find ourselves, at a point such as A.

Why do we find ourselves on an uphill rather than a downhill slope, as at B? In another paper, Boltzmann offers a remarkable proposal to explain this, too.[12] Perhaps our perception of past and future depends on the entropy gradient, in such a way that we are bound to regard the future as lying 'uphill'. Thus the perceived direction of time would not be objective, but a product of our own orientation in time. Creatures at point B would see the future as lying in the other direction, and there's no objective sense in which they are wrong and we are right, or vice versa. Boltzmann compares this to the discovery that spatial up and down are not absolute directions, the same for all observers everywhere.

For present purposes, what matters about the Boltzmann-Schuetz hypothesis is that it offers an explanation of the local asymmetry of thermodynamics in terms which are symmetric on a larger scale. So it is a no-asymmetry solution—the puzzle of the thermodynamic asymmetry simply vanishes on the large scale.

7　The big problem

Unfortunately, however, this clever proposal has a sting in its tail, a sting so serious that it now seems almost impossible to take the hypothesis seriously. The problem flows directly from Boltzmann's own link between entropy and probability. In Figure 1, the vertical axis is a logarithmic probability scale. For every downward increment, dips in the curve of the corresponding depth are exponentially more improbable. So a dip of the depth of point A or point B is much more likely to occur in the form shown at point C—where the given depth is very close to the minimum of the fluctuation—than in association with a much bigger dip, as at A and B. Hence if our own region has a past of even lower entropy, it is much more improbable than it needs to be, given its present entropy. So far, this point seems to have been appreciated already in the 1890s, in exchanges between Boltzmann and Zermelo. What doesn't seem to have appreciated is its devastating consequence, namely, that according to the Boltzmann measure it is much easier to produce fake records and memories, than to produce the real events of which they purport to be records.

Why does this consequence follow? Well, imagine that the universe is vast enough to contain many separate fluctuations, each containing everything that we see around us, including the complete works of Shakespeare, in all their twenty-first century editions. Now imagine choosing one of these fluctuations at random. It is vastly more likely that we'll select a case in which the Shakespearean texts are a product of a spontaneous recent fluctuation, than one in which they were really written four hundred years earlier by a poet called William Shakespeare. Why? Simply because entropy is much higher now than it was in the sixteenth century (as we normally assume that century to have been). Recall that according to Boltzmann, probability increases exponentially with entropy. Fluctuations like our twenty-first century—'Shakespearian' texts and all—thus occur much more often in typical world-histories than fluctuations like the lower-entropy sixteenth century. So almost all fluctuations including the former don't include the latter. The same goes for the rest of history—all our 'records' and 'memories' are almost certainly misleading.

To make this conclusion vivid we can take advantage of the fact that in the Boltzmann picture, there isn't an objective direction of time. So we can equally well think about the question of 'what it takes' to produce what we see around us from the reverse of the normal temporal perspective. Think of starting in what we call the future, and moving in the direction we call towards the past. Think of all the apparently miraculous accidents it takes to produce the kind of world we see around us. Among other things, our bodies themselves, and our editions of Shakespeare, have to 'undecompose', at random, from (what we normally think of as) their future decay products. That's obviously extremely unlikely, but the fact that we're here shows that it happens. But now think of what it takes to get even further back, to a sixteenth century containing Shakespeare himself. The same kind of near-miracle needs to happen many more times. Among other things, there are several billion intervening humans to 'undecompose' spontaneously from dust.

So the Boltzmann-Schuetz hypothesis implies that our apparent historical evidence is almost certainly unreliable. So far as I know, this point is first made in print by von Weizsäcker in 1939.[13] Von Weizsäcker notes that 'improbable states can count as documents [i.e., records of the past] only if we presuppose that still less probable states preceded them.' He concludes that 'the most probable situation by far would be that the present moment represents the entropy minimum, while the past, which we infer from the available documents, is an illusion.'

Von Weizsäcker also notes that there's another problem of a similar kind. The Boltzmann-Schuetz hypothesis implies that as we look further out into space, we should expect to find no more order than we already have reason to believe in. But we can now observe vastly more of the universe than was possible in Boltzmann's day, and there seems to be low entropy all the way out.

So the Boltzmann-Schuetz hypothesis faces some profound objections. Fortunately, as we're about to see, modern cosmology goes at least some way to providing us with an alternative.

8 Initial smoothness

We have seen that the observed thermodynamic asymmetry requires that entropy was low in the past. Low entropy requires concentrations of energy in useable forms, and presumably there are many ways such concentrations could exist in the universe. On the face of it, we seem to have no reason to expect any particularly neat or simple story about how it works in the real world—about where the particular concentrations of energy we depend on happen to originate. Remarkably, however, modern cosmology suggests that all the observed low entropy is associated with a single characteristic of the early universe, soon after the big bang. The crucial thing is that matter is distributed extremely smoothly in the early universe. This provides a vast reservoir of low entropy, on which everything else depends. In particular, smoothness is necessary for galaxy and star formation, and most familiar irreversible phenomena depend on the sun.

Why does a smooth arrangement of matter amount to a low entropy state? Because in a system dominated by an attractive force such as gravity, a uniform distribution of matter is highly unstable (and provides a highly useable supply of potential energy). However, about 10^5 years after the big bang, matter seems to have been distributed smoothly to very high accuracy.

One way to get a sense how surprising this is, is to recall that we've found no reason to disagree with Boltzmann's suggestion that there's no objective distinction between past and future—no sense in which things really happen in the direction we think of as past-to-future. Without such a distinction, there's no objective sense in which the big bang is not equally the end point of a gravitational collapse. Somehow that collapse is coordinated with astounding accuracy, so that the matter involved manages to avoid forming large agglomerations (in fact, black holes), and instead spreads itself out very evenly across the universe. (By calculating the entropy of black holes with comparable mass, Penrose[14] has estimated the probability of such a smooth arrangement of matter at $10^{-10^{123}}$.)

In my view, this discovery about the cosmological origins of low entropy is one of the great achievements of late twentieth century physics. It is a remarkable discovery in two quite distinct ways, in fact. First, it is the only anomaly necessary to account for the low entropy we find in the universe, at least so far as we know. So it is a remarkable theoretical achievement—it wraps up the entire puzzle of the thermodynamic asymmetry into a single package, in effect. Second, it is astounding that it happens at all, according to existing theories of how gravitating matter should behave (which suggests, surely, that there is something very important missing from those theories).[b]

9 Open questions

Why is the universe smooth soon after the big bang? This is a major puzzle, but—if we accept that the one-asymmetry model—it is

the only question we need to answer, to solve the puzzle of the thermodynamic arrow. So we have an answer to the question with which we began. What has happened to the puzzle noticed by those nineteenth century giants? It has been transformed by some of their twentieth century successors into a puzzle for cosmology, a puzzle about the early universe.

It is far from clear how this remaining cosmological puzzle is to be explained. Indeed, there are some who doubt whether it needs explaining.[15,16,17] But these issues are beyond the scope of this paper. I want to close by calling attention to some open questions associated with this understanding of the origins of the thermodynamic asymmetry, and by making a case for an unusually sceptical attitude

---

[b]True, it is easy to fail to see how astounding the smooth early universe is, by failing to see that the big bang can quite properly be regarded as the end point of a gravitational collapse. But anyone inclined to deny the validity of this way of viewing the big bang faces a perhaps even more daunting challenge: to explain what is meant by, and what is the evidence for, the claim that time has an objective direction!

to the second law.

One fascinating question is whether whatever explains why the universe is smooth after the big bang would also imply that the universe would be smooth before the big crunch, if the universe eventually recollapses. In other words, would entropy would eventually decrease, in a recollapsing universe? This possibility was first suggested by Thomas Gold some forty years ago.[18] It has often been dismissed on the grounds that a smooth recollapse would require an incredibly unlikely 'conspiracy' among the components parts of the universe, to ensure that the recollapsing matter did not clump into black holes. However, as we have already noted, this incredible conspiracy is precisely what happens towards (what we usually term) the big bang, if we regard that end of the universe as a product of a gravitational collapse. The statistics themselves are time-symmetric. If something overrides them at one end of the universe, what right do we have to assume that the same does not happen at the other? Until we understand more about the origins of the smooth early universe, then, it seems best to keep an open mind about a smooth late universe.

Another interesting and open question is whether a future low entropy boundary condition would have effects *now*. Events at the present era provide us with evidence of a low entropy past. Could there also be evidence of a low entropy future? The answer depends on our temporal distance from such a future boundary condition, in relation to the relaxation time of cosmological processes. It has been argued that a symmetric time-reversing universe would require more radiation in the present era than we actually observe—radiation which in the reversed time sense originates in the stars and galaxies of the opposite end of the universe.[19] But because of its anti-thermodynamic character, from our point of view, it is doubtful whether this radiation would be detectable, at least by standard means.[2]

Some people dismiss the question whether entropy would reverse in a recollapsing universe on the grounds that the current evidence suggests that the universe will not recollapse. However, it seems reasonable to expect that when we find out why the universe is smooth near the big bang, we'll be able to ask a theoretical question about what that reason would imply in the case of universe which did recollapse. Moreover, as a number of writers have pointed out,[20,21] much the same question arises if just a bit of the universe recollapses—e.g., a galaxy, collapsing into a black hole. This process seems to be a miniature version of the gravitational collapse of a whole universe, and so it makes sense to ask whether whatever constrains the big bang also constrains such partial collapses.

10 Scepticism about the second law

In my view, the moral of these considerations is that until we know more about why entropy is low in the past, it is sensible to keep an open mind about whether it might be low in the future. The appropriate attitude is a kind of healthy scepticism about the universality of the second law of thermodynamics.

The case for scepticism goes like this. What we've learnt about why entropy increases in our region is that it does so because it is very low in the past (for some reason we don't yet know), and the increase we observe is the most likely outcome consistent with that restriction. As noted, however, the statistics underpinning this reasoning are time-symmetric, and hence the predictions we make about the future depend implicitly on the assumption that there is no corresponding low entropy boundary condition in that direction. Thus the Boltzmann probabilities don't enable us to predict without qualification that entropy is unlikely to decrease, but only that it is unlikely to decrease, *unless there is the kind of boundary condition in the future that makes entropy low in the past.* In other words, the second law is likely to continue to hold so long as there isn't a low entropy boundary condition in the future. But it can't be used to exclude this possibility—even probabilistically!

Sceptics about the second law are unusual in the history of thermodynamics, and I would like to finish by giving some long-overdue credit to one of the rare exceptions. Samuel Hawksley Burbury (1831–1911) was not one of the true giants of thermodynamics. However, he made an important contribution to the identification of the puzzle of the time-asymmetry of thermodynamic phenomena. And he was more insightful than any of his contemporaries—and most writers since, for that matter—in being commendably cautious about declaring the puzzle solved.

Burbury was an English barrister. He read mathematics at Cambridge as an undergraduate, but his major work in mathematical physics came late in life, when deafness curtailed his career at the Bar. In his sixties and seventies, he thus played an important part in discussions about the nature and origins of the second law. In a review of Burbury's monograph *The Kinetic Theory of Gases* for *Science* in 1899, the reviewer describes his contribution as follows:

> [I]n that very interesting discussion of the Kinetic Theory which was begun at the Oxford meeting of the British Association in 1894 and continued for months afterwards in *Nature,* Mr. Burbury took a conspicuous part, appearing as the expounder and defender of Boltzmann's H-theorem in answer to the question which so many [had] asked in secret, and which Mr. Culverwell asked in print, '*What is the H-theorem and what does it prove?*' Thanks to this discussion, and to the more recent publication of Boltzmann's *Vorlesungen über Gas-theorie,* and finally to this treatise by Burbury, the question is not so difficult to answer as it was a few years ago.[22]

It is a little misleading to call Burbury a defender of the H-theorem. The crucial issue in the debate referred to here was the source of the time-asymmetry of the H-theorem, and while Burbury was the first to put his finger on the role of assumption of molecular chaos, he himself regarded this assumption with considerable suspicion. Here's how he puts it in 1904:

> Does not the theory of a general tendency of entropy to diminish [*sic*][c] take too much for granted? To a certain extent it is supported by experimental evidence. We must accept such evidence as far as it goes and no further. We have no right to supplement it by a large draft of the scientific imagination.[23]

---

[c]Burbury is apparently referring to Boltzmann's quantity H, which does decrease as entropy increases.

Burbury's reasons for scepticism are not precisely those which seem appropriate today. Burbury's concern might be put like this. To see that the dynamical processes routinely fail to produce entropy increases towards the past is to see that it takes an extra ingredient to ensure that they do so towards the future. We're then surely right to wonder whether that extra ingredient is sufficiently universal, even towards the future, to guarantee that the second law will always hold. As the first clearly to identify the source of the time-asymmetry in the H-theorem, Burbury was perhaps more sensitive to this concern than any of his contemporaries.

At the same time, however, Burbury seems never to have distanced himself sufficiently from the H-theorem to see that the real puzzle of the thermodynamic asymmetry lies elsewhere. The interesting question is not whether there is a good dynamical argument to show that entropy will alway increase towards the future. It is why entropy steadily *decreases* towards the past—in the face, note, of such things as the effects of collisions and external influences, which are 'happening' in that direction as much as in the other! As we've seen, this re-orientation provides a new reason for being cautious about proclaiming the universal validity of the second law. Once we regard the fact that entropy decreases towards the past as itself a puzzle, as something in need of explanation, then it ought to occur to us that whatever explains it might be non-unique—and thus that in principle, there might be a low entropy boundary condition in the future, as well as in the past.

References

1. L. Sklar in *Time's Arrows Today,* ed. S. Savitt (Cambridge University Press, Cambridge, 1995).
2. H. Price, *Time's Arrow and Archimedes' Point* (Oxford University Press, New York, 1996).
3. H. Price, *British Journal for the Philosophy of Science* **53**, 83 (2002).
4. H. Price in *Time, Reality and Experience,* ed. C. Callender (Cambridge University Press, Cambridge, 2002).
5. T.M. Ridderbos and M. Redhead, *Foundations of Physics* **28**, 1237 (1998).
6. S.H. Burbury, *Nature* **51**, 78 (1894).
7. S.H. Burbury, *Nature* **51**, 320 (1895).
8. D.Z. Albert, *British Journal for the Philosophy of Science* **45**, 669 (1994).
9. D.Z. Albert, *Time and Chance* (Harvard University Press, Cambridge, 2000).
10. L. Boltzmann, *Sitzungsberichte de kaiserlichen Akademie der Wissenschaften, Wien* **75**, 67 (1877).
11. L. Boltzmann, *Nature,* **51**, 413 (1895).
12. L. Boltzmann, *Annalen der Physik* **60**, 392 (1897).
13. C. von Weizsäcker, *Annalen der Physik (5 Folge)* **36**, 275 (1939), reprinted in translation in *The Unity of Nature* (Farrar Straus Giroux, New York, 1980).
14. R. Penrose, *The Emperor's New Mind* (Oxford University Press, Oxford, 1989).
15. C. Callender, *Metascience* **11**, 68 (1997).
16. C. Callender, *British Journal for the Philosophy of Science* **49**, 135 (1998).
17. L. Sklar, *Physics and Chance* (Cambridge University Press, Cambridge, 1993).
18. T. Gold, *American Journal of Physics* **30**, 403 (1962).

19. M. Gell-Mann and J. Hartle in *Physical Origins of Time Asymmetry,* ed. J. Halliwell, J. Perez-Mercader and W. Zurek (Cambridge University Press, Cambridge, 1994).
20. S.W. Hawking, *Physical Review, D* **33**, 2489 (1985).
21. R. Penrose in *General Relativity: an Einstein Centenary,* ed. S.W. Hawking and W. Israel (Cambridge University Press, Cambridge, 1979).
22. E.H. Hall, *Science, New Series* **10**, 685 (1899).
23. S.H. Burbury, *Philosophical Magazine, Series 6,* **8**, 43 (1904).

ARROW OF TIME FROM TIMELESS QUANTUM GRAVITY

CLAUS KIEFER

Institut für Theoretische Physik, Universität zu Köln, Zülpicher Str. 77, 50937 Köln,
Germany
E-mail: kiefer@thp.uni-koeln.de

Although most of the fundamental laws of nature do not distinguish between past
and future, most observed phenomena are actually irreversible. These 'arrows of
time' seem to have one common root – the expansion of the universe. To describe
the origin of the universe, a theory of quantum gravity seems to be necessary.
Such a theory does not contain any time parameter. I discuss how the standard
concept of time may emerge as an approximation and how the observed arrows of
time could be understood from a simple boundary condition in timeless quantum
gravity.

1   Arrows of time

The fundamental laws of nature do not contain a distinguished direction of time.
By fundamental laws I refer to the theories of the fundamental interactions – gravitational, strong, and electroweak interaction. There is one exception from this
rule: the CP violation connected with the weak interaction. Due to the validity
of the CPT theorem, CP violation occurs together with T violation, i.e. the violation of time reversal. This is an important process which is discussed in other
contributions to this volume. In my contribution, the focus is on other processes
which have been called *arrows of time* by Arthur Eddington. These correspond to
genuine irreversible phenomena, distinguished from CP violation by the fact that
they *cannot* be compensated by a symmetry transformation to give a non-violated
symmetry. In the CP-case one can compensate T-asymmetry by CP-asymmetry
to get CPT-symmetry. The latter is a general property of local relativistic field
theories.

There are many classes of phenomena which exhibit an arrow of time. This
means that their time-reversed version is under ordinary conditions never observed.
The most important ones are the following (see [1] for a detailed exposition and
references):

- Radiation arrow (advanced versus retarded radiation);

- Second Law of thermodynamics (increase of entropy);

- Quantum theory (measurement process and emergence of classical properties);

- Gravitational phenomena (expansion of the universe and emergence of structure by gravitational condensation).

The expansion of the universe is distinguished because it does not refer to a class
of phenomena; it is a single process. It has therefore been suggested that it is the
common root for all other arrows of time – the 'master arrow'. We shall see in the
course of our discussion that this seems indeed to be the case. But first we shall
consider in more detail the various arrows of time.

225

The *radiation arrow* is distinguished by the fact that fields interacting with local sources are usually described by *retarded* solutions, which in general lead to a damping of the source. Advanced solutions are excluded. They would describe the reversed process, during which the field propagates coherently towards its source, leading to its excitation instead of damping. This holds, in fact, for all wave phenomena. In electrodynamics, a solution of Maxwell's equations can be described by

$$A^\mu = \text{source term plus boundary term}$$
$$= A^\mu_{\text{ret}} + A^\mu_{\text{in}}$$
$$= A^\mu_{\text{adv}} + A^\mu_{\text{out}} \, ,$$

where A^μ is the vector potential. The important question is then why the observed phenomena obey $A^\mu \approx A^\mu_{\text{ret}}$ or, in other words, why

$$A^\mu_{\text{in}} \approx 0 \tag{1}$$

holds instead of $A^\mu_{\text{out}} \approx 0$. Eq. (1) is called a 'Sommerfeld radiation condition'. In a famous debate, Einstein and Ritz discussed whether (1) is a new law of nature (in which case the field would not possess independent degrees of freedom) or whether it can be derived from the Second Law of thermodynamics. Ritz voted for the former, Einstein for the latter. They wrote (see [2])

> " ... Ritz considers the restriction to the form of the retarded potential as one of the roots of the Second Law, while Einstein believes that the irreversibility is exclusively based on reasons of probability."

Today the opinion is on Einstein's side: due to the absorption properties of the material which constitutes the walls of the laboratory in which electrodynamic experiments are being performed, ingoing fields will be absorbed within a very short time and (1) will be fulfilled. For the thermal properties of absorbers the Second Law of thermodynamics is responsible.

The condition (1) also seems to hold for the universe as a whole ('darkness of the night sky'). The so-called Olbers' paradox can be solved by noting that the universe is, in fact, not static, but has a finite age and is much too young to have enough stars for a bright night sky. This is of course not yet sufficient to understand the validity of (1) for the universe as a whole. In an early stage the universe was a hot plasma in thermal equilibrium. Only the expansion of the universe and the ensuing redshift of the radiation are responsible for the fact that radiation has decoupled from matter and cooled to its present value of about three Kelvin – the temperature of the approximately isotropic cosmic background radiation with which the night sky 'glows'. During the expansion a strong thermal non-equilibrium could develop, which enabled the formation of structure.

The second arrow concerns the *Second Law* of thermodynamics: for a closed system entropy does not decrease. The total change of entropy is given by

$$\frac{dS}{dt} = \underbrace{\left(\frac{dS}{dt}\right)_{\text{ext}}}_{dS_{\text{ext}} = \delta Q/T} + \underbrace{\left(\frac{dS}{dt}\right)_{\text{int}}}_{\geq 0} \, ,$$

so that according to the Second Law the second term is non-negative. As the increase of entropy is also relevant for physiological processes, the Second Law is responsible for the subjective experience of irreversibility, in particular for the aging process. If applied to the universe as a whole, it would predict the increase of its total entropy, which would seem to lead to its 'heat death' (*Wärmetod*).

The laws of thermodynamics are based on microscopic statistical laws which are time-symmetric. How can the Second Law be derived from such laws? Already in the 19th century objections were formulated against a statistical foundation of the Second Law. These were, in particular,

- Loschmidt's reversibility objection (*Umkehreinwand*), and

- Zermelo's recurrence objection (*Wiederkehreinwand*).

Loschmidt's objection states that reversible dynamics must lead to an equal amount of transitions from an improbable to a probable state and from a probable to an improbable state. With overwhelming probability the system should be in its most probable state, i.e. in thermal equilibrium. Zermelo's objection is based on a theorem by Poincaré, according to which every system comes arbitrarily close to its initial state (and therefore to its initial entropy) after a finite amount of time. This objection is irrelevant, since the corresponding 'Poincaré times' are bigger than the age of the universe already for systems with few particles. The reversibility objection can only be avoided if a special boundary condition of low entropy holds for the early universe. Therefore, for the derivation of the Second Law

> one needs a special *boundary condition*!

Such a boundary condition must either be postulated or derived from a fundamental theory. The formal description of entropy increase from such a boundary condition is done by *master equations* [3]. These are equations for the 'relevant' (coarse-grained) part of the system. In an open system, the entropy can of course decrease, provided the entropy capacity of the environment is large enough to at least compensate this entropy decrease. This is crucial for the existence of life, and a particular efficient process in this respect is photosynthesis. The huge entropy capacity of the environment comes in this case from the high temperature gradient between the hot Sun and the cold empty space: few high-energy photons (with small entropy) arrive on Earth, while many low-energy photons (with high entropy) leave it. Thus, also the thermodynamic arrow of time points towards cosmology: how can gravitationally condensed objects like the Sun can come from in the first place?

Another important arrow of time is the quantum-mechanical arrow. The Schrödinger equation is time-reversal invariant, but the measurement process, either through

- a dynamical *collapse* of the wave function, or

- an Everett *branching*

distinguishes a direction. Growing entanglement with other degrees of freedom leads to *decoherence*: the *irreversible* emergence of classical properties for a local system through interaction with its environment [3]. A particle trajectory in a bubble chamber, for example, emerges through the interaction with the atoms in the chamber. Independent of the existence of a real collapse of the wave function, decoherence readily explains the occurrence of an 'apparent' collapse. Information about quantum-mechanical superpositions is delocalised into correlations with the inaccessible environment and no longer available at the system itself. Decoherence is described quantitatively through the dynamics of a reduced density matrix which is obtained from the total system by integrating out the environmental degrees of freedom. It obeys a master equation. The local entropy increases during the emergence of entanglement with the environment. It is, therefore, a measure of entanglement and corresponds to the size of the apparent ensemble of wave functions into which the density matrix can be decomposed. Again, decoherence only works if a special initial condition – a condition of weak entanglement – holds. But where can this come from?

The last of the main arrows is the gravitational arrow of time. Although the Einstein field equations are time-reversal invariant, gravitational systems in Nature distinguish a certain direction: the universe as a whole *expands*, while local systems such as stars form by *contraction*, e.g. from gas clouds. It is by this gravitational contraction that the high temperature gradients between stars such as the Sun and the empty space arise. Because of the negative heat capacity for gravitational systems, homogeneous states possess a low entropy, whereas inhomogeneous states possess a high entropy – just the opposite than for non-gravitational systems.

An extreme case of gravitational collapse is the formation of black holes. According to general relativity, their gravitational field is so strong that nothing, not even light, can escape. In spite of this, they possess fundamental thermodynamical properties (temperature and entropy) which are connected with their event horizon. The time-reversed version of black holes, the so-called white holes, do not seem to exist in our universe. Temperature and entropy of a black hole exhibit their meaning if quantum theory is taken into account: according to the Hawking effect (see e.g. [4]) a black hole radiates with a temperature that is in the spherically-symmetric case given by the 'Hawking temperature',

$$T_{\mathrm{BH}} = \frac{\hbar}{8\pi G k_{\mathrm{B}} M} \approx 6.2 \times 10^{-8} \frac{M_\odot}{M} \; \mathrm{K} \; . \tag{2}$$

Connected with this temperature is the 'Bekenstein-Hawking entropy'

$$S_{\mathrm{BH}} = \frac{k_{\mathrm{B}} A}{4 G \hbar} , \tag{3}$$

where A denotes the surface area of the black-hole horizon. This entropy is much bigger than the entropy of the object from which the black hole has formed. The entropy of the Sun, for example, is $S_\odot \approx 10^{57}$, but the entropy of a solar-mass black hole is $S_{\mathrm{BH}} \approx 10^{77}$, i.e. twenty orders of magnitudes larger (all entropies are measured in units of k_{B}). If all matter in the observable universe were in a single gigantic black hole, its entropy would be $S_{\mathrm{BH}} \approx 10^{123}$. Black holes thus seem to be the most efficient objects for swallowing information. A 'generic' universe would thus basically consist of black holes. Since this is not the case, our universe must have been started with a very special initial condition. Can this be analysed further? Close to the big bang the classical theory of general relativity breaks down. It is most likely that a quantum theory of gravity must be invoked to describe this early phase. This is discussed in the next section.

2 Quantum gravity

There does not yet exist a consensus about a final theory of quantum gravity. Still, there are many reasons why one could expect such a theory to replace general relativity at the most fundamental level. Among them are (cf. [5]):

- *Singularity theorems of general relativity*: Under very general conditions, the occurrence of a singularity, and therefore the breakdown of the theory, is unavoidable. A more fundamental theory is therefore needed to overcome these shortcomings. Motivated by history (the quantum cure of singularities in the electrodynamic description of atoms), the general expectation is that this fundamental theory is a quantum theory of gravity.

- *Unification*: Apart from general relativity, all known fundamental theories are *quantum* theories. It would thus seem awkward if gravity, which couples universally to all other fields, should remain the only classical entity in a fundamental description. Moreover, it seems that classical fields cannot be coupled to quantum fields without leading to inconsistencies (Bohr-Rosenfeld type of analysis).

- *Gravity as a regulator*: Many models indicate that the consistent inclusion of gravity in a quantum framework automatically eliminates the divergences that plague ordinary quantum field theory.

- *Problem of time*: In ordinary quantum theory, the presence of an external time parameter t is crucial for the interpretation of the theory: 'Measurements' take place at a certain time, matrix elements are evaluated at fixed times, and the norm of the wave function is conserved *in* time. In general relativity, on the other hand, time as part of spacetime is a dynamical quantity. Both concepts of time must therefore be modified at a fundamental level.

The task of quantising gravity has not yet been accomplished, but approaches exist within which sensible questions can be asked. Two approaches are at the centre of current research: *Superstring theory* (or *M-theory*) and *canonical quantum gravity*. Superstring theory is much more ambitious and aims at a unification of all interactions within a single quantum framework. Canonical quantum gravity, on the other hand, attempts to construct a consistent, non-perturbative, quantum theory of the gravitational field on its own. This is done through the application of standard quantisation rules to the general theory of relativity. A recent review of these approaches is [6].

The fundamental length scales that are connected with these theories are the Planck length, $l_P = \sqrt{G\hbar/c^3}$, or the string length, l_s. It is generally assumed that the string length is somewhat larger than the Planck length. Although not fully established in quantitative detail, canonical quantum gravity should follow as an approximation from superstring theory for scales $l \gg l_s > l_P$. It seems therefore sufficient to base the following discussion on canonical quantum gravity, although I want to emphasise that the same conceptual issues arise in superstring theory.

Depending on the choice of the canonical variables, the canonical theory can be subdivided into the following approaches:

- *Quantum geometrodynamics*: This is the traditional approach that uses the three-dimensional metric as its configuration variable.

- *Quantum connection dynamics*: The configuration variable is a non-Abelian connection that has many similarities to gauge theories.

- *Quantum loop dynamics*: The configuration variable is the trace of a holonomy with respect to loops, analogous to Wilson loops.

There exists a connection between the last two approaches, whereas their connection to the first approach is less clear. For the above reason one should, however, expect that a relation between all approaches exists at least on a semiclassical level. The following discussion is done within quantum geometrodynamics, since this seems to be the most appropriate language for the conceptual issues related to time.

Which are the central equations of quantum geometrodynamics? Since the classical theory (general relativity) possesses certain invariance properties (invariance with respect to coordinate transformations), it contains *constraints*. These are equations without second time derivatives. Because one has the freedom to choose four coordinates locally, there are four constraints at each space point. I shall not go into any details here (see e.g. [5] and [6]). The most important equation is the 'Wheeler-DeWitt equation'. It has the form of a zero-energy Schrödinger equation,

$$\boxed{\hat{H}\Psi[h_{ab}, \varphi] = 0} \, , \tag{4}$$

where $\hat{H}$ denotes the total Hamilton operator of both gravitational (the three-metric h_{ab}) and non-gravitational (here represented by a scalar field φ) degrees of freedom. The Wheeler-DeWitt equation has certain characteristic features:

- It depends on the *three*-dimensional metric, but is invariant under coordinate transformations, so it depends in fact only on the three-dimensional geometry.

- It does not contain any time parameter ('timeless nature').

- It follows from any theory which classically is time-reparametrisation invariant.

- It is (locally) hyberbolic and thereby defines an 'intrinsic time'.

Spacetime as a classical concept has disappeared. How can this be understood? In classical canonical gravity, a spacetime can be represented as a 'trajectory' in configuration space – the space of all three-metrics. Although time coordinates have no intrinsic meaning in classical general relativity either, they can nevertheless be used to parametrise this trajectory in an essentially arbitrary way. Since no trajectories exist anymore in quantum theory, no spacetime exists at the most fundamental, and therefore also no time coordinates to parametrise any trajectory. A simple analogy is provided by the relativistic particle: in the classical theory there is a trajectory which can be parametrised by some essentially arbitrary parameter, e.g. the proper time. Reparametrisation invariance leads to one constraint, $p^2 + m^2 = 0$. In the quantum theory, no trajectory exists anymore, the wave function obeys the Klein-Gordon equation as an analogue of (4), and any trace of a classical time parameter is lost (although, of course, for the relativistic particle the background Minkowski spacetime is present, which is not the case for gravity).

The Wheeler-DeWitt equation (4) is in general very complicated. It assumes a simple form only in simple models. One example is cosmology where all gravitational degrees of freedom are frozen out, except global degrees of freedom like the scale factor ('radius') of the universe. In the case of a Friedmann universe with scale factor $a \equiv e^\alpha$ and massive scalar field ϕ, (4) reads

$$\hat{H}\psi \equiv \left(\hbar^2 \frac{\partial^2}{\partial\alpha^2} - \hbar^2 \frac{\partial^2}{\partial\phi^2} + m^2\phi^2 e^{6\alpha} - e^{4\alpha} \right) \psi(\alpha,\phi) = 0 \ . \tag{5}$$

One recognises explicitly the hyperbolic nature ('wave nature') of this equation. The role of intrinsic time is played by α; this becomes evident if further degrees of freedom are added: they all come with the sign of the kinetic term for ϕ.

Since no time parameter t is contained in (4), one cannot pose any initial conditions with respect to t. Instead, one can specify the wave function (and its derivative) – in the example (5) – at a fixed value of α. This is the natural boundary condition for a hyperbolic equation. It has drastic consequences if one wants to describe a universe that classically expands, reaches a maximum and recollapses again [1,7,8]. Both big bang and big crunch correspond to the same region in configuration space – the region of $\alpha \to -\infty$. They are thus intrinsically indistinguishable. The Wheeler-DeWitt equation connects larger scale factors with smaller scale factors, but not two ends of a classical trajectory. If one wants to mimick the classical trajectory by a 'recollapsing' wave packet, one has to include both the 'initial' and the 'final' wave packet into one initial condition with respect to α. If one of the two packets were lacking, one would not be able to recover the classical trajectory as an approximation.

There is another interesting feature in the case of recollapsing universes: it is in general not possible to construct from (5) a wave packet that follows as a narrow tube the classical trajectory [7,1]. Therefore, a semiclassical approximation is not valid all along the trajectory and quantum effects can play a role even far away from the Planck scale – e.g. at the turning point of the classical universe.

3   Recovery of time and its direction

The Wheeler-DeWitt equation (4) is timeless. To derive an arrow of time from such an equation, two steps must be performed:

1. A time parameter t must be recovered as an approximate concept, and increasing t should be correlated with increasing scale factor.

2. A natural boundary condition must be invoked to derive entropy increase with increasing scale factor (and therefore with increasing t).

The first step involves a Born-Oppenheimer type of approximation scheme with respect to the Planck mass $m_P \equiv \hbar/l_{PC}$. The situation is formally similar to molecular physics where the heavy nuclei move slowly, followed adiabatically by the light electrons. In situations where the relevant scales are much smaller than the Planck mass, the gravitational kinetic term can be neglected in a first approximation. One makes for solutions of (4) the ansatz

$$\Psi[h_{ab}, \varphi] = e^{im_P^2 S[h_{ab}]} \Phi[h_{ab}, \varphi] \ . \tag{6}$$

Inserting this into (4) and and making an expansion with respect to m_P, one finds that $S[h_{ab}]$ obeys the gravitational Hamilton-Jacobi equation. This is known to be equivalent to Einstein's field equations. In this sense the classical background spacetime emerges as an approximation (such as geometrical optics emerges as a limit from wave optics). One can now pick out one classical spacetime from the many classical solutions (spacetimes) that are described by $S[h_{ab}]$. The 'matter wave functional' $\Phi[h_{ab}, \varphi]$ can then be evaluated on this particular spacetime described by $h_{ab}(\mathbf{x}, t)$ and can therefore shortly be labelled $\Phi(t, \varphi)$. If other semiclassical variables are present (such as the homogeneous field ϕ in (5)), they are included in S. The time parameter t is defined from $S[h_{ab}]$ as parametrising the classical trajectory (spacetime) running orthogonally to $S[h_{ab}] = $ const. in the space of three-geometries. In the special case (5) of the Friedmann universe, t is defined by the scale factor $a(t)$ and the homogeneous scalar field $\phi(t)$. It can be shown from (4) that the time evolution of the state Φ,

$$\frac{\partial}{\partial t} \Phi(t, \varphi) = \int d^3 x \, \dot{h}_{ab}(\mathbf{x}, t) \, \frac{\delta}{\delta h_{ab}(\mathbf{x}, t)} \Phi[h_{ab}(\mathbf{x}, t), \varphi] \ . \tag{7}$$

is given by a functional Schrödinger equation in the external classical spacetime found from $S[h_{ab}]$,

$$i\hbar \frac{\partial}{\partial t} \Phi(t, \varphi) = \hat{H}^{\mathrm{mat}} \Phi(t, \varphi) \ , \tag{8}$$

where $\hat{H}^{\mathrm{mat}}$ is the matter field Hamiltonian in the Schrödinger picture, parametrically depending on (generally nonstatic) metric coefficients of the curved spacetime background. In this way, the Schrödinger equation for non-gravitational fields has been recovered from quantum gravity as an approximation. A derivation similar to the above can already be performed within ordinary quantum mechanics if one assumes that the total system is in a 'timeless' energy eigenstate, see [9]. In fact, Mott had already considered in 1931 a time-independent Schrödinger equation for a total system consisting of an α-particle and an atom [10]. If the state of the α-particle can be described by a plane wave (corresponding in this case to high velocities), one can make an ansatz similar to (6) and derive a time-dependent Schrödinger equation for the atom alone, in which time is *defined* by the α-particle.

The ansatz (6) is already special, since it is a product of a pure phase part depending on gravity with a matter wave function. The i in the Schrödinger equation (8) has its origin in the choice of this phase. Can this be justified? The answer is yes, and the process of decoherence (see above) plays a crucial role in this demonstration. Tracing out irrelevant degrees of freedom (in the present context these can be density fluctuations or gravitational waves), one can see that states of the form (6) are most robust against environmental influence and that the variables contained in $S[h_{ab}]$ assume quasiclassical properties [1,5].

The first of the above-mentioned two steps has thus been achieved: in certain regions of configuration space where the wave functional is of the approximate form (6), a time parameter can be introduced as an approximate notion. This time parameter is defined through macroscopic (decohered) quantities to which in particular the global size of the universe belongs. But there has not yet emerged any direction of time. How can this be achieved?

It is already recognisable from (5) that the potential term vanishes near the 'big bang' $\alpha \to -\infty$. This property is robust against the inclusion of (small) perturbations, i.e. degrees of freedom describing density fluctuations or gravitational waves. Denoting these variables ('modes') symbolically by $\{x_i\}$, one has for the total Hamiltonian in the Wheeler-DeWitt equation an expression of the form (cf. [1])

$$\hat{H} = \frac{\partial^2}{\partial \alpha^2} + \sum_i \left(-\frac{\partial^2}{\partial x_i^2} + V_i(\alpha, x_i) \right) + V_{\text{int}}(\alpha, \{x_i\}) \, , \tag{9}$$

where the last term describes the interaction between the modes (assumed to be small), and the V_i describe the interaction of the mode x_i with the scale factor α. Both terms have, in fact, the property that they vanish for $\alpha \to -\infty$. It is therefore possible to impose in this limit a separating solution of $\hat{H}\Psi = 0$,

$$\Psi \stackrel{\alpha \to -\infty}{\longrightarrow} f(\alpha) \prod_i \chi_i(x_i) \, , \tag{10}$$

i.e. a solution of lacking entanglement. If this is taken as an 'initial condition', the Wheeler-DeWitt equation automatically – through the occurrence of the potentials in (9) – leads to a wave function which for increasing α becomes *entangled* between α and all modes. This, then, leads to an increase of local entropy, i.e. an increase of the entropy which is connected with the subset of 'relevant' degrees of freedom. Calling the latter $\{y_i\}$, one has

$$S(\alpha, \{y_i\}) = -k_{\text{B}} \text{tr}(\rho \ln \rho) \, , \tag{11}$$

where ρ is the reduced density matrix corresponding to α and $\{y_i\}$ and which is obtained by tracing out all irrelevant degrees of freedom in the full wave function. Entropy thus increases with increasing scale factor – this is the gravitational arrow of time. It is also the arrow of time that is connected with decoherence. It is therefore the root for both the quantum mechanical and the thermodynamical arrow of time. Quantum gravity could thus really yield the master arrow, the formal reason being the asymmetric appearance of α in the Wheeler-DeWitt equation: the potential goes to zero near the big bang, but becomes highly non-trivial for increasing size of the universe. It is imaginable that a full understanding of quantum gravity necessarily requires a boundary condition of the type (10), see in this context [11].

In the case of a classically recollapsing universe the boundary condition (10) has interesting consequences: since it is formulated at $\alpha \to \infty$, increasing entropy is always correlated with increasing α, i.e. increasing size of the universe. Consequently, the arrow of time formally reverses near the classical turning point [8]. It turns out that this region is fully quantum, so no paradox arises; it just means that there are many quasi-classical components of the wave function, each describing a universe that is experienced from within as expanding. All these components interfere destructively near the classical turning point, which would then constitute the 'end' of evolution (cf. the remarks at the end of Sect. 2).

Needless to say that the above remarks about the quantum-gravitational origin of the arrow of time are still speculative. However, the equations on which the discussion is based are straightforward consequences of unifying two established physical theories – quantum theory and general relativity. One might therefore hope that even a fundamental theory such as string theory would not spoil these conclusions. Quantum gravity can thus provide for the first time the necessary conceptual means for an understanding of entropy increase in our universe.

Acknowledgements

I would like to thank the organisers of this meeting, Ikaros Bigi and Martin Fäßler, for inviting me to such an inspiring conference amidst beautiful surroundings.

References

1. H. D. Zeh, *The physical basis of the direction of time*, 4th edn (Springer, Berlin, 2001).
2. A. Einstein and W. Ritz, *Phys. Z.* **10**, 323 (1909).
3. E. Joos, H. D. Zeh, C. Kiefer, D. Giulini, J. Kupsch, and I.-O. Stamatescu, *Decoherence and the appearance of a classical world in quantum theory*, 2nd edn (Springer, Berlin, 2003).
4. C. Kiefer, in *Classical and quantum black holes*, ed. P. Fré *et al* (Institute of Physics Publishing, Bristol, 1999).
5. C. Kiefer, in *Towards quantum gravity*, ed. J. Kowalski-Glikman (Springer, Berlin, 2000).
6. D. Giulini, C. Kiefer, and C. Lämmerzahl (ed.), *Aspects of quantum gravity*, Lecture Notes in Physics (Springer, Berlin, 2003).
7. C. Kiefer, *Phys. Rev.* *D* **38**, 1761 (1988).
8. C. Kiefer and H. D. Zeh, *Phys. Rev.* *D* **51**, 4145 (1995).
9. J. S. Briggs and J. M. Rost, *Found. Phys.* **31**, 693 (2001).
10. N. F. Mott, *Proc. Cambridge Phil. Soc.* **27**, 553 (1931).
11. B. S. DeWitt, *Phys. Rev.* **160**, 1113 (1967).

THE EVOLUTION OF THE UNIVERSE

JUAN GARCIA-BELLIDO

Theory Division CERN, CH-1211 Genève, Switzerland
E-mail: bellido@mail.cern.ch

With the recent measurements of temperature and polarization anisotropies in the microwave background by WMAP, we have entered a new era of precision cosmology, with the cosmological parameters of a Standard Cosmological Model determined to 1%. This Standard Model is based on the Big Bang theory and the inflationary paradigm, a period of exponential expansion in the early universe responsible for the large-scale homogeneity and spatial flatness of our observable patch of the Universe. The spectrum of metric perturbations, seen in the microwave background as temperature anisotropies, were produced during inflation from quantum fluctuations that were stretched to cosmological size by the expansion, and later gave rise, via gravitational collapse, to the observed large-scale structure of clusters and superclusters of galaxies. Furthermore, the same theory predicts that all the matter and radiation in the universe today originated at the end of inflation from an explosive production of particles that could also have been the origin of the present baryon asymmetry, before the universe reached thermal equilibrium at a very large temperature. From there on, the universe cooled down as it expanded, in the way described by the standard hot Big Bang model.

1   Introduction

Our present understanding of the universe is based upon the successful hot Big Bang theory, which explains its evolution from the first fraction of a second to our present age, around 13 billion years later. This theory rests upon four strong pillars, a theoretical framework based on general relativity, as put forward by Albert Einstein and Alexander A. Friedmann in the 1920s, and three strong observational facts. First, the expansion of the universe, discovered by Edwin P. Hubble in the 1930s, as a recession of galaxies at a speed proportional to their distance from us. Second, the relative abundance of light elements, explained by George Gamow in the 1940s, mainly that of helium, deuterium and lithium, which were cooked from the nuclear reactions that took place at around a second to a few minutes after the Big Bang, when the universe was a hundred times hotter than the core of the sun. Third, the cosmic microwave background (CMB), the afterglow of the Big Bang, discovered in 1965 by Arno A. Penzias and Robert W. Wilson as a very isotropic blackbody radiation at a temperature of about 3 degrees Kelvin, emitted when the universe was cold enough to form neutral atoms, and photons decoupled from matter, 380 000 years after the Big Bang. Today, these observations are confirmed to within a few percent accuracy, and have helped establish the hot Big Bang as the preferred model of the universe.

The Big Bang theory could not explain, however, the origin of matter and structure in the universe; that is, the origin of the matter–antimatter asymmetry, without which the universe today would be filled by a uniform radiation continuosly expanding and cooling, with no traces of matter, and thus without the possibility to form gravitationally bound systems like galaxies, stars and planets that could sustain life. Moreover, the standard Big Bang theory assumes, but cannot explain, the origin of the extraordinary smoothness and flatness of the universe on the very large scales seen by the microwave background probes and the largest galaxy catalogs. It can neither explain the origin of the primordial density perturbations that gave rise, via gravitational collapse, to cosmic structures like galaxies, clusters and superclusters; nor the nature of the dark matter and dark energy that we believe permeates the universe; nor the origin of the Big Bang itself.

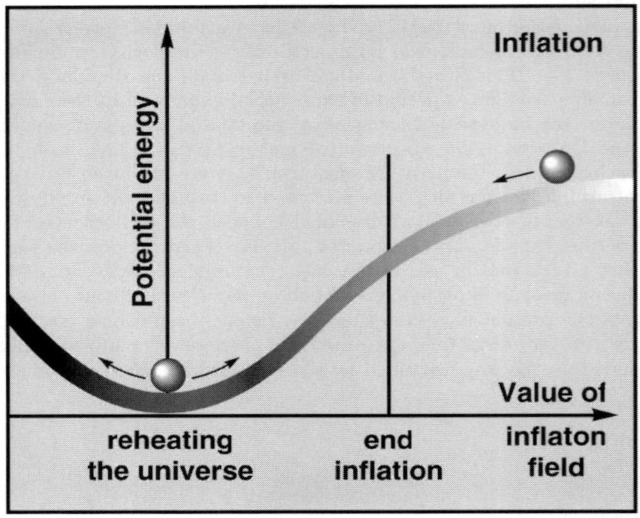

Figure 1. The inflaton field can be represented as a ball rolling down a hill. During inflation, the energy density is approximately constant, driving the tremendous expansion of the universe. When the ball starts to oscillate around the bottom of the hill, inflation ends and the inflaton energy decays into particles. In certain cases, the coherent oscillations of the inflaton could generate a resonant production of particles which soon thermalize, reheating the universe.

In the 1980s, a new paradigm, deeply rooted in fundamental physics, was put forward by Alan H. Guth, Andrei D. Linde and others, to address these fundamental questions. According to the inflationary paradigm, the early universe went through a period of exponential expansion, driven by the approximately constant energy density of a scalar field called the inflaton. In modern physics, elementary particles are represented by quantum fields, i.e. a function of space and time whose quantum oscillations can be interpreted as particles. For instance, the photon is the particle associated with the electromagnetic field. In our case, the inflaton field has, associated with it, a large potential energy density, which drives the exponential expansion during inflation, see Fig. 1. We know from general relativity that the density of matter determines the expansion of the universe, but a constant energy density acts in a very peculiar way: as a repulsive force that makes any two points in space separate at exponentially large speeds. (This does not violate the laws of causality because there is no information carried along in the expansion, it is simply the stretching of space-time.)

This superluminal expansion is capable of explaining the large scale homogeneity of our observable universe and, in particular, why the microwave background looks so isotropic: regions separated today by more than $1°$ in the sky were, in fact, in causal contact before inflation, but were stretched to cosmological distances by the expansion, see Fig. 2. Any inhomogeneities present before the tremendous expansion would be washed out. Moreover, in the usual Big Bang scenario a flat universe, one in which the gravitational attraction of matter is exactly balanced by the cosmic expansion, is unstable under perturbations: a small deviation from flatness is amplified and soon produces either an empty universe or a collapsed one. For the universe to be nearly flat today, it must have been extremely flat at nucleosynthesis for example, deviations not exceeding more than one part in 10^{15}. This extreme fine tuning of initial conditions was also solved by the inflationary paradigm, see Fig. 3. Thus inflation is an extremely elegant hypothesis that explains how a region much, much greater than our own observable universe could have become smooth and flat without recourse to *ad hoc* initial conditions.

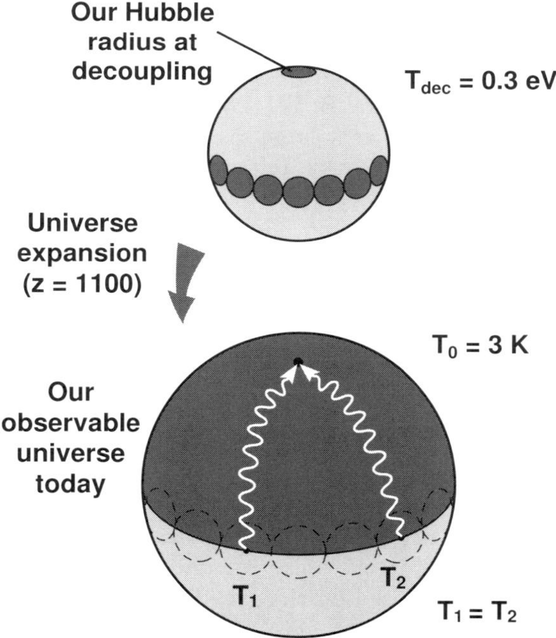

Figure 2. Perhaps the most acute problem of the Big Bang model is explaining the extraordinary homogeneity and isotropy of the microwave background. Information cannot travel faster than the speed of light, so the causal region (so-called horizon or Hubble radius) at the time of photon decoupling could not be larger than 300 000 light years across, or about $1°$ projected in the sky today. So why should regions that are separated by more than $1°$ in the sky have the same temperature, when the photons that come from those two distant regions could not have been in causal contact when they were emitted? This constitutes the so-called horizon problem, which is spectacularly solved by inflation.

2   The origin of structure in the universe

If cosmological inflation made the universe so extremely flat and homogeneous, where did the galaxies and clusters of galaxies come from? One of the most astonishing predictions of inflation, one that was not even expected, is that quantum fluctuations of the inflaton field are stretched by the exponential expansion and generate large-scale perturbations in the metric. Inflaton fluctuations are small wave packets of energy that, according to general relativity, modify the space-time fabric, creating a whole spectrum of curvature perturbations. The use of the word spectrum here is closely related to the case of light waves propagating in a medium: a spectrum characterizes the amplitude of each given wavelength. In the case of inflation, the inflaton fluctuations induce waves in the space-time metric that can be decomposed into different wavelengths, all with approximately the same amplitude, that is, corresponding to a scale-invariant spectrum. These patterns of perturbations in the metric are like fingerprints that unequivocally characterize a period of inflation. When matter fell in the troughs of these waves, it created density perturbations that collapsed gravitationally to form galaxies, clusters and superclusters of galaxies, with a spectrum that is also scale invariant. Such a type of spectrum was proposed in the early 1970s (before inflation) by Edward R. Harrison, and independently by the Russian cosmologist Yakov B. Zel'dovich, to explain the distribution of galaxies and clusters of galaxies on very large scales in our observable universe.

Various telescopes – like the Hubble Space Telescope, the twin Keck telescopes in Hawaii and the European Southern Observatory telescopes in Chile – are exploring the most distant regions of the universe and discovering the first galaxies at large distances. According to the Big Bang theory, the further the galaxy is, the larger its recession velocity, and the larger the shift towards the red of the spectrum of light from that galaxy. Astronomers thus measure distances in units of red-shift z. The furthest galaxies observed so far are at redshifts of $z \simeq 7$, or 13 billion light years from the Earth, whose light was emitted when the universe had only about 2% of its present age. Only a few galaxies are known at those redshifts, but there are at present various catalogs like the IRAS PSCz and Las Campanas redshift survey, that study the spatial distribution of hundreds of thousands of galaxies up to distances of a billion light years, or $z < 0.1$, that recede from us at speeds of tens of thousands of kilometres per second. These catalogs are telling us about the evolution of clusters of galaxies in the universe, and already put constraints on the theory of structure formation based on the gravitational collapse of the small inhomogeneities produced during inflation. From these observations one can infer that most galaxies formed at redshifts of the order of $2 - 4$; clusters of galaxies formed at redshifts of order 1, and superclusters are forming now. That is, cosmic structure formed from the bottom up: from galaxies to clusters to superclusters, and not the other way around.

This fundamental difference is an indication of the type of matter that gave rise to structure. We know from primordial nucleosynthesis that all the baryons in the universe cannot account for the observed amount of matter, so there must be some extra matter (dark since we don't see it) to account for its gravitational pull. Whether it is relativistic (hot) or non-relativistic (cold) could be inferred

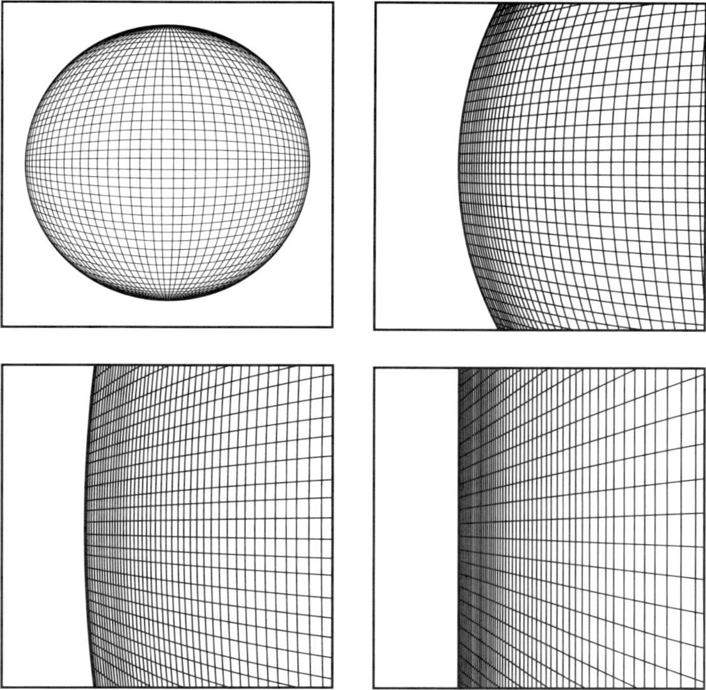

Figure 3. The exponential expansion during inflation made the radius of curvature of the universe so large that our observable patch of the universe today appears essentialy flat, analogous (in three dimensions) to how the surface of a balloon appears flatter and flatter as we inflate it to enormous sizes. This is a crucial prediction of cosmological inflation that will be tested to extraordinary accuracy in the next few years.

from observations: relativistic particles tend to diffuse from one concentration of matter to another, thus transferring energy among them and preventing the growth of structure on small scales. This is excluded by observations, so we conclude that most of the matter responsible for structure formation must be cold. How much there is is a matter of debate at the moment. Some recent analyses suggest that there is not enough cold dark matter to reach the critical density required to make the universe flat. If we want to make sense of the present observations, we must conclude that some other form of energy permeates the universe. In order to resolve this issue, even deeper galaxy redshift catalogs are underway, looking at millions of galaxies, like the Sloan Digital Sky Survey (SDSS) and the Anglo-Australian two degree field Galaxy Redshift Survey, which are at this moment taking data, up to redshifts of $z < 3$, or several hundred billion light years away, over a large region of the sky. These important observations will help astronomers determine the nature of the dark matter and test the validity of the models of structure formation.

However, if galaxies did indeed form from gravitational collapse of density per-

Table 1. **The parameters of the standard cosmological model**. The standard model of cosmology has about 20 different parameters, needed to describe the background space-time, the matter content and the spectrum of metric perturbations. We include here the present range of the most relevant parameters (with 1σ errors), as recently determined by MAP, and the error with which the Planck satellite will be able to determine them in the near future. The rate of expansion is written in units of $H = 100\,h$ km/s/Mpc

physical quantity	symbol	MAP	Planck
total density	Ω_0	1.02 ± 0.02	0.7%
baryonic matter	Ω_B	0.044 ± 0.004	0.6%
cosmological constant	Ω_Λ	0.73 ± 0.04	0.5%
cold dark matter	Ω_M	0.23 ± 0.04	0.6%
hot dark matter	$\Omega_\nu h^2$	< 0.0076 (95% c.l.)	1%
sum of neutrino masses	$\sum m_\nu$ (eV)	< 0.23 (95% c.l.)	1%
CMB temperature	$T_0\ (K)$	2.725 ± 0.002	0.1%
baryon to photon ratio	$\eta \times 10^{10}$	6.1 ± 0.3	0.5%
baryon to matter ratio	$\Omega_\mathrm{B}/\Omega_\mathrm{M}$	0.17 ± 0.01	1%
spatial curvature	Ω_K	< 0.02 (95% c.l.)	0.5%
rate of expansion	h	0.71 ± 0.03	0.8%
age of the universe	t_0 (Gyr)	13.7 ± 0.2	0.1%
age at decoupling	t_dec (kyr)	379 ± 8	0.5%
age at reionization	t_r (Myr)	180 ± 100	5%
spectral amplitude	A	0.833 ± 0.085	0.1%
spectral tilt (at $k_0 = 0.05$ Mpc^{-1})	n_s	0.93 ± 0.03	0.2%
spectral tilt variation	$dn_\mathrm{s}/d\ln k$	-0.031 ± 0.017	0.5%
tensor-scalar ratio	r	< 0.71 (95% c.l.)	5%
reionization optical depth	τ	0.17 ± 0.04	5%
redshift of matter-energy equality	z_eq	3233 ± 200	5%
redshift of decoupling	z_dec	1089 ± 1	0.1%
width of decoupling	Δz_dec	195 ± 2	1%
redshift of reionization	z_r	20 ± 10	2%

turbations produced during inflation, one should also expect to see such ripples in the metric as temperature anisotropies in the cosmic microwave background, that is, minute deviations in the temperature of the blackbody spectrum when we look at different directions in the sky. Such anisotropies had been looked for ever since Penzias and Wilson's discovery of the CMB, but had eluded all detection, until NASA's Cosmic Background Explorer (COBE) satellite discovered them in 1992. The reason why they took so long to be discovered was that they appear as perturbations in temperature of only one part in 100 000. There is, in fact, a dipolar anisotropy of one part in 1000, in the direction of the Virgo cluster, but that is interpreted consistently as our relative motion with respect to the microwave background due to the local distribution of mass, which attracts us gravitationally towards the Virgo cluster. When subtracted, we are left with a whole spectrum of anisotropies in the higher multipoles (quadrupole, octopole, etc.), see Fig. 4. Soon after COBE, other groups quickly confirmed the detection of temperature anisotropies at around $30\,\mu$K, at higher multipole numbers or smaller angular scales.

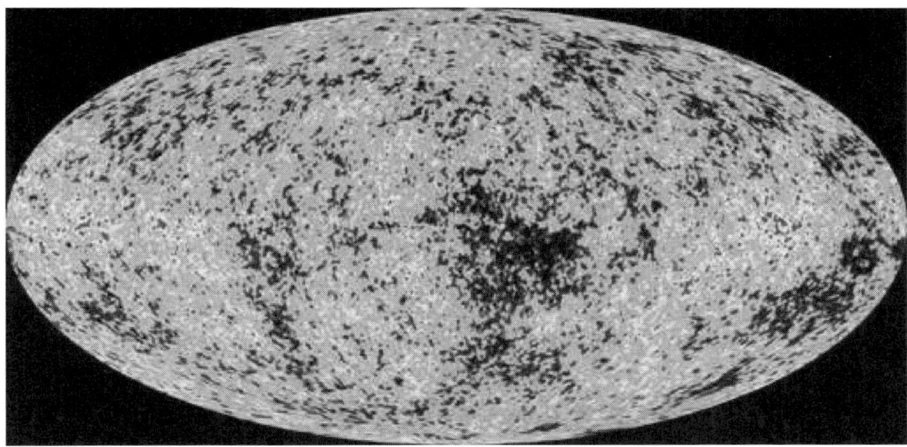

Figure 4. The microwave background sky as seen by WMAP, with 10 arc minute resolution. It shows the intrinsic CMB anisotropies, corresponding to the quadrupole and higher multipoles, at the level of a few parts in 10^5. The galaxy is a foreground and has been subtracted.

There are at this moment dozens of ground and balloon-borne experiments analysing the anisotropies in the microwave background with angular resolutions from $7°$ to a few arc minutes in the sky. The physics of the CMB anisotropies is relatively simple: photons scatter off charged particles (protons and electrons), and carry energy, so they feel the gravitational potential associated with the perturbations imprinted in the metric during inflation. An overdensity of baryons (protons and neutrons) does not collapse under the effect of gravity until it enters the causal Hubble radius. The perturbation continues to grow until radiation pressure opposes gravity and sets up acoustic oscillations in the plasma, very similar to sound waves. Since overdensities of the same size will enter the Hubble radius at the same time, they will oscillate in phase. Moreover, since photons scatter off these baryons, the acoustic oscillations occur also in the photon field and induces a pattern of peaks in the temperature anisotropies in the sky, at different angular scales, see Fig. 5. The larger the amount of baryons, the higher the peaks. The first peak in the photon distribution corresponds to overdensities that have undergone half an oscillation, that is, a compression, and appear at a scale associated with the size of the sonic horizon at last scattering (when the photons decoupled) or about $1°$ in the sky. Other peaks occur at harmonics of this, corresponding to smaller angular scales. Since the amplitude and position of the primary and secondary peaks are directly determined by the sound speed (and, hence, the equation of state) and by the geometry and expansion of the universe, they can be used as a powerful test of the density of baryons and dark matter, and other cosmological parameters.

By looking at these patterns in the anisotropies of the microwave background, cosmologists can determine not only the cosmological parameters but also the primordial spectrum of metric perturbations produced during inflation. It turns out that the observed temperature anisotropies are compatible with a scale-invariant spectrum, as predicted by inflation. This is remarkable, and gives very strong support to the idea that inflation may indeed be responsible for both the CMB anisotropies and the large-scale structure of the universe. Different models of inflation have different specific predictions for the fine details associated with the spectrum generated during inflation. It is these minute differences that will allow cosmologists to differentiate bewteen alternative models of inflation and discard those that do not agree with observations. However, most importantly, perhaps, the pattern of anisotropies predicted by inflation is completely different from those predicted by alternative models of structure formation, like cosmic defects: strings, vortices, textures, etc. These are complicated networks of energy density concentrations left over from an early universe phase transition, analogous to the defects formed in the laboratory in certain kinds of liquid crystals when they go through a phase transition. The cosmological defects have spectral properties very different from those generated by inflation. That is why it is so important to launch more sensitive instruments, and with better angular resolution, to determine the properties of the CMB temperature and polarization anisotropies. With the recent observations of these anisotropies by the Microwave Anisotropy Probe (MAP) satellite, launched by NASA in 2000, we can now discard topological defects as the source of structure in the universe at more than ten standard deviations. The full sky coverage of MAP and its extraordinary angular resolution (10 arcminutes) allows cosmologists to determine today a handful of cosmological parameters at the 1% level, see table 1. We have thus entered the era of precision cosmology and we can now speak of a truly Standard Model of Cosmology.

In the next few years, a third generation satellite – the Planck Surveyor, due to be launched by the European Space Agency in 2007 – will measure those temperature anisotropies with 10 times better angular resolution and 10 times better sensitivity than MAP, and thus allow cosmologists to determine the parameters of the standard cosmological model with 1 per mil accuracy. What makes the microwave background observations particularly powerful is the absence of large systematic errors that plague other cosmological measurements. As we have discussed above, the physics of the microwave background is relatively simple, compared to, say, the physics of supernova explosions, and computations can be done consistently within perturbation theory. Thus, most of the systematic errors are theoretical in nature, due to our ignorance about the primordial spectrum of metric perturbations from inflation. There is a great effort at the moment in trying to cover a large region in the parameter space of models of inflation, to ensure that we have considered all possible alternatives, like isocurvature or pressure perturbations, non scale invariant or tilted spectra and non-Gaussian density perturbations.

In particular, inflation also predicts a spectrum of gravitational waves. Their amplitude is directly proportional to the total energy density during inflation, and thus its detection would immediately tell us about the energy scale (and, therefore, the epoch in the early universe) at which inflation occurred. If the period of inflation responsible for the observed CMB anisotropies is associated with the Grand

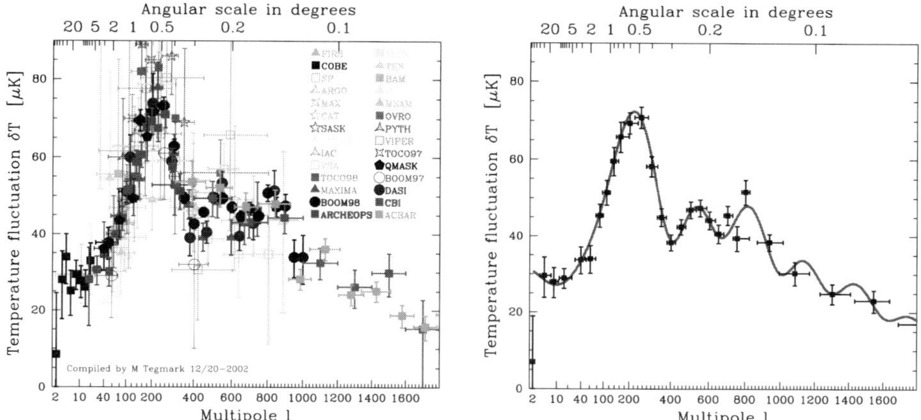

Figure 5. There are at present about thirty experiments (in satellites, from the ground and balloon-borne) looking at the microwave background temperature anisotropies with angular resolutions from $7°$ to a few arc minutes in the sky, corresponding to multipole numbers $l = 2 - 3000$. The right panel shows the l-binned spectrum. Present observations suggest the existence of a series of acoustic peaks in the angular distribution, as predicted by inflation. The theoretical curve (red thick line) illustrates the concordance Λ-CDM model which fits the data.

Unification scale, 12 orders of magnitude above the electroweak scale, when the strong and electroweak interactions are supposed to unify, then there is a chance that we might see the effect of gravitational waves in the future satellite measurements, specially from the analysis of polarization anisotropies in the microwave background maps.

Moreover, the stochastic background of gravitational waves generated during inflation could eventually be observed by ground-based laser interferometers like LIGO and VIRGO, which will start taking data as gravitational wave observatories in the next few years. These are extremely sensitive devices that could distinguish minute spatial variations, of one part in 10^{23} or better, induced when a gravitational wave from a distant source passes through the Earth and distorts the space-time metric. Gravitational waves moving at the speed of light are a fundamental prediction of general relativity. Their existence was indirectly confirmed by Russell A. Hulse and Joseph H. Taylor, through the precise observations of the decay in the orbital period of the pulsar PSR1913+16, due to the emission of gravitational radiation. In the near future, observations of gravitational waves with laser interferometers will open a completely new window onto the universe. It will allow us to observe with a very different probe (that of the gravitational interaction) a huge range of phenomena, from the most violent processes in our galaxy and beyond, like supernova explosions, neutron star collisions, quasars, gamma ray bursts, etc., to the origin of the universe. Moreover, NASA and ESA have joined efforts to construct LISA, an interferometer in space, with satellites millions of kilometers apart, whose sensitivity is good enough to detect the minutest perturbations in space-time induced by the stochastic background of gravitational waves coming from inflation.

In our quest for the parameters of the standard cosmological model, various groups are searching for distant astrophysical objects that can serve as standard candles to determine the distance to the object from their observed apparent luminosity. A candidate that has recently been exploited with great success is a certain type of supernova explosions at large redshifts. These are stars at the end of their life cycle that become unstable and violently explode in a natural thermonuclear explosion that out-shines their progenitor galaxy. The intensity of the distant flash varies in time, it takes about three weeks to reach its maximum brightness and then it declines over a period of months. Although the maximum luminosity varies from one supernova to another, depending on their original mass, their environment, etc., there is a pattern: brighter explosions last longer than fainter ones. By studying the light curves of a reasonably large statistical sample, cosmologists from two competing groups, the Supernova Cosmology Project and the High-redshift Supernova Project, are confident that they can use this type of supernova as a standard candle. Since the light coming from some of these rare explosions has travelled for a large fraction of the size of the universe, one expects to be able to infer from their distribution the spatial curvature and the rate of expansion of the universe. One of the surprises revealed by these observations is that the universe appears to be accelerating instead of decelerating, as was expected from the general attraction of matter; something seems to be acting as a repulsive force on very large scales. The most natural explanation for this is the existence of a cosmological constant, a diffuse vacuum energy that permeates all space and, as explained above, gives the universe an acceleration that tends to separate gravitationally bound systems from each other. The origin of such a vacuum energy is one of the biggests problems of modern physics. Its observed value is 120 orders of magnitude smaller than predicted by quantum mechanics. If confirmed, it will pose a real challenge to theoretical physics, one that may affect its most basic foundations.

3 The origin of matter in the universe

Cosmological inflation may be responsible for the metric perturbations that later gave rise to the large scale structures we see in the universe, but where did all the matter in the universe come from? Why isn't all the energy in photons, which would have inevitably redshifted away in a cold universe devoid of life? How did we end up being matter dominated? Everything we see in the universe, from planets and stars, to galaxies and clusters of galaxies, is made out of matter, so where did the antimatter in the universe go? Is this the result of an accident, a happy chance occurrence during the evolution of the universe, or is it an inevitable consequence of some asymmetry in the laws of nature? Theorists believe that the excess of matter over antimatter comes from fundamental differences in their interactions soon after the end of inflation.

Inflation is an extremely efficient mechanism in diluting any particle species or fluctuations. At the end of inflation, the universe is empty and extremely cold, dominated by the homogeneous coherent mode of the inflaton. Its potential energy density is converted into particles, as the inflaton field oscillates coherently around the minimum of its potential, see Fig 1. These particles are initially very far from equilibrium, but they strongly interact among themselves and soon reach thermal equilibrium at a very large temperature. From there on, the universe expanded isoentropically, cooling down as it expanded, in the way described by the standard hot Big Bang model. Thus the origin of the Big Bang itself, and the matter and energy we observe in the universe today, can be traced back to the epoch in which the inflaton energy density decayed into particles. Such a process is called reheating of the universe.

Recent developments in the theory of reheating suggest that the decay of the inflaton energy could be explosive due to the coherent oscillations of the inflaton, which induce its stimulated decay. The result is a resonant production of particles in just a few inflaton oscillations, an effect very similar to the stimulated emission of a laser beam of photons. The number of particles produced this way is exponentially large, which may explain the extraordinarily large entropy, of order 10^{89} particles, in our observable patch of the universe today. However, the inflaton is supposed to be a neutral scalar field, and thus its interactions cannot differentiate between particles and antiparticles. How did we end up with more matter than antimatter? The study of this cosmological asymmetry goes by the name of baryogenesis since baryons (mainly protons and neutrons) are the fundamental constituents of matter in planets, stars and galaxies in the universe today. So, what are the conditions for baryogenesis?

Everything we know about the properties of elementary particles is included in the standard model of particle physics. It describes more than 100 observed particles and their interactions in terms of a few fundamental constituents: six quarks and six leptons, and their antiparticles. The standard model describes three types of interactions: the electromagnetic force, the strong and the weak nuclear forces. These forces are transmitted by the corresponding particles: the photon, the gluon and the W and Z bosons. The theory also requires a scalar particle, the Higgs particle, responsible for the masses of quarks and leptons and the breaking of the electroweak symmetry at an energy scale 1000 times the mass of the proton. The Higgs is believed to lie behind most of the mysteries of the standard model, including possibly also the asymmetry between matter and antimatter.

In 1967, the Russian physicist Andrei Sakharov pointed out the three necessary conditions for the baryon asymmetry of the universe to develop [16]. First, we need interactions that do not conserve baryon number B, otherwise no asymmetry could be produced in the first place. Second, C and CP symmetry must be violated, in order to differentiate between matter and antimatter, otherwise B non-conserving interactions would produce baryons and antibaryons at the same rate, thus maintaining zero net baryon number. Third, these processes should occur out of thermal equilibrium, otherwise particles and antiparticles would be produced at the same rate. The standard model is baryon symmetric at the classical level, but violates B at the quantum level, through the chiral anomaly. Electroweak interactions violate C and CP, but the magnitude of the latter is clearly insufficient to account for the observed baryon asymmetry [16,17]. This failure suggests that there must be other sources of CP violation in nature, and thus the standard model of particle physics is probably incomplete.

One of the most popular extensions of the standard model includes a new symmetry called supersymmetry, which relates bosons (particles that mediate interactions) with fermions (the constituents of matter). Those extensions generically predict other sources of CP violation coming from new interactions at scales above 1000 times the mass of the proton. Such scales will soon be explored by particle colliders like the Large Hadron Collider (LHC) at CERN (the European Centre for Particle Physics) and by the Tevatron at Fermilab. The mechanism for baryon production in the early universe in these models relies on the strength of the electroweak phase transition, as the universe cooled and the symmetry was broken. Only for strongly first-order phase transitions is the universe sufficiently far from equilibrium to produce enough baryon asymmetry. Unfortunately, the phase transition in these models is typically too weak to account for the observed asymmetry, so some other mechanism is needed.

If reheating after inflation occurred in an explosive way, via the resonant production of particles from the inflaton decay, as recent developments suggest, then the universe has actually gone through a very non-linear, non-perturbative and very far from equilibrium stage, before thermalizing via particle interactions. Electroweak baryogenesis could then take place during that epoch, soon after the end of low energy inflation at the electroweak scale. Such models can be constructed but require a specially flat direction (a very small mass for the inflaton) during inflation, in order to satisfy the constraints from the amplitude of temperature anisotropies seen by COBE. Such flat directions are generic in supersymmetric extensions of the standard model. After inflation, the inflaton acquires a large mass from its interaction with the Higgs field.

The crucial ingredient of departure from equilibrium, necessary for the excess production of baryons over antibaryons, is strongly present in this new scenario of baryogenesis, as the universe develops from a zero-temperature and zero-entropy state, at the end of inflation, to a thermal state with exponentially large numbers of particles, the origin of the standard hot Big Bang. If, during this stage, fundamental or effective interactions that are B, C and CP violating were fast enough compared to the rate of expansion, the universe could have ended with the observed baryon asymmetry of one part in 10^{10}, or one baryon per 10^9 photons today, as deduced from observations of the light element abundances. Recent calculations suggest than indeed, the required asymmetry could be produced as long as some new physics, just above the electroweak symmetry breaking scale, induces a new effective CP violating interaction.

These new phenomena necessarily involve an interaction between the Higgs particle, responsible for the electroweak symmetry breaking, and the inflaton field, responsible for the period of cosmological inflation. Therefore, for this scenario to work, it is expected that both the Higgs and the inflaton particles be discovered at the future particle physics colliders like the LHC and the Next Linear Collider (NLC). Furthermore, this new physics would necessarily involve new interactions in the quark sector, for example inducing CP violations in the B meson (a bound state composed of a bottom quark and an antidown quark) system. Such violations are the main research objective of the B factory at SLAC in California and at KEK, the High Energy Accelerator Research Organization in Tsukuba, Japan. These experiments have already been collecting data for a couple years, and for the moment are in perfect agreement with the Standard Model of particle physics [16].

However, perhaps in the near future they may detect a deviation which could give us a clue to the origin of CP, and thus to the matter–antimatter asymmetry of the Universe and, possibly, to baryogenesis from reheating after inflation.

4   Conclusions

We have entered a new era in cosmology, were a host of high-precision measurements are already posing challenges to our understanding of the universe: the density of ordinary matter and the total amount of energy in the universe; the microwave background anisotropies on a fine-scale resolution; primordial deuterium abundance from quasar absorption lines; the acceleration parameter of the universe from high-redshift supernovae observations; the rate of expansion from gravitational lensing; large scale structure measurements of the distribution of galaxies and their evolution; and many more, which already put constraints on the parameter space of cosmological models. However, these are only the forerunners of the precision era in cosmology that will dominate this millennium, and will make cosmology a science in its own right.

It is important to bear in mind that all physical theories are approximations of reality that can fail if pushed too far. Physical science advances by incorporating earlier theories that are experimentally supported into larger, more encompassing frameworks. The standard Big Bang theory is supported by a wealth of evidence, nobody really doubts its validity anymore. However, in the last decade it has been incorporated into the larger picture of cosmological inflation, which has become the new standard cosmological model. All cosmological issues are now formulated in the context of the inflationary paradigm. It is the best explanation we have at the moment for the increasing set of cosmological observations.

In the next few years we will have an even larger set of high-quality observations that will test inflation and the cold dark matter paradigm of structure formation, and determine most of the 20 or more parameters of the standard cosmological model to a few per mil accuracy, see Table 1. It may seem that with such a large number of parameters one can fit almost anything. However, that is not the case when there is enough quantity and quality of data. An illustrative example is the standard model of particle physics, with around 21 parameters and a host of precise measurements from particle accelerators all over the world. This model is, nowadays, rigurously tested, and its parameters measured to a precision of better than 1% in most cases. It is clear that high-precision measurements will make the standard model of cosmology as robust as that of particle physics. This is definitely a very healthy field, but there is still a lot to do. With the advent of better and larger precision experiments, cosmology is becoming a mature science, where speculation has given way to phenomenology.

However, there are still many unanswered fundamental questions in this emerging picture of cosmology. For instance, we still do not know the nature of the inflaton field, is it some new fundamental scalar field in the electroweak symmetry breaking sector, or is it just some effective description of a more fundamental high energy interaction? Hopefully, in the near future, experiments in particle physics might give us a clue to its nature. Inflation had its original inspiration in the Higgs field, the scalar field supposed to be responsible for the masses of elementary particles (quarks and leptons) and the breaking of the electroweak symmetry. Such a field has not been found yet, and its discovery at the future particle colliders may help understand one of the truly fundamental problems in physics, the origin of masses. If the experiments discover something completely new and unexpected, it would automatically affect inflation at a fundamental level.

One of the most difficult challenges that the new cosmology will have to face is understanding the origin and nature of the cosmological constant. Ever since Einstein introduced it as a way to counteract gravitational attraction, it has haunted cosmologists and particle physicists. We still do not have a mechanism to explain its extraordinarily small value, 120 orders of magnitude below what is predicted by quantum physics. For several decades there has been the reasonable speculation that this fundamental problem may be related to the quantization of gravity. General relativity is a classical theory of space-time, and it has proved particularly difficult to construct a consistent quantum theory of gravity, since it involves fundamental issues like causality and the nature of space-time itself.

The value of the cosmological constant predicted by quantum physics is related to our lack of understanding of gravity at the microscopic level. However, its effect is dominant at the very largest scales of clusters or superclusters of galaxies, on truly macroscopic scales. We can speculate that perhaps general relativity is not the correct description of gravity on the very largest scales. In fact, it is only in the last few billion years that the observable universe has become large enough that these global effects could be noticeable. In its infancy, the universe was much smaller than it is now, and, presumably, general relativity gave a correct description of its evolution, as confirmed by the successes of the standard Big Bang theory. As it expanded, larger and larger regions were encompassed, and, therefore, deviations from general relativity would slowly become important. It may well be that the recent determination of a cosmological constant from observations of supernovae at high redshifts is hinting at a fundamental misunderstanding of gravity on the very large scales. If this were indeed the case, we should expect that the new generation of precise cosmological observations will not only affect our cosmological model of the universe but also a more fundamental description of nature.

References

1. Abbott, L. 1988 The mystery of the cosmological constant. *Scientific Am.* May, pp. 106–113.
2. Brush, S. G. 1992 How cosmology became a science. *Scientific Am.* August, pp. 34–40.
3. Freedman, W. L. 1992 The expansion rate and size of the universe. *Scientific Am.* November, pp. 54–60.
4. Guth, A. H. 1997 *The inflationary universe.* Reading: Perseus Books.
5. Henry J. P., Briel U. G. & Böhringer, H. 1998 The evolution of galaxy clusters. *Scientific Am.* December, pp. 52–56.
6. Hogan, C. J., Kirshner, R. P. & Suntzeff, N. B. 1999 Surveying space-time with supernovae. *Scientific Am.* January, pp. 46–51.

7. Hu, W. 1997 The physics of the microwave background anisotropies.
   `http://background.uchicago.edu/~whu/physics/physics.html`

8. Jeffries, D. A., Saulson, P. R., Spero, R. E. & Zucker, M. E. 1983 Gravitational wave observatories. *Scientific Am.* December, pp. 50–58.

9. Krauss, L. M. 1999 Cosmological Antigravity. *Scientific Am.* January, pp. 53–59.

10. Linde, A. D. 1994 The self-reproducing inflationary universe. *Scientific Am.* November, pp. 32–39.

11. Linde, A. D. 1990 *Particle Physics and Inflationary Cosmology.* New York: Harwood Academic Press.

12. Peebles, P. J. E., Schramm, D. N., Turner, E. L. & Kron, R. G. 1994 The evolution of the universe. *Scientific Am.* October, pp. 53–57.

13. Tegmark, M. 1999 The microwave background anisotropies experiments.
   `http://www.hep.upenn.edu/~max/cmb/experiments.html`

14. Quinn, H. R. & Witherell, M. S. 1998 The asymmetry between matter and antimatter. *Scientific Am.* October, pp. 76–81.

15. WMAP Collaboration 2003. `http://map.gsfc.nasa.gov`

16. For more details see the article by I.I. Bigi in these proceedings.

17. For more details see the article by E. Blucher in these proceedings.

DID TIME AND ITS ARROW HAVE A BEGINNING?

G. VENEZIANO

Theory Division, CERN, CH-1211 Geneva 23, Switzerland

I will discuss recent ideas on the issues appearing in the title in the context of some unconventional cosmologies suggested by string theory.

1 Foreword

In this talk I will try to address two very important questions for cosmology. The first one concerns the beginning of time, the second the beginning of its arrow, i.e. of a clear physical distinction between past and future. I hasten to point out that, while I have given much thought to the first question, particularly within the context of string theory, I still feel quite uneasy at addressing the second one. When I received the invitation to come and talk at this Conference I gladly accepted, both for the attraction of such a beautiful place, and with the hope to make progress on the time-arrow issue. Well, it will be for you to judge how much my hope has been fulfilled ...

The plan of the talk will be as follows: I will start from the title and provide what I believe to be the standard answers to the questions it contains. After criticizing the standard attitude as one putting the answers in the premises, I will turn my attention to string theory, to some of its general properties, and to how it should provide a new point of view concerning the beginning-of-time issue. This will lead me to illustrate – in a necessarily synthetic way – some string-inspired cosmologies based on the duality symmetries of the field equations and, in particular, the so-called pre-big-bang scenario. Finally, I will discuss how arguments related to the concept of holography apply to this scenario and may ultimately offer a new handle on the arrow-of-time issue.

2 Title and standard answers

The title of this talk contains two distinct questions. Let us discuss them in turn and provide some "standard" answers.

- Did time have a beginning?

  If the big-bang singularity (conventionally defined to occur at $t = 0$) is real, as it is the case in classical general relativity, the answer to this question is a clear yes. No meaning can be given to $t < 0$! In St. Augustin's words, our world and time were created together.

- Did the arrow of time have a beginning?

  In order to answer this question let us associate the arrow of time with the direction, if any, in which entropy increases. If, in the only meaningful time interval $t \geq 0$, $t \sim 0$ represents a state of low (i.e. less than maximal) entropy, then indeed entropy must increase in the direction of growing t. This direction *defines* the future. Roger Penrose [1] suggested that low-entropy singularities (with a divergent Ricci-, but a finite Weyl-tensor) should characterize "initial" (big bang), as opposed to "final" (big crunch), singularities. And indeed, if the big bang singularity was of the type needed to explain today's homogeneity, isotropy and flatness, then it had to have low entropy and the arrow of time points away from it since the very beginning.

- The standard answers thus seem to be: "yes, yes" with two coinciding beginnings.

However, the standard hot-big bang model has its (in)famous fine-tuning problems which are indeed related to assuming it had low-entropy. In other words, the standard hot big bang model, rather than explaining the arrow of time, has the origin of it as one of its major puzzles. These problems are supposedly "solved" by the inflationary paradigm [2]. In chaotic inflation [3], for instance, initial conditions have to be given in a very quantum regime (the so-called Planck era) where it is not clear that one can define an arrow at all. This is the situation prevailing almost all the time and almost everywhere. An arrow may emerge, however, within each lucky patch that underwent inflation, became smooth and flat, and finally re-heated giving a sort of homogeneous "small bang". Each one of these mini-bangs will have its own arrow of time.

Thus, in standard inflation, the answers to our two questions appear to be:

- Probably, time did have a beginning at a big bang singularity ... but

- clear arrows of time have only emerged locally, i.e. within those patches that underwent inflation and reheating. Elsewhere it is hard, if not meaningless, to distinguish past and future.

3   String theory and singularities

We have just argued that a beginning of time is forced upon us by the cosmological, big bang singularity. But is such a singularity really unavoidable? General theorems[4] tell us that such is the general case in Einstein's relativity. However, classical general relativity is certainly inadequate for describing what happens around or before the so-called Planck time, $t_P \equiv \sqrt{G_N \hbar / c^5} \sim 10^{-43}$ s after the big bang. A consistent theory of quantum gravity is needed and, so far, only one such theory is known: string theory (see [5] for two recent – but very different – reviews of the subject).

Quantum string theory (QST), unlike quantum field theory (QFT) or classical string theory, introduces a fundamental length in physics, λ_s: it is nothing but the length scale needed to convert the (area) action of a string into a pure number in order to go over to its quantum theory. And indeed, λ_s is related, through the uncertainty principle, to the characteristic (minimal) size of a quantum string. From all we know λ_s plays many roles simultaneously in string theory [6]: that of Planck's constant (replacing $\hbar$ [6]), that of an ultraviolet cutoff, that of a minimal radius for compactified dimensions, that of a maximal temperature and of a minimal black-hole size, etc. We thus also expect λ_s to eliminate the singularities of CGR, and, in particular, the one associated with our prejudices about the beginning of time. In any case, in string theory, the standard analysis becomes totally inadequate for $t < \lambda_s/c$, a time scale expected to be a few orders of magnitude larger than t_P.

Unfortunately, string theory is only understood, at present, as a perturbative series in *two* expansion parameters:

- $g_s^2 = e^\phi$. This is like the loop expansion parameter of Quantum Field Theory (say α for gauge theories) with the noteworthy difference that the coupling has been promoted to a scalar field, the so-called dilaton ϕ .

- $\lambda^2 = \lambda_s^2 \Delta$, where Δ stands for two derivatives or, if we prefer, two powers of momentum. The Quantum Field Theory limit of Quantum String Theory is obtained for $\lambda \to 0$ (i.e. $\lambda_s \to 0$, or the low-energy limit).

Typical of string theory is the fact that the dilaton gives, at the same time, the value of the gauge couplings and that of Newton's constant in string units, best expressed (in four dimensions and up to a numerical constant) as

$$e^\phi = g_s^2 \sim \alpha_{GUT} \sim l_P^2 \lambda_s^{-2} \ , \quad l_P = c\, t_P \ , \tag{1}$$

a relation that embodies the unification of gravity with all other forces at the string scale $M_s = \lambda_s^{-1}$.

Modulo very few lucky exceptions, our non-perturbative understanding of QST is, so far, very limited and therefore:

- We are not able to describe what happens at strong coupling and, as a result, we cannot solve the problem of perturbative vacuum degeneracy, of SUSY breaking, of dilaton and moduli stabilization, etc.

- We are not able to describe what happens at strong curvatures, $(R \sim \lambda_s^{-2})$. We can only make the educated guess that the big bang singularity of CGR gets replaced by a strong-curvature phase that can only be described in string-theoretic (rather than field-theoretic) terms.

What do we know instead from perturbation theory? Actually quite a lot. There are just five perturbative superstring theories, all living, for consistency, in (9+1) space-time dimensions. Thus a firm prediction of string theory is the existence of extra dimensions of space, possibly so tiny that we cannot *directly* see them. These five theories are connected, through a web of dualities, both among themselves and to (10+1)-dimensional supergravity, a very elegant and constrained field theory much in the spotlight in the 70's and 80's. Actually, these six theories appear to correspond just to six different limits of one and the same "Mother" (or M) theory, whose exact definition remains unclear to this date.

For each one of the six theories we can find the massless excitations/particles. These always include a graviton, a dilaton-axion pair (with one caveat, see below), non-abelian gauge fields, and their fermionic superpartners. The actual gauge group, the matter content, the Yukawa couplings, etc. depend on the specific vacuum around which we are studying the theory, e.g. on the way six(seven) of the nine(ten) dimensions are assumed to be curled up. One amusing point is that, while the five 10-dimensional theories always have a scalar-pseudoscalar pair of partners to the graviton (the already mentioned dilaton-axion pair), 11-D supergravity does not. However, the role of the dilaton in this latter theory is played by the size of the 10th spatial dimension (usually a small circle or a segment). At weak coupling ($\phi \to -\infty$) the size of the 10th dimension goes to zero and that's why the five perturbative theories appear as 9-dimensional! As the coupling grows, the 10th dimension opens up. As long as we are at low-energy (small curvatures) we can study what happens by appealing to 11-D supergravity but, as we shall see, this will not be quite sufficient for tackling some of the most crucial issues in string cosmology.

4 Duality and string-inspired cosmologies

If the big-bang (BB) singularity is replaced by a large (but finite)-curvature phase we should be able to continue time to the past of such a phase and define a pre-bangian era. This phase provides extra time for things to happen, hence it can be used to cure the fine-tuning problems of standard, non-inflationary cosmology by producing a good big bang as the result of a preceding evolution. Two guesses on the nature of that phase inspired by the peculiar symmetries of string cosmology are represented in Figs. 1 and 2.

In both figures the vertical axis represents the expansion rate (or Hubble) parameter. The horizontal axis measures instead a convenient combination of the Hubble parameter and of the rate of change of the dilaton. Also indicated is the line $\dot{\phi} = 0$ separating regions of growing coupling from those of decreasing coupling. Trajectories in this plane represent some possible (homogeneous and spatially flat) cosmologies. The standard symmetry under $t \to -t$ corresponds to reflections with respect to the origin, while an extra symmetry of string cosmology (known as scale-factor duality [7]) corresponds to reflections around the x-axis. Obviously, the product of the two transformations give reflections about the y-axis.

The first string-inspired cosmology I wish to present is shown in Fig. 1 and is usually referred to as pre-big bang (PBB) cosmology [8]. Since the presence of a pre-bangian phase is common to both models, I have added a "doc" (for "dé nomination d'origine controllée") for the original scenario. The trajectory starts near the origin in the *first* quadrant. Thus, as the arrow indicates, it is an evolution from weak coupling and weak curvature towards strong coupling and strong curvature. This phase has been called dilaton-driven inflation (DDI). Without effects due to the finite string size λ_s such a phase terminates, after a finite proper time, in a singularity. A basic assumption of the PBB scenario is that the singularity is avoided, thanks to string-size and strong coupling effects, and connects smoothly (following the curved trajectory) to the (quasi) Friedman-Robertson-Walker (FRW) phase indicated by an arrow in the second quadrant. Eventually, through entropy production and dilaton stabilization, the evolution relaxes to a $\phi = $ constant genuine FRW cosmology. I will discuss in the next Section the way such a scenario can be put in motion. Needless to say, the least obvious part of it has been, and still is, the transition between the DDI and FRW phases since this is believed to occur in a strong-coupling, high-curvature regime on which very little is known.

The second string-inspired cosmology is a simpler version [9] of the original "ekpy-

PBB doc

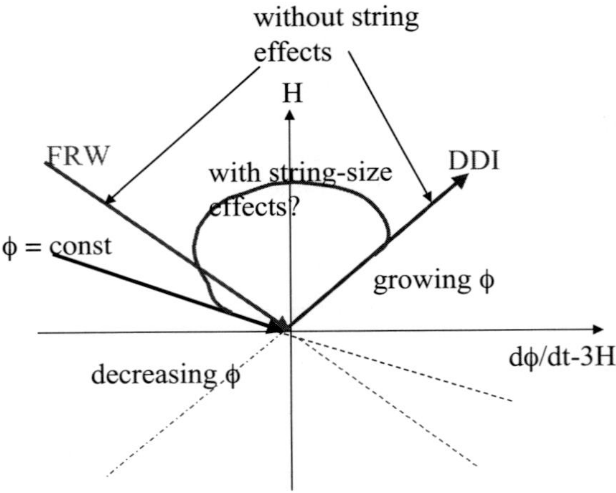

Figure 1.

rotic" scenario, first proposed in [10]. It is illustrated in Fig. 2. The cosmological trajectory starts once more near the origin but in the fourth, rather than in the first, quadrant. Thus the initial evolution represents now an accelerated contraction with decreasing coupling. In other words, the initial state is at strong coupling and weak curvature, a regime where, as we have explained, D=10 superstring theory, is well represented by D=11 supergravity. Thus, in this scenario, the initial state can be seen as consisting of two 10-dimensional membranes (one of which representing our world) coming against each other. The transition between the pre and post big bang phase occurs now at strong curvature and weak (essentially vanishing) coupling, i.e. it corresponds to the collision of the two branes. Such a collision should then give rise to release of heat and entropy and to the beginning of the FRW era. In a very schematic way, PBB and ekpyrotic models correspond, respectively, to anti-clock and clock-wise trajectories in the phase diagram of Figures 1 and 2.

From Big Crunch to Big Bang
(KOSST)

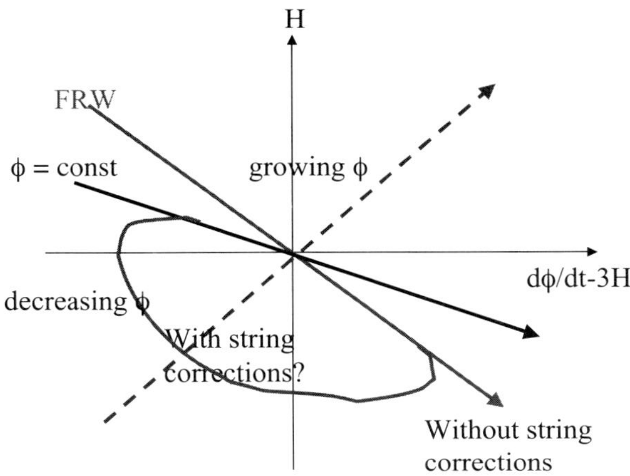

Figure 2.

5 The standard PBB scenario

But how did it all start? And, finally, how can we go back to the question of the arrow of time? A particular realization of the PBB scenario, that we may call "chaotic pre-big bang" was proposed in [11] and may help clarify these issues.

One imagines most of space to be, for most of the time, in the weak-coupling small-curvature regime. The generic solution of the field equations in this perturbative region is a superposition of massless waves of all kinds, travelling in all directions with all kinds of wave-numbers, something best described as the (9+1)-dimensional analog of a chaotic sea or sky: apparently, a highly entropic, chaotic state where the arrow of time is ill-defined. The crucial (and non-trivial) point is that the perturbative solutions I have just mentioned are unstable against gravitational collapse. In the presence of gravity the usual intuition about what maximizes entropy (a homogeneous state in thermal equilibrium) fails. The system tends to form inhomogeneities and these, rather than being washed out, tend to clump further and further until black holes are formed. Well, string-inspired cosmologies thrive precisely on such an instability.

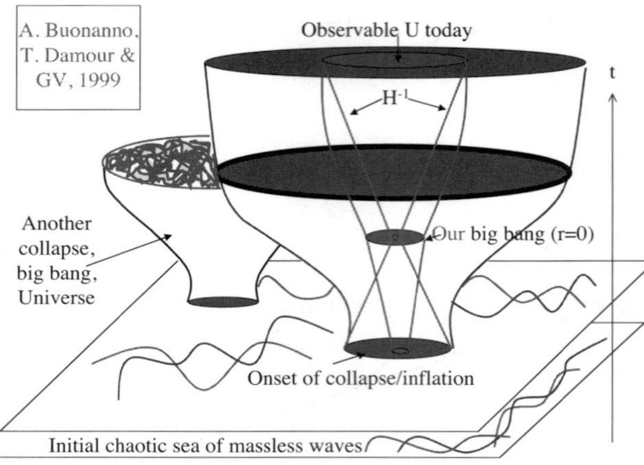

Figure 3.

How does DDI set in, for instance? If some (scale-invariant, λ_s-independent) conditions are met a closed trapped surface develops within which a black-hole is formed. According to CGR, a singularity is lying inside the horizon and the direction pointing towards it is time-like. In other words, space-time inside the horizon is of a cosmological, big-crunch type with a clear distinction between past and future. The phenomenon is very similar to the one by which black holes are formed in General Relativity. However, the presence of the additional dilaton field ϕ changes drastically the physical interpretation. While in Classical General Relativity the region of space lying inside the black-hole horizon undergoes a contraction, or a crunch, this is only the case here, if distances are measured with the Planck stick, i.e. in Planck-length (l_P) units. However, in string theory, l_P is actually a derived quantity (see Eq. (1)), the true scale of string physics being rather the string length λ_s. At very weak coupling (assumed to be the case initially) $l_P \ll \lambda_s$ but, as the coupling grows, the ratio l_P/λ_s also grows. This is why what looks like an implosion (or big crunch) in Planck units becomes an explosion (or big bang) in string units.

An artistic (?) view of the formation of two black holes and of the corresponding FRW Universes is displayed in Fig. 3.

6 Entropy bounds, holography, and the arrow of time

Let us finally come to our main concern, that of the arrow of time. We have already said that, in the presence of gravity, entropy is increased through gravitational collapse. It would thus look natural to imagine that, in the PBB picture illustrated in the previous section, a state of maximal entropy gets formed inside the black hole. If this were the case there would be little hope to find an arrow of time emerge "after" the collapse/inflation supposedly leading to the FRW era. The subtle point here is that, at each given time, actual entropy should be compared to its maximal possible value. Until that bound is reached there is an arrow of time in the direction of saturating it. But what is that bound? This has been the subject of much recent (and not so recent) debate, that I will now briefly summarize.

More than 20 years ago Bekenstein [12] suggested a universal bound on the entropy of a weak-gravity system of energy E and size R (assumed to be larger than its gravitational radius $R_g \sim G_N E$):

$$S \leq S_{Bek} = ER/\hbar c = RR_g l_P^{-2} \, , \tag{2}$$

where numerical factors are omitted hereafter. A black hole saturates the bound (if R is identified with its Schwarzschild radius R_g) since its entropy is given by the Bekenstein-Hawking formula:

$$S_{BH} = (R_g/l_P)^2 \, . \tag{3}$$

A more recent approach to entropy bounds is based on the so-called "holography" idea [13] saying that entropy for an isolated system of size R is limited by the area of a surface surrounding it (in Planck units):

$$S \leq S_{hol} = (R/l_P)^2 \, . \tag{4}$$

This bound appears to be very tight since it says that entropy, usually believed to be extensive and thus to grow like the volume of the system, is actually bound by a quantity proportional the area. However, for systems larger than their gravitational radius, $S_{hol} > S_{Bek}$, and therefore the holography bound is actually weaker than Bekenstein's.

The problem is how to extend the bound to non isolated systems and, in particular, to a cosmological setup. Recently a generalization has emerged [14,15,16] whereby entropy is maximized by allowing (roughly) one black-hole per Hubble volume H^{-3}

$$S < S_{Hubble} = n_{BH} S_{BH} = (V H^3)(H^{-2} l_P^{-2}) = V H l_P^{-2} \, . \tag{5}$$

It is not hard to check [14] that the criterion for the onset of gravitational collapse (i.e. the formation of a closed trapped surface) roughly coincides with saturation of this bound within the collapsing region. Once the collapse starts it cannot stop. One can follow how the bound (5) evolves for the region inside the newly formed black hole and discovers that it stays constant. There is therefore no threat to the bound as long as no strong non-adiabatic process takes place. The only such process we know about is particle production in time-dependent backgrounds, the same phenomenon supposedly responsible for the formation of structure in inflationary cosmology. Since the rate of entropy production due to this effect is proportional to (some powers of) the Hubble parameter and the coupling, it turns out [14] that the bound (5) is only in danger when a combination of the Hubble parameter and the string coupling reaches a critical value. This also corresponds to the point at which back-reaction effects from particle production is expected to become significant. In fact, if the curvature singularity or the unlimited growth of the coupling are not avoided, the bound is inevitably violated.

We thus have found, on the basis of entropy bounds, an argument, paralleling an old one by Bekenstein himself [17] for the spurious nature of the cosmological singularity . Moreover, if the singularity is indeed avoided through a bounce leading to standard FRW cosmology, the bound is just saturated at the bounce. After the bounce the bound grows so fast that the arrow of time gets clearly established away from the (fake) big bang[14].

7 Conclusions

String theory, through its fundamental scale λ_s, appears to call for a deep revision of our traditional ideas about the origin of time and of its arrow.

Most likely the BB singularity will be proven to be absent (or to be harmless) in string theory, allowing us to ask questions about what was there before.

In string-inspired scenarios most likely time has always existed (and will continue existing forever) but did not always (and does not everywhere) have a well-defined arrow.

In the particular string-cosmology known as the Pre-Big Bang scenario, such an arrow will naturally arise, inside the horizon of each black hole, as the direction in which gravitational collapse proceeds.

Regions inside each horizon acquire relatively low entropy; if the BH singularity is avoided, each one becomes a distinct Universe with its own, well defined, arrow of time.

It is my great pleasure to thank Martin Faessler and Ikaros Bigi for having invited me to such a unique Conference in such a unique spot.

References

1. R. Penrose, *The Emperor's New Mind*, Oxford University Press, New York, 1989.
2. E. W. Kolb and M. S. Turner, *The Early Universe*, Addison-Wesley, Redwood City, CA, 1990;
   A. D. Linde, *Particle Physics and Inflationary Cosmology*, Harwood, New York, 1990.
3. A. D. Linde *Phys. Lett.* B **129**, 177 (1983).
4. S. W. Hawking and R. Penrose, *Proc. Roy. Soc. Lond.* **A 314**, 529 (1970).
5. J. Polchinski, *String Theory*, Cambridge University Press, 1998;
   B. Greene, *The Elegant Universe*, Norton W. W. and Co., New York & London, 1999.
6. G. Veneziano, *Europhys. Lett.* **2**, 133 (1986).
7. G. Veneziano, *Phys. Lett.* B **265**, 287 (1991).
8. J. E. Lidsey, D. Wands and E. J. Copeland, *Phys. Rep.* **337**, 343 (2000);
   M. Gasperini and G. Veneziano, *Phys. Rep.* **373**, 1 (2003).
9. J. Khouri, B.A. Ovrut, N. Seiberg, P.J. Steinhardt, and N. Turok, *Phys. Rev.* D **65**, 086007 (2002).
10. J. Khouri, B.A. Ovrut, P.J. Steinhardt and N. Turok, *Phys. Rev.* D **64**, 123522 (2001).
11. A. Buonanno, T. Damour and G. Veneziano, *Nucl. Phys.* B **543**, 275 (1999).
12. J. Bekenstein , *Phys. Rev.* D **23**, 287 (1981).

13. G. 't Hooft , *Abdus Salam Festschrift, A Collection of Talks*, World Scientific, Singapore, 1993;
L. Susskind, *J. Math. Phys.* **36**, 6377 (1995).
14. G. Veneziano, *Phys. Lett.* B **454**, 22 (1999).
15. R. Easther and D. A. Lowe, *Phys. Rev. Lett.* **82**, 4967 (1999);
D. Bak and S. Rey, *Class. Quant. Grav.* **17**, L1 (2000).
16. R. Brustein and G. Veneziano, *Phys. Rev. Lett.* **84**, 5695 (2000).
17. J. Bekenstein, *Int. J. Theor. Phys.* **28**, 967 (1989).

THE WORMHOLE HAZARD

S. KRASNIKOV

The Central Astronomical Observatory at Pulkovo,
St. Petersburg, 196140, Russia.
Email: redish@gao.spb.ru

To predict the outcome of (almost) any experiment we have to assume that our
spacetime is globally hyperbolic. The wormholes, if they exist, cast doubt on the
validity of this assumption. At the same time, no evidence has been found so far
(either observational, or theoretical) that the possibility of their existence can be
safely neglected.

1 Introduction

According to a widespread belief general relativity is the science of gravitational
force. Which means, in fact, that it is important only in cosmology, or in extremely
subtle effects involving tiny post-Newtonian corrections.

However, this point of view is, perhaps, somewhat simplistic. Being concerned
with the structure of spacetime itself, relativity sometimes poses problems vital to
the whole of physics. Two best known examples are singularities and time machines.
In this talk I discuss another, a little less known, but, in my belief, equally important
problem (closely related to the preceding two). In a nutshell it can be formulated
in the form of two question: What principle must be added to general relativity to
provide it (and all other physics along with it) with predictive power? Does not
the (hypothetical) existence of wormholes endanger that (hypothetical) principle?

2 Global hyperbolicity and predictive power

2.1 Globally hyperbolic spacetimes

The globally hyperbolic spacetimes are the spacetimes with especially simple and
benign causal structure, the spacetimes where we can use physical theories in the
same manner as is customary in the Minkowski space.

1. Definition. A spacetime M is *globally hyperbolic* if it contains a subset S
(called a *Cauchy surface*) which is met exactly once by every inextendible causal
curve in M.

2. Remark We shall not need the concept of a globally hyperbolic *subset* of a
spacetime, so below, whenever I call a subset U of M globally hyperbolic[a], I mean
that U is globally hyperbolic, when considered as an (extendible) spacetime by
itself, not as a part of M.

Topologically globally hyperbolic spacetimes are 'dull' [1] in the following sense.

3. Property. All Cauchy surfaces of a given globally hyperbolic spacetime M are
homeomorphic. Moreover,

$$M = S \times \mathbb{R}^1,$$

---

[a]A more rigorous term would be *intrinsically globally hyperbolic.* [7]

where $\mathcal{S}$ is a three-dimensional manifold, and $\mathcal{S} \times \{x\}$ are Cauchy surfaces of M for all $x \in \mathbb{R}^1$.

The causal structure of globally hyperbolic spacetimes is also nice.

4. Property. Globally hyperbolic spacetimes do not contain closed causal curves (for, if such a curve ℓ meets a set $\mathcal{P}$ once, then the causal curve $\ell \circ \ell$ meets $\mathcal{P}$ twice).

5. Examples. The Minkowski and Friedmann spacetimes are both globally hyperbolic. And a 'conical' spacetime

$$\mathrm{d}s^2 = -\mathrm{d}t^2 + \mathrm{d}x^2 + \mathrm{d}r^2 + r^2\mathrm{d}\varphi^2, \qquad r > 0, \; \varphi \in [0, \varphi_0 < 2\pi),$$

which describes a cosmic string, is not. Indeed, consider a ball $B = \{p \colon r(p) \leqslant \pi/2\}$. By definition a Cauchy surface must be *achronal* (i. e. no its points can be connected by a timelike curve). But the t-coordinates of all points of $\mathcal{P} \cap B$, where $\mathcal{P}$ is achronal and passes through the origin of the coordinate system, are bounded from above by some $t_0 > 0$. So, the inextendible causal curve

$$x = \phi = 0, \quad t = 2t_0 + \tan r$$

will never intersect $\mathcal{P}$. Whence $\mathcal{P}$ is not a Cauchy surface.

Note that global hyperbolicity is not directly related to the absence of singularities. For example, the (two-dimensional) anti-de Sitter spacetime, which is a strip

$$\mathrm{d}s^2 = \sec^2 x \left(-\mathrm{d}t^2 + \mathrm{d}x^2\right), \qquad -\pi/2 < x < \pi/2$$

(see Fig. 1a), is non-globally hyperbolic, though it is singularity free (the lines $x = \pm\pi/2$ are infinities rather than singularities — it would take infinitely large time for an observer moving with a finite acceleration to reach them). At the same time the Schwarzschild spacetime (see Fig. 1b) *is* globally hyperbolic in spite of the singularities.

Let us call a system (i. e. a set of fields, and/or particles) *predictable* if there is a spacelike surface such that the state of the system on that surface uniquely determines its state at any future point. A theory is *prognostic* if all systems described by that theory are predictable. A non-prognostic theory, i. e. a theory unable to predict the outcome of an experiment, is of doubtful practical value.

The importance of globally hyperbolic spacetimes lies in the fact that all usual relativistic theories (mechanics, electrodynamics, etc.) are prognostic in such spacetimes. If the knowledge of everything that takes place at $\mathcal{S}$ is insufficient for determining the state of the system at p, it means that the necessary additional information either comes to p just out of nowhere (and so the corresponding theory is indeterministic), or reaches p after propagating along spacelike curves (only such curves can avoid meeting $\mathcal{S}$). In the latter case the theory admits superluminal signalling.

The reverse is also true: in a prognostic theory information does not propagate faster than light. Indeed, take a point p' slightly to the future from p and consider the set $I^-(p')$ of all points reachable from p' by past-directed timelike curves. It turns out that $I^-(p')$ is globally hyperbolic, if so is the whole spacetime (this is not obvious from Definition 1, but can be easily established from another definition of global hyperbolicity, [1] the equivalence of which to ours is a highly non-trivial fact).

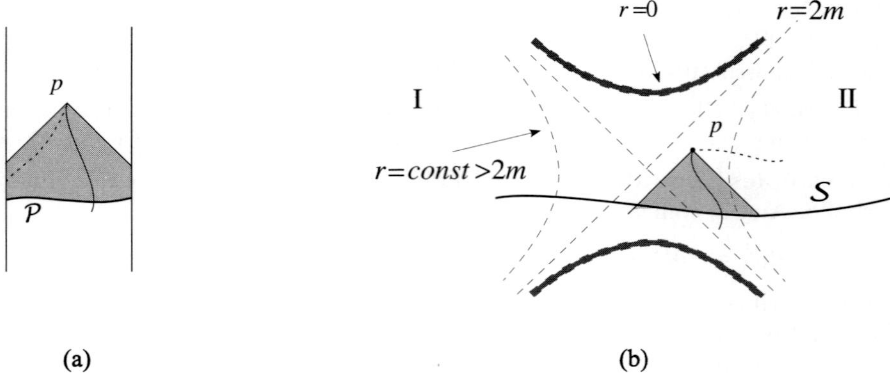

Figure 1. *a*. In the anti-de Sitter space some 'unexpected' (i. e. unknown at $\mathcal{P}$) information can come to p from infinity along the (timelike) dashed curve. *b*. In the Schwarzschild spacetime everything, that reaches p, either originates at $\mathcal{S}$, or moves with a superluminal speed.

Thus in a prognostic theory a state of the system in p is uniquely determined by what takes place in $I^-(p')$ (in fact, even by those events that lie on its Cauchy surface). In this sense no information from $M - I^-(p')$ is available to an observer in p. Our assertion then follows by continuity.

6. Example. Consider a field obeying the equation

$$(\Box + \mu)\phi = 0.$$

This is a second-order linear hyperbolic equation. The Cauchy problem for such equations is known to be well-posed in any globally hyperbolic spacetime. [1] So, ϕ is predictable and hence, in particular, is unsuitable for superluminal signaling (at least on the classical level). Which is noteworthy, because, when $\mu > 0$, such a field is traditionally referred to as 'tachyonic'.

When a spacetime is globally hyperbolic and all material fields are predictable[b], general relativity itself is also a prognostic theory: the metric of a spacetime is uniquely determined by the Einstein equations, if it is known on a Cauchy surface.

2.2 *The general case*

If the condition of global hyperbolicity is relaxed the situation with predictability is *much worse*. Let us illustrate it by a few simple examples.

Remove the points ($t = \pm 1$, $x = \pm 1$) from the Minkowski plane. Next, make the cuts along the two spacelike segments $\{t = \pm 1, \; -1 < x < 1\}$ connecting them (see Fig. 2a). Now, preserving their orientation, glue the banks of the cuts — the upper bank of the lower cut to the lower bank of the upper cut and vice

---

[b]Actually, for a rigorous proof a few additional mild assumptions are necessary, see chapter 7 in Ref. [1].

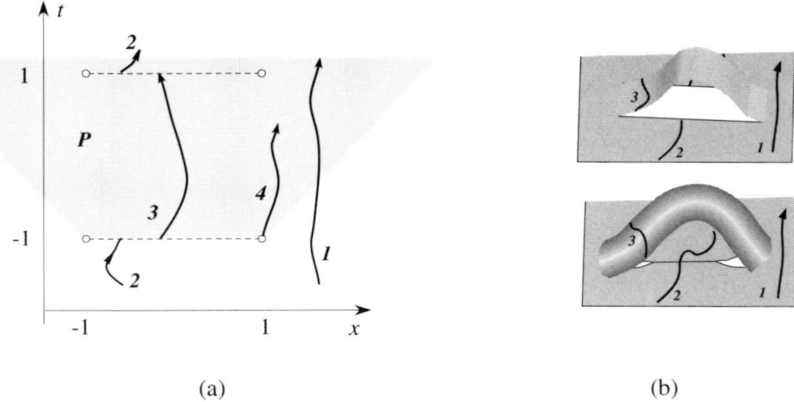

Figure 2. The banks of the cuts (which are depicted by the dashed lines) are glued so that curve 2 is actually continuous and curve 3 — closed.

versa. The obtained spacetime M — called the Deutsch–Politzer (DP) space — is a combination of a cylinder and a plane with two holes (see Fig. 2b). The holes result from excising the 'corner' points (we cannot glue them back into the spacetime) and form singularities of a somewhat unusual kind[c]: the spacetime is flat and thus the singularities are not associated with the divergence of curvature, or its derivatives. In addition to singularities the spacetime contains closed causal curves, which makes it one of the simplest models of the time machine [2,3]. These are the images of the causal curves which in the original (Minkowski) space start from the lower segment and terminate at the upper.

7. Remark. We described the DP space as a result of some 'cut-and-paste' procedure. This, however, gives *no* grounds at all to consider it as something defective and unphysical. In fact, it is exactly as 'physical' as the Minkowski space, being the solution of the Einstein equations with exactly the same (zero) source and exactly the same initial conditions (see below). If desired, one could start from the Deutsch–Politzer spacetime and obtain the Minkowski space by a similar surgery.

The *whole* spacetime M is not globally hyperbolic (by property 4). However, all its 'pathologies', i. e. singularities and time loops, are confined to a region P (in Fig. 2a it is shown gray) bounded by null geodesics emerging from the lower 'holes'. The remaining part $N \equiv M - \overline{P}$ of M is a 'nice' spacetime. Specifically, N, considered as a spacetime (see remark 2), *is* globally hyperbolic, the line $\mathcal{S}_N = \{p\colon t(p) = -2\}$ being one of its Cauchy surfaces. We have thus a spacetime that 'loses' its global hyperbolicity in the course of evolution. This loss, as can be easily seen, is fatal for the predictability of material systems.

8. Example. Suppose our subject matter are pointlike sterile particles, which

---

[c]In many respects they resemble the conical singularity from Example 1.

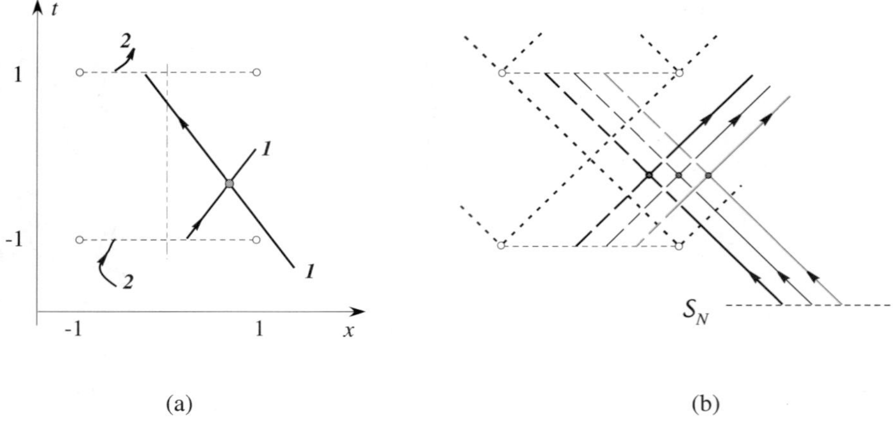

Figure 3. *a.* The twisted Deutsch–Politzer space. The banks of the horizontal cuts are glued so that both curves (1 and 2) are smooth. *b.* The time travel paradox. The trajectories of all possible (given there are only 3 particles at $\mathcal{S}_N$) particles are shown. The world lines of the 'initial' particles cannot be continued in agreement with the laws of interaction.

is, probably, the simplest possible subject. Let us fix the simplest possible initial data — no particles at all at $\mathcal{S}_N$. It is easy to check that neither these initial data, nor the equations of motion[d] give us any clue about what will happen later. Possibly the spacetime will always remain empty. But exactly as well later there may appear some particles. Such 'new' particles ('lions' in terminology of [4]) may have the world lines emerging from a singularity (as with particle 4), or rolled into a circle (particle 3).

Moreover, it may happen that quite an innocent Cauchy problem in a non-globally hyperbolic spacetime does not have *any solution* whatsoever. This fact is known as the 'time travel paradox'. [4]

9. Example. Consider a spacetime which is constructed exactly as the Deutsch–Politzer space, except that before gluing the cuts, one of them is twisted by 180° (see Fig. 3a). Such a spacetime — called the twisted DP space — is especially convenient for constructing paradoxes, because the world lines even of free falling particles have self-intersections. Our concern is with the behaviour of massless particles whose dynamics is determined by the following rules. Each particle is characterized by two integer parameters — we shall call them 'color' and 'flavor' — that take three and two values, respectively. The particles are assumed to be sterile with a *single* exception: if two identical (i. e. with the same values of both parameters) particles meet, they both change their flavor. Suppose now that at some moment $t < -1$ there are three particles of different colors, all moving to the left, as shown in Fig. 3b. If the spacetime remained globally hyperbolic

---

[d]Nor, of course, any their consequences, like the energy conservation.

at later times, these initial conditions would just determine the evolution of the particles, but not in this case. In the twisted DP space *no* evolution (governed by the formulated laws of interaction) is consistent with the initial data. [4]

The situation with the evolution of the spacetime geometry is also bad in the general case. Note, for example, that any sufficiently small region of the DP space — including the vicinity of $\mathcal{S}_N$ — is isometric to the corresponding region of the Minkowski space. Which means, in particular, that if the Minkowski space is a solution of some Cauchy problem (i. e. of a system {some local — Einstein's, say — equations + conditions on $\mathcal{S}_N$}), then the DP space is also a solution of the same Cauchy problem. And this is always the case: whenever a Cauchy problem in general relativity has a globally hyperbolic solution M, it also has infinitely many non-globally hyperbolic solutions (remove from M a two-sphere $\mathbb{S}$ lying to the future from $\mathcal{S}$; the double covering of $M - \mathbb{S}$ is an example of such 'spurious' solution). Within classical relativity (complemented, if desired, with any additional local laws) all these solutions are equipollent.

3 Cosmic censorship hypothesis

Summing up the preceding we can state that physics is a nice (prognostic) theory if it is known somehow that our spacetime is globally hyperbolic. Otherwise, not so much can be said about anything whatsoever. Strictly speaking, in the general case we cannot give a substantiated answer even to the question: Where a car, moving at 60 km per hour, will find itself in 20 minutes? The honest answer would be: If the car is lucky enough it will pass 20 km, but exactly as well it may happen that (after 10 minutes, say) a singularity will appear out of thin air (as in the DP case) and the car will vanish in it, or will be attacked by a monster that will have emerged from the singularity, etc. All such possibilities are in perfect agreement with both the initial data and the known physical laws.

In everyday life people cope with this difficulty by assuming (*implicitly* as a rule) that the spacetime, indeed, is globally hyperbolic. So, it seems tempting to solve the whole problem by just explicitly adopting global hyperbolicity as an additional *postulate*. The spacetimes of the DP type would then be ruled out and general relativity would consider only globally hyperbolic maximal solutions of the Einstein equations as appropriate models of the universe.

Such a program, however, immediately meets two problems. The first is of a philosophical nature: it is hard to justify such a *non-local* postulate (indeed, as we saw in example 9, our ability to perform some experiments *now* may depend on whether or not a causal loop will appear somewhere in the *future*). The second problem is more serious: this new postulate can come into contradiction with the 'old' ones. Spacetimes are conceivable (see below) where the loss of global hyperbolicity is an *inevitable* consequence of the Einstein equations (for specific initial conditions, of course, and with maximality required). So, the best one can hope is that such situations are impossible in 'realistic' circumstances. And it is this hope — known as the Cosmic censorship hypothesis — that is endangered by the wormholes.[e]

---

[e]If one discovered how to create a timelike singularity, it also would violate the Cosmic censorship

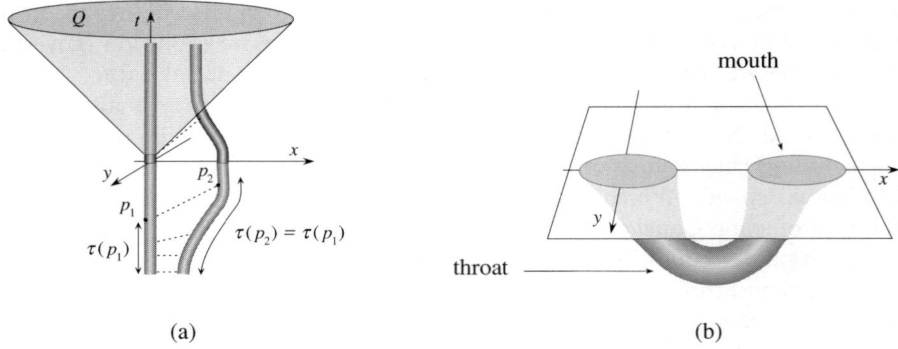

Figure 4. When the identification is performed, the dashed lines become circles. Note that as time goes these lines get more and more tilted. Finally, one of them (the one lying on the boundary of the gray cone) becomes null and the first closed causal curve appears.

4 Wormholes

Pick in Minkowski space two close cylinders C_1 and C_2, of which the second has a bend between, say, $t = -1$ and $t = 1$. Except for that bend both cylinders must be parallel to the t-axis (see Fig. 4a). On the boundaries of the cylinders $\mathcal{B}_{1,2} \equiv \mathrm{Bd}\, C_{1,2}$ define a function τ as follows: $\tau(p)$ is the length of the longest (recall that the metric is Lorentzian) timelike curve that lies in $\mathcal{B}_{1,2}$ and connects p with the surface $t = -2$. Clearly, $\tau(p) = t(p)$ for all $p \in \mathcal{B}_1$, but not for $p \in \mathcal{B}_2$. Now remove the interiors of the cylinders and identify the boundaries $\mathcal{B}_1$ and $\mathcal{B}_2$ so that

$$p_1 = p_2 \quad \Rightarrow \quad \tau(p_1) = \tau(p_2), \qquad p_{1,2} \in \mathcal{B}_{1,2}. \tag{$*$}$$

Finally, smooth out the junction by curving appropriately a close vicinity of $\mathcal{B}_1 = \mathcal{B}_2$ (to remove the discontinuities in derivatives of the metric).

To see what thus obtained spacetime M_W presents, consider its section $\mathcal{S}_N = \{p:\ t(p) = -2\}$. According to the procedure described above, $\mathcal{S}_N$ is obtained from the Euclidean space $\mathbb{E}^3$ by removing two open balls, identifying the boundaries of the holes, and smoothing out the junction. So, $\mathcal{S}_N$ is a wormhole (in the two-dimensional case, see Fig. 4b, we would call it a handle). The former holes are called *mouths* and the 'conduit' connecting them the *throat*. The form of the throat depends on just how we have smoothed out the junction, but — and this is important — it can be made (almost) constant (that is what the condition ($*$) was imposed for). So, M_W describes a wormhole one of whose mouths is somehow pushed away from the other (without changing the length and the form of the throat) and then returned back. [6]

---

hypothesis. This possibility, however, is much better known (see, for example, in. [5]) and I shall not discuss it here.

This spacetime, proposed by Morris, Thorne, and Yurtsever (MTY) [6], received much attention, because at one time it was believed to describe the creation of a time machine. Indeed, the identified points of $\mathcal{B}_1$ and $\mathcal{B}_2$ initially (i. e. at $t = -2$) have the same t-coordinate, but later, according to (*), each $p_1 \in \mathcal{B}_1$ is identified with a $p_2 \in \mathcal{B}_2$ such that $t(p_2) > t(p_1)$. As soon as $t(p_2) - t(p_1)$ becomes greater than $x(p_2) - x(p_1)$ the identified points turn out to be causally connected in the 'initial' (i. e. Minkowski) space. Which means that M_W contains closed causal curves.

As with the DP space, the causal loops are confined to a region Q (the gray cone in Fig. 4a) that lies to the future of a globally hyperbolic (when considered in itself) spacetime $L \equiv M_W - \overline{Q}$. So, we again have a spacetime loosing its global hyperbolicity in the course of evolution. However, there is also a striking difference with the previous case. The additional postulate — to the effect that the whole (i. e. maximal) spacetime must be globally hyperbolic — only required that N would evolve into the Minkowski (and not Deutsch–Politzer, say) space, something that intuitively is quite acceptable and even appealing. But L *does not* have globally hyperbolic extensions and therefore makes us choose between maximality and global hyperbolicity.

Thus the wormholes (if they exist, if they are stable, if their mouths can be separately moved, etc.) refute the Cosmic censorship hypothesis. Just take a wormhole, push one of its mouths, and pull it back. Whatever results[f], it will be in any case non-globally hyperbolic.

5 Global hyperbolicity protection

How serious is the wormhole hazard and do wormholes exist in nature, in the first place? The wormholes are often considered as 'a marginal idea' [8], as something 'too exotic', that is, essentially, as something that unlikely exists. The reasons for such a belief are not always clear, but, at any rate, they are neither experimental, nor theoretical.

The idea at the heart of general relativity is that the spacetime we live in is a curved four-dimensional manifold. Which immediately provokes a question: What is so special in $\mathbb{R}^4$, that one would believe it to be the only possible topology of the universe? The answer (as of today, at least) is obvious — nothing. One hardly would be surprised, for example, if it turn out that we live in a spatially closed universe.

The argument: "We haven't ever seen any wormholes, so they don't exist" does not, of course, stand up. The wormholes do not shine like stars and are not supposed to be 'seen'. The presence of a wormhole surely would strongly affect the surrounding matter, but to what observable effects it must lead, is yet to be learned. Some progress in this direction has been achieved by Cramer *et al.* [9], who noticed that the gravitational lensing of wormholes may differ from that of stars. By Birkhoff's theorem the gravitational field of a static spherically symmetric wormhole (in the empty region around it) is that of a pointlike massive source. The

---

[f] As has been proved recently, it need not be a time machine: among the possible extensions of L there always are causal ones. [7]

value m of the corresponding mass depends on what wormhole is considered and at present there are no reasons to regard any m as 'more realistic'. It is important, however, that *in particular* m may be negative. Such wormholes would act more like diverging lenses in contrast to the stars with their positive masses. These considerations enabled Torres *et al.* [10] to find some (though not too restrictive) bounds on the possible abundance of wormholes with negative m.

In a globally hyperbolic spacetime only primordial wormholes may exist (see property 3). So, the most direct way to support the Cosmic censorship hypothesis would be to find a mechanism excluding their existence at the onset of classical physics (i. e. at the end of the Planck era). Needless to say that no such mechanism is known yet.

Another way out would be to prove that realistic wormholes cannot be *traversable* [11], i. e. large and long-lived enough to pass a macroscopic object through the throat[g]. Indeed, the Einstein equations ensure [12] that to be traversable a wormhole must be maintained by the 'exotic matter' (i. e. the matter whose energy density is negative in some points). Which means, in fact, that at the classical level the traversable wormholes *are* prohibited (except for a few rather exotic possibilities such as the classical scalar field, [13] or ghost radiation [14] as a source). So, if it turns out that quantum effects (which are known to produce negative energy densities as, for example, in the Casimir effect [15]) also fail to support macroscopic wormholes, the problem would be solved. Actually, however, the quantum fields seem to be well suited for the task. In particular, a wormhole was found such that the zero point fluctuations of the electro-magnetic or neutrino fields in its throat produce just enough exotic matter to sustain the wormhole. [16]

Lastly, it may happen that traversable wormholes do exist, but, nevertheless, the MTY scenario does not work because the mouths cannot be moved appropriately. The point is that in the spacetimes of that type there always exist 'almost closed' null geodesics. In Fig. 4a such a geodesic goes from the left mouth, enters the right one, comes again from the left, etc. Though always remaining in L, a photon with this world line makes infinitely many trips between the mouths getting more and more blue. [17] This may indicate that the process is unstable in the sense that an occasional photon can prevent one from bringing the mouths close enough — the closer are the mouths the stronger is the resistance offered by the photon. Of course, classically such an instability can be cancelled by just placing an opaque screen between the mouths, but one does not expect the *quantum* modes to be counteracted as easily. Which suggests that the MTY process may suffer quantum instability. To verify this hypothesis it is instructive to study the behavior of the vacuum expectation of the stress-energy tensor near $\mathrm{Bd}\,L$ and to check whether or not it is bounded. [18] This has been done in a few simplest (two-dimensional) cases and no evidence for the quantum instability was found — exactly as with the Minkowski space the energy density in some cases blows up [19] and in other cases [20,21] does not.

---

[g]A standard example of a *non*-traversable wormhole is the Schwarzschild solution. Even though there are two asymptotically flat regions (I and II in Fig. 1b) connected with a 'bridge', a traveler cannot get from one of them to the other because the bridge collapses too fast.

References

1. S. W. Hawking and G. F. R. Ellis *The Large Scale Structure of Spacetime* (Cambridge University Press, Cambridge, England 1973).
2. D. Deutsch, *Phys. Rev. D* **44**, 3197 (1991).
3. H. D. Politzer, *Phys. Rev. D* **46**, 4470 (1992).
4. S. Krasnikov, *Phys. Rev. D* **65**, 064013 (2002).
5. P. R. Brady, I. G. Moss, and R. C. Myers, *Phys. Rev. Lett.* **80**, 3432 (1998).
6. M. S. Morris, K. S. Thorne, and U. Yurtsever, *Phys. Rev. Letters* **61**, 1446 (1988).
7. S. Krasnikov, *Class. Quantum Grav.* **19**, 4109 (2002).
8. P. C. V. Davies, *www.edge.org/documents/archive/edge77.html*.
9. J. G. Cramer, R. L. Forward, M. S. Morris, M. Visser, G. Benford, and G. A. Landis, *Phys. Rev. D* **51**, 3117 (1995).
10. D. F. Torres, G. E. Romero, and L. A. Anchordoqui, *Phys. Rev. D* **58**, 123001 (1998).
11. M. S. Morris and K. S. Thorne, *Am. J. Phys.* **56**, 395 (1988).
12. J. L. Friedman, K. Schleich, and D. M. Witt, *Phys. Rev. Lett.* **71**, 1486 (1993).
13. C. Barceló and M. Visser, *Phys. Lett. B* **466**, 127 (1999).
14. S. A. Hayward, *Phys. Rev. D* **43**, 3878 (1991).
15. N. D. Birrel and P. C. V. Davies *Quantum Fields in Curved Space* (Cambridge University Press, Cambridge, England 1982).
16. S. Krasnikov, *Phys. Rev. D* **62**, 084028 (2000).
17. S. V. Krasnikov, *Class. Quantum Grav.* **11**, 2755 (1994).
18. V. P. Frolov, *Phys. Rev. D* **43**, 3878 (1991).
19. U. Yurtsever, *Class. Quantum Grav.* **8**, 1127 (1991).
20. S. V. Krasnikov, *Phys. Rev. D* **54**, 7322 (1996).
21. S. V. Sushkov, *Class. Quantum Grav.* **14**, 523 (1997).

ELEMENTARY PARTICLES AS BLACK HOLES

G. 'T HOOFT

Institute f. Theor. Physics, Utrecht University, P.O. Box 80006, NL-3508 TA Utrecht, the Netherlands
E-mail: G.tHooft@phys.uu.nl

There is no write-up of this talk. However the reader can find very similar material discussed in: G. 't Hooft, *Int.J.Mod.Phys.* **A11** (1996) 4623, gr-qc/9607022; gr-qc/9711053.

COUNTER-EXAMPLE WHERE COSMIC TIME KEEPS ITS ORIGINAL ROLE IN QUANTUM COSMOLOGY

E. I. GUENDELMAN AND A. B. KAGANOVICH

Ben Gurion University, Beer Sheva 84105, Israel
E-mail: guendel@bgumail.bgu.ac.il and alexk@bgumail.bgu.ac.il

In the minisuperspace models of quantum cosmology, the absence of time in the Wheeler-DeWitt (constraint) equation, is the main point leading to the generally accepted conclusion that in the quantum cosmology there is no possibility to describe the evolution of the universe procceding in the cosmic time (the time usually used in classical cosmology). We show that in spite of the constraint, under the specific circumstances, the averaging of some of the Heisenberg equations can give nontrivial additional information about *explicit time dependence of the expectation values of certain dynamical variables in quantum cosmology*. This idea is realized explicitly in a higher dimensional model with a negative cosmological constant and dust as the sources of gravity. When there is an anisotropy in the evolution of the universe, the above phenomenon (i.e. explicit cosmic time dependence of certain expectation values) appears and we find the new quantum effect which consists in "quantum inflationary phase" for some dimensions and simultaneous "quantum deflationary contraction" for the remaining dimensions. The expectation value of the "volume" of the universe remains constant during this quantum "inflation-deflation" process.

1 Introduction

In the quantization of generally covariant systems, like General Relativity (in any number of dimensions), one has to take into account a fundamental constraint which basically tell us that the total Hamiltonian of the classical system equals zero[1], $H \approx 0$. The sign $\approx$ is used[2] instead of $H = 0$ to emphasize the fact that although H is zero along the classical trajectories, it is still non trivial in a sense that it may have non zero Poisson brackets with other dynamical variables.

In the quantized version of the theory, the physical states have to satisfy the constraint equation $H\Psi = 0$ which means that those states should be time independent. One believes usually that the expectation values of the Heisenberg equations for operator of *any* dynamical variable is always zero just by virtue of the constraint equation, i.e. expectation values of all the dynamical variables must be time independent too. Such a situation is interpreted in the literature as the statement (having almost the power of a theorem) that time disappears from quantum gravity[1,3,4]. In particular, in the context of quantum cosmology, this statement is formulated as the generally accepted conclusion that there is no possibility to describe the evolution of the universe procceding in the cosmic time.

We will see here that the above conclusions can be premature. This will be done by presentation of an explicit counter-example where cosmic time keeps its original role in quantum cosmology: the expectation values of certain variables have non trivial dependence on the same cosmic time which enters, for instance, in the classical equations of the Friedmann cosmology.

The reason for the recovery of a non trivial time dependence in quantum cosmology, in spite of the fact that H equals zero when applied on physical states, lies

272

on the two complementary facts: a) The Hamiltonian is still a nontrivial operator since it must have non trivial commutators in order to reproduce the Heisenberg equations. This means that there must be non physical states $|N.P.\rangle$ for which $H|N.P.\rangle \neq 0$. b) In our specific model, the physical states Ψ, for which $H|\Psi\rangle = 0$ is satisfied, are found to be non normalizable states. This forces us to consider in any case the "non physical" states of a) in order to define expectation values of relevant operators with the help of a limit process $|N.P.\rangle \rightarrow |\Psi\rangle$. Since for the non physical states $H|N.P.\rangle \neq 0$, non trivial time dependence can appear in the expectation values of some of the Heisenberg equations.

We will display this interesting phenomenon of the appearance of nontrivial time dependence in quantum cosmology in the context of a Kaluza-Klein model which allows for anisotropic evolution: expansion of 3 dimensions and contraction of the extra dimensions for a physically attractive scenario. In this model, a negative cosmological constant does not let the total volume of the universe grow, while quantum effects stabilize the volume against collapse.

It is important to point out that the appearance of time we display in our model, *does not* rely on some WKB approximation or on the use of some field variable as time, rather, it is the *genuine, original cosmic time* which fulfils its natural duty to be the time parameter of the theory even at the intrinsic quantum level. A somewhat related approach which also gives cosmic time dependence for averages of certain dynamical variables in the presence of anisotropy was developed by Kheyfets and Miller in Ref.[5]. For the generalization of this approach see Ref.[6].

2 Description of the model and its classical dynamics

We start from studying a higher dimensional homogeneous totally anisotropic spatially flat cosmological model

$$ds^2 = -dt^2 + \sum_{l=1}^{D} a_l^2(t)dx_l^2 \tag{1}$$

which is assumed to be toroidally compact: $0 \leq x_l \leq L_l$. The scalar curvature corresponding to Eq.(1) is given by

$$R = \frac{2}{V}\frac{d^2V}{dt^2} - \frac{D-1}{D}\left(\frac{d(\ln V)}{dt}\right)^2 + \frac{1}{D}\sum_{l<m}\left(\frac{d(\ln a_l)}{dt} - \frac{d(\ln a_m)}{dt}\right)^2 \tag{2}$$

where $V = \prod_{l=1}^{D} a_l \equiv$ "volume" of the universe. We will assume that the only sources of gravity are a negative cosmological constant $\Lambda < 0$ and dust. The gravitational action can be written then as

$$S_{gr} = \frac{1}{\kappa}\int d^Dx dt\sqrt{-g}(R - 2\Lambda) = -\frac{1}{\kappa}\prod_{l=1}^{D}L_l\int L dt \tag{3}$$

where $\kappa = 16\pi G$ and up to a total derivative term, the Lagrangian L is given by

$$L = \frac{1}{V}\frac{D-1}{D}\left(\frac{dV}{dt}\right)^2 - \frac{V}{D}\sum_{l<m}\left(\frac{d(\ln a_l)}{dt} - \frac{d\ln(a_m)}{dt}\right)^2 + 2\Lambda V \tag{4}$$

For simplicity, we choose units where $\frac{1}{\kappa}\prod_{i=1}^{D}L_l = 1$ so that

$$S_{gr} = -\int L\,dt \tag{5}$$

In addition to the equations derived from (4) and (5) we have to impose the constraint saying that the Hamiltonian H is zero (the statement which coincides with the $0-0$ component of Einstein's equations). The presence of dust affects only the constraint equation which reads

$$H = \frac{1}{V}\frac{D-1}{D}\left(\frac{dV}{dt}\right)^2 - \frac{V}{D}\sum_{l<m}\left(\frac{d(\ln a_l)}{dt} - \frac{d(\ln a_m)}{dt}\right)^2 - 2\Lambda V - \mu = 0, \tag{6}$$

where $\mu > 0$ is a constant which has the interpretation of the dust energy density times the volume V of the universe.

The form of the Lagrangian and the constraint may be simplified to a marked degree if one uses the following parametrization for $a_l(t)$

$$a_l(t) = [V(t)]^{1/D}e^{\theta_l(t)}. \tag{7}$$

Since $\sum_{l=1}^{D}\theta_l \equiv 0$ one can exclude $\theta_D = -\sum_{i=1}^{D-1}\theta_i$ and proceed with the D independent variables: V and $\theta_i,\; i = 1, 2, ... D-1$. Finally, one can see that the Lagrangian and the constraint take the diagonalized and normalized form

$$L = \left(\frac{d\rho}{dt}\right)^2 - \rho^2\sum_{i=1}^{D-1}\left(\frac{dz^i}{dt}\right)^2 - \omega^2\rho^2, \tag{8}$$

$$H = \left(\frac{d\rho}{dt}\right)^2 - \rho^2\sum_{i=1}^{D-1}\left(\frac{dz^i}{dt}\right)^2 + \omega^2\rho^2 - \mu = 0, \tag{9}$$

where

$$\omega^2 = -\frac{D}{2(D-1)}\Lambda; \quad \rho^2 = \frac{4(D-1)}{D}V \tag{10}$$

and

$$z^i = \frac{1}{1+\sqrt{D}}\sqrt{\frac{D}{2(D-1)}}[\theta_1 + \theta_2 + ... + (2+\sqrt{D})\theta_i + ... + \theta_{D-1}],$$

$$i = 1, 2, ..., D-1 \tag{11}$$

Notice that the Lagrangian (8) and the constraint (9) are invariant under $\frac{D(D-1)}{2}$ dimensional symmetry group of translations and rotations of a $D-1$ dimensional Euclidean space. In particular, the translational symmetry $z^i \rightarrow z^i + b^i$ with constants $b^i,\; i = 1, 2, ..., D-1$, gives rise to $D-1$ conserved quantities

$$F_i = -2\rho^2\frac{dz^i}{dt}, \quad i = 1, 2, ..., D-1 \tag{12}$$

As we will see in what follows, it is very important that the set of quantities $\frac{dz^i}{dt}$ (which are the linear combinations of $\frac{d\theta^i}{dt}$) measure anisotropy of the evolution of the universe. In fact, since $\sum_{i=1}^{D} \theta_i \equiv 0$, all θ_i's can have the same time dependence only if such time dependence is the trivial one, i.e., all of the θ_i's are constants. Therefore, the set of $D-1$ conserved independent quantities F_i measure anisotropy of the evolution of the universe.

It is interesting to note that the constraint may be represented now in the following form

$$H = \left(\frac{d\rho}{dt}\right)^2 + U_{eff}^{(class)}(\rho) - \mu = 0 \tag{13}$$

where the classical effective volume dependent potential appears:

$$U_{eff}^{(class)}(\rho) \equiv -\frac{F^2}{4\rho^2} + \omega^2 \rho^2, \quad F^2 \equiv \sum_{i=1}^{D-1} F_i^2 \tag{14}$$

All classical solutions exhibit cosmological singularities. This feature, as we will see below, can be avoided in quantum cosmology in the presence of a negative cosmological constant and dust.

3 Minisuperspace quantization and solutions of the Wheeler-DeWitt equation

To produce a quantum theory from the classical one, we have to postulate canonical commutation relations and in addition, the constraint equation (9) has to be imposed as a condition on the wave function of the universe Ψ. The resulting Wheeler-DeWitt equation $H\Psi = 0$ is the fundamental equation governing the quantum cosmology.

It is convenient to rewrite the Lagrangian (8) and the Hamiltonian (9) in a geometrical fashion

$$L = f_{\alpha\beta}(q)\frac{dq^\alpha}{dt}\frac{dq^\beta}{dt} - \omega^2 \rho^2, \tag{15}$$

$$H = f_{\alpha\beta}(q)\frac{dq^\alpha}{dt}\frac{dq^\beta}{dt} + \omega^2 \rho^2 - \mu = \frac{1}{4}f^{\alpha\beta}\pi_\alpha\pi_\beta + \omega^2 \rho^2 - \mu \tag{16}$$

where

$$q^\alpha = (\rho, z^i), \quad i = 1, 2, ..., D-1; \quad f_{\alpha\beta}(q) = diag(1, -\rho^2, ..., -\rho^2) \tag{17}$$

are the coordinates and metric of our minisuperspace, respectively, and π_α are the momenta canonically conjugate to q^α.

The ambiguity due to the operator ordering problem can be taken into account[7] by adding the "nonminimal" term $\xi\mathcal{R}_f$, where $\mathcal{R}_f$ is the scalar curvature corresponding to the metric $f_{\alpha\beta}(q)$ and ξ is an arbitrary real constant depending on the operator ordering used. The resulting Wheeler-DeWitt equation reads[8]

$$\frac{1}{\rho^{D-1}}\frac{\partial}{\partial\rho}\left(\rho^{D-1}\frac{\partial\Psi}{\partial\rho}\right) - \frac{1}{\rho^2}\sum_{i=1}^{D-1}\frac{\partial^2\Psi}{\partial z^{i2}} + \left[\xi\frac{(D-1)(D-2)}{\rho^2} - 4\omega^2\rho^2 + 4\mu\right]\Psi = 0, \tag{18}$$

where we have used that the scalar curvature of the minisuperspace metric $f_{\alpha\beta}$ is
$\mathcal{R}_f = (D-1)(D-2)\rho^{-2}$.

The inner product for sufficiently regular wavefunctions Ψ and Φ is defined in the geometric form

$$(\Psi, \Phi) = \int d^D q \sqrt{|det(f_{\alpha\beta})|}\, \Psi^*(q)\Phi(q). \tag{19}$$

To solve Eq.(18), let us note that the Hamiltonian commutes with the generators $-i\frac{\partial}{\partial z^i}$ of the symmetry $z^i \to z^i + b^i$ with $D-1$ arbitrary real numbers b^i. It is therefore possible to take the solutions of (18) as eigenstates of the generators $-i\frac{\partial}{\partial z^i}$

$$\Psi(\rho, z^i) = \frac{1}{(2\pi)^{(D-1)/2}} R(\rho) \exp\left(i \sum_{i=1}^{D-1} F_i z^i \right), \tag{20}$$

where F_i is an eigenvalue of the operator $-i\frac{\partial}{\partial z^i}$, which is the quantum version of the conserved quantity defined by Eq.(12). The function $R(\rho)$ is then determined as the solution of the equation

$$\frac{d^2 R}{d\rho^2} + \frac{D-1}{\rho}\frac{dR}{d\rho} + \frac{K}{\rho^2}R - 4\omega^2\rho^2 R = -4\mu R, \tag{21}$$

where

$$K \equiv F^2 + \xi(D-1)(D-2) \tag{22}$$

and F^2 is defined in Eq.(14). It is possible to regard Eq.(21) as a stationary Schrö dinger equation in a D-dimensional space (with mass of particle $= 1/2$) with an effective potential

$$U_{eff}^{(quant)}(\rho) \equiv -\frac{K}{\rho^2} + 4\omega^2\rho^2. \tag{23}$$

Depending on whether (i) $K > 0$, (ii) $K = 0$ or (iii) $K < 0$ the effective potential has an attractive core, is exactly a harmonic potential or has a repulsive core respectively. Then 4μ in Eq.(21) plays the role of the "energy" eigenvalue. Since the region $\rho \to \infty$ is classically forbidden, a physically acceptable solution of Eq(21) has to vanish in this limit. This leads to the quantization of the eigenvalues of the linear operator corresponding to Eq.(21). However, in our situation, it is not very appealing to quantize the energy of dust μ because this is not a dynamical variable. In contrast to this, the conserved quantity F^2, which is a measure of the anisotropy of the cosmological evolution, can be regarded as the quantized dynamical variable. Using the fact that F^2 appears linearly in Eq.(21), for a *given* μ, we select the appropriate values of F^2 such that the operator corresponding to the l.h.s. of Eq.(21) has as its eigenvalue the number 4μ. Therefore we have different values of F^2 specifying uniquely the eigenfunctions.

The solution of Eq.(21) is given by

$$R = R_n(\rho) = N_n \rho^{2s_n} e^{-|\omega|\rho^2} \Phi\left(-n, \frac{\mu}{|\omega|} - 2n, 2|\omega|\rho^2\right), \tag{24}$$

where N_n is a normalization factor, Φ is a confluent hypergeometric function, n is a non-negative integer and

$$s_n = \frac{1}{4}[-(D-2) + \sqrt{(D-2)^2 - 4K_n}] = \frac{1}{2}\left(\frac{\mu}{|\omega|} - 2n - \frac{D}{2}\right). \qquad (25)$$

The corresponding quantized values of the length of the vector F_i are

$$F^2 = F_n^2 = 4\left\{(D-2)\left[\frac{1}{4}(D-2) - \xi(D-1)\right] - \left[\frac{\mu}{|\omega|} - 1 - 2n\right]^2\right\}, \qquad (26)$$

and K_n in (25) is determined by F_n^2 via Eq.(22). Notice that the direction of F_i remains arbitrary.

Avoidance of the cosmological singularity in the context of quantum cosmology can be defined as a statement that the amplitude $\Psi \to 0$ as $\rho \propto \sqrt{V} \to 0$. In our case, this is possible if $s_n > 0$, i.e.

$$\mu > (2n + \frac{D}{2})|\omega|. \qquad (27)$$

We see that the presence of enough dust is a necessary condition for the avoidance of the cosmological singularity. It follows from Eq.(25) that condition $s_n > 0$ implies also that $K_n < 0$, that is $F_n^2 < -\xi(D-1)(D-2)$. Assuming $D > 2$, this can be achieved for some $F_n^2 > 0$ only if $\xi < 0$. Notice that $K_n < 0$ means that the quantum effective potential (23) has a repulsive core.

4 The Dynamics of Quantum Cosmology Proceeding in Cosmic Time

4.1 Heisenberg picture

Before studying the problem of defining the role of time in quantum cosmology processes in the framework of our model, it is important to understand some aspects of the classical dynamics. This is of importance since the quantum behavior has to satisfy, in some way, the correspondence principle. In particular, the time dependence of z^i is determined by Eq.(12) which implies

$$\frac{dz^i}{dt} = -\frac{F_i}{2\rho^2}, \quad i = 1, 2, ..., D-1. \qquad (28)$$

Eq.(28) can be obtained not only from the Euler-Lagrange equations but also from the Hamiltonian formalism, i.e.

$$\frac{dz^i}{dt} = [z^i, H]_{PB}, \quad \frac{d\pi_{z^i}}{dt} = -\frac{d}{dt}\left(2\rho^2 \frac{dz^i}{dt}\right) = [\pi_{z^i}, H]_{PB} = 0 \qquad (29)$$

where π_{z^i} is the momenta canonically conjugated to z^i. Note that *although $H = 0$ along the classical trajectories, H has nontrivial Poisson brackets with z^i. This* allows a nontrivial (cosmic) time dependence through Eqs.(29), which provides *information not contained in the constraint equation (9).*

We will now see that a similar situation appears in the quantum version of the theory, where so far the only thing we have cared about has been the Wheeler-DeWitt (quantum constraint) equation (18). Normally, when quantizing a theory without constraints, the Poisson bracket $[A, B]_{PB}$ is replaced by $i[A, B]$, where $[A, B]$ is the quantum mechanical commutator of A and B. For any operator Q which does not depend explicitly on time, the Heisenberg equations hold

$$\frac{dQ}{dt} = i[H, Q]. \tag{30}$$

For our purpose, it is convenient to work in the Heisenberg picture, where operators satisfy Eq.(30), but states are taken to be time independent, that is

$$\frac{\partial \Psi}{\partial t} = 0. \tag{31}$$

We are interested in a subspace of this space of functions, the so-called "physical subspace", in which the constraint equation (18) holds[2]. The consistency requirement that the constraint be preserved in time, is here trivially satisfied because the constraint function H coincides with the Hamiltonian. Solution of Eq.(18) has been presented in the previous section, and it is now our purpose to look at the consequences of the Heisenberg equations that determine *the evolution of operators defined in the whole space of functions* (and not just states satisfying the constraint $H\Psi = 0$).

The operator equations that govern the cosmic time dependence of z^i ($i = 1, 2, ..., D-1$) are

$$\frac{dz^i}{dt} = i[H, z^i] = \frac{i}{4\rho^2} \sum_{k=1}^{D-1} \left[\frac{\partial^2}{\partial z^{k2}}, z^i \right] = \frac{i}{2\rho^2} \frac{\partial}{\partial z^i}. \tag{32}$$

4.2   Quantum Mechanical Averaging

We are going now to evaluate the averages of $\frac{dz^i}{dt}$ and other dynamical variables. However, one should be careful on how averages are defined for wavefunctions (20) since these are normalizable in a continuous way only.

To see that naive manipulations in this case can lead to wrong results, let us consider Eq.(30) together with the subsidiary condition $H\Psi = 0$. If we assume H to be an Hermitian operator, we would get (using the inner product (19)) that $\frac{d}{dt}\langle \Psi | Q | \Psi \rangle = \langle \Psi | \frac{dQ}{dt} | \Psi \rangle = \langle \Psi | i[H, Q] | \Psi \rangle = 0$, from where it seems to follow that all averages are time independent. However, the above manipulations are not correct, since the assumption that H is an Hermitian operator holds only if the functions considered are not pathological. In fact, the latter is not always true since if we consider, for example, the case $Q = z^i$, the state $z^i | \Psi \rangle$ (that enters in the expression $\langle \Psi | i[H, Q] | \Psi \rangle$), where $| \Psi \rangle$ is given by (20), is very singular in the limit $|z^i| \to \infty$. So, the conclusion about the time independence of the expectation values of *all* canonical operators $Q(q, \pi)$, i.e. that $\frac{d}{dt}\langle \Psi | Q | \Psi \rangle = 0$ is always true, has been based on a wrong assumption.

Notice that the z^i-dependence of the wavefunction (20) (resulting from the symmetry $z^i \to z^i + b^i$) resembles another more familiar example in physics, namely the momentum eigenstate $|\vec{p}\rangle$ of a free nonrelativistic particle governed by the Hamiltonian $H = \frac{\hat{p}^2}{2m}$. We have in the Heisenberg picture $\frac{d}{dt}\langle\vec{p}|\vec{x}|\vec{p}\rangle = \langle\vec{p}|\frac{d}{dt}\vec{x}|\vec{p}\rangle = i\langle\vec{p}|[H,\vec{x}]|\vec{p}\rangle = i\langle\vec{p}|(H\vec{x} - \vec{x}H)|\vec{p}\rangle$. Taking into account that $H|\vec{p}\rangle = \frac{\vec{p}^2}{2m}$ and *assuming* H is hermitian, we get $\frac{d}{dt}\langle\vec{p}|\vec{x}|\vec{p}\rangle? =?i(\frac{\vec{p}^2}{2m} - \frac{\vec{p}^2}{2m})\langle\vec{p}|\vec{x}|\vec{p}\rangle = 0$. The dubious step is actually based on wrongly assuming H to be hermitian. For that property no longer holds in this case because of the singular behavior of the state $\vec{x}|\vec{p}\rangle$ at large distance, and this is the reason why the result is wrong. In fact, we know that the right answer is obtained by replacing $i[H,\vec{x}]$ by $\hat{\vec{p}}/m$, which leads to the conclusion $(\langle\vec{p}|\vec{p}\rangle)^{-1}\frac{d}{dt}\langle\vec{p}|\vec{x}|\vec{p}\rangle = \vec{p}/m$. Although the last argument leads to the right result, the method of calculation is not totally satisfactory, since it involves carelessly manipulating infinite factors like $\langle\vec{p}|\vec{p}\rangle$, etc. The same conclusion can however be obtained more rigorously by defining

$$\frac{1}{\langle\vec{p}|\vec{p}\rangle}\frac{d}{dt}\langle\vec{p}|\vec{x}|\vec{p}\rangle \equiv \lim_{|\varphi\rangle\to|\vec{p}\rangle}\lim_{|\chi\rangle\to|\vec{p}\rangle}\left[\frac{1}{\langle\varphi|\chi\rangle}\frac{d}{dt}\langle\varphi|\vec{x}|\chi\rangle\right] \tag{33}$$

provided $|\varphi\rangle$ and $|\chi\rangle$ are states in the Heisenberg representation which have good enough large distance behavior (e.g. wave packets) so that the integrals appearing in the numerator and denominator in the r.h.s. of (33) are well defined (it means that the states $|\varphi\rangle$ and $|\chi\rangle$ are not eigenstates of the operators $\hat{\vec{p}}$ and H).

Proceeding in the same fashion in our quantum cosmology model, we define in general

$$\langle Q\rangle \equiv \lim_{|\varphi\rangle\to|\Psi\rangle}\lim_{|\chi\rangle\to|\Psi\rangle}\left[\frac{1}{\langle\varphi|\chi\rangle}\langle\varphi|Q|\chi\rangle\right], \tag{34}$$

where $|\Psi\rangle = |F_i, n\rangle$ is the eigenstate of the operator $-i\frac{\partial}{\partial z^i}$ with eigenvalue F_i and the corresponding wavefunction $\langle\rho, z^i|\Psi\rangle = \Psi(\rho, z^i)$ is the solution (20), (24) of the constraint $H\Psi = 0$; $|\varphi\rangle$ and $|\chi\rangle$ are states in the Heisenberg representation (and therefore static) which have good enough behavior as $|z^i| \to \infty$, so that the integrals appearing in the numerator and denominator of (34) are well defined. This necessarily implies that these states do not satisfy the constraint, i.e. $H\varphi \neq 0$, $H\chi \neq 0$ although for a given point (ρ, z^i), the wavefunctions $\langle\rho, z^i|\varphi\rangle$ and $\langle\rho, z^i|\chi\rangle$ approach the wavefunction $\Psi(\rho, z^i)$ in the limit process appearing in (34).

In particular, when $Q = \frac{dz^i}{dt}$, we find

$$\langle\frac{dz^i}{dt}\rangle = \frac{d\langle z^i\rangle}{dt} \equiv \lim_{|\varphi\rangle\to|\Psi\rangle}\lim_{|\chi\rangle\to|\Psi\rangle}\left[\frac{1}{\langle\varphi|\chi\rangle}\langle\varphi|\frac{dz^i}{dt}|\chi\rangle\right], \ where \ |\Psi\rangle = |F_i, n\rangle. \tag{35}$$

Using the definition (35) to evaluate the average of (32) in a state of the form given by Eq.(20), we get the result

$$\frac{d\langle z^i\rangle}{dt} = -\langle\frac{1}{2\rho^2}\rangle_n F_i, \tag{36}$$

which corresponds exactly with the classical result (28).

In the same way one can proceed to evaluate other averages. In particular, this way of proceeding gives for functions of ρ alone, $Q(\rho)$:

$$\langle Q(\rho)\rangle_n = \frac{\int Q(\rho)R_n^2(\rho)\rho^{D-1}d\rho}{\int R_n^2(\rho)\rho^{D-1}d\rho}, \tag{37}$$

where $R_n(\rho)$ is determined by Eq.(24). The convergence of $\frac{d\langle z^i \rangle}{dt}$ implies the convergence of $\langle \frac{1}{\rho^2} \rangle$. One can show that $\langle \frac{1}{\rho^2} \rangle_n$ is finite provided $\mu > (2n+1)|\omega|$ and

$$\langle \frac{1}{\rho^2} \rangle_n = \frac{2\omega^2}{\mu - (2n+1)|\omega|}. \tag{38}$$

It is interesting to look at the average of $\frac{d^2 z^i}{dt^2}$ which is given (using Heisenberg equations) by

$$\frac{d^2 z^i}{dt^2} = -\left(\frac{D-4}{\rho^4} + \frac{2}{\rho^3} \frac{\partial}{\partial \rho} \right) \frac{\partial}{\partial z^i}. \tag{39}$$

It turns out that the average of $\frac{d^2 z^i}{dt^2}$ in a state given by Eqs.(20) and (24) is identically zero provided

$$\mu > 2(n+1)|\omega| \tag{40}$$

One can show that the restriction (40) on the amount of dust is a necessary condition to provide the consistency of the main three items of our above analyzes: the constraint equation $H\Psi = 0$, the Heisenberg equations and definition of averages.

Recall that the condition (27) is needed if we require that the universe have a zero probability amplitude of having zero volume. Comparing (40) with the condition (27), we conclude that for $D \geq 4$, the quantum cosmology problem under consideration has a satisfactory solution if the condition (27) is satisfied. We will assume it in what follows.

It is very important also that the condition (27) is a stronger restriction on the amount of dust than is actually needed to provide that *the average of the "volume" of the universe* $\langle V \rangle_n = \frac{D}{4(D-1)} \langle \rho^2 \rangle_n$ is finite and it turns out to be *time independent*:

$$\langle V \rangle_n = \frac{D}{4(D-1)} \frac{\int R_n^2(\rho) \rho^{D+1} d\rho}{\int R_n^2(\rho) \rho^{D-1} d\rho} = \frac{D}{16(D-1)\omega^2} \frac{\mu^2 - (2n+1)^2 \omega^2}{\mu - (2n-1)|\omega|} \tag{41}$$

In particular, for the ground state, $n = 0$, $\langle V \rangle_{n=0} = \frac{D}{16(D-1)\omega^2}(\mu - |\omega|)$. The integrals appearing in (41) when they converge of course give a positive result as it must be from the definition of $\langle V \rangle_n$.

4.3 The Cosmic Time Dependence of the Expectation Values of the Cosmological Quantities: Some General Results.

To see more clearly the above results, let us now represent them in terms of usual cosmological quantities. In the totally anisotropic model of Eq.(1), the corresponding classical variables are the "scale factors" $a_l(t)$ $(l = 1, 2, ..., D)$ which have been parametrized by means of Eq.(7) in terms of the "volume" of the universe $V(t)$ and D functions $\theta_l(t)$ $(l = 1, 2, ..., D)$. One should note that due to the identity $\sum_{l=1}^{D} \theta_l \equiv 0$, only $D - 1$ of the θ_l's are independent . Relations between variables V and θ_i $(i = 1, 2, ..., D-1)$ and those ρ and z^i are given by Eqs.(10) and (11).

In quantum cosmology, the average of z^i, even with the improved definition (34), does not exist and therefore (34) has nothing to say on whether $\langle z_i \rangle$ is time dependent or time independent. In contrast to this, the average of $\langle \frac{dz^i}{dt} \rangle$ is well defined and in the present of an anisotropy ($F_i \neq 0$), it is a nonzero finite constant determined by Eqs.(36) and (38). Then we find that the time evolution of $\langle z_i \rangle$ is of the form $\langle z_i \rangle = \langle \frac{dz^i}{dt} \rangle t + c^i = \left(-\langle \frac{1}{2\rho^2} \rangle_n F_i \right) t + c^i$, where c^i are undetermined constants. This yields the following (cosmic) time dependence of the θ_i variables

$$\langle \theta_i \rangle = \alpha_i t + \gamma_i, \tag{42}$$

where γ_i are integration constants and for $i = 1, 2, ..., D - 1$

$$\alpha_i = \frac{\sqrt{D-1}}{D(\sqrt{D}+1)} \frac{2\omega^2}{\mu - (2n+1)|\omega|} \left[(D + \sqrt{D} - 1)F_i - \sum_{j \neq i} F_j \right] \tag{43}$$

From the identity $\sum_{l=1}^{D} \theta_l \equiv 0$ we have for $\langle \theta_D \rangle$

$$\langle \theta_D \rangle = \alpha_D t + \gamma_D \quad where \quad \alpha_D = -\sum_{i=1}^{D-1} \alpha_i, \quad \gamma_D = -\sum_{i=1}^{D-1} \gamma_i \tag{44}$$

Besides, in the classical cosmology one can define D expansion parameters $H_l \equiv \frac{1}{a_l(t)} \frac{da_l(t)}{dt}$. In the quantum version of the theory, we have of course to define the ordering of the operators a_l and $\frac{da_l}{dt}$. For example, $a_l^{-1} \frac{da_l}{dt}$ or $a_l^{-1/2} \frac{da_l}{dt} a_l^{-1/2}$, etc. give quantum mechanically distinct definitions for H_l. We will choose a definition

$$H_l \equiv \frac{d}{dt}(\ln a_l). \tag{45}$$

Using this definition, the parametrization (7) and results of the previous subsection one can evaluate *the expectation values of the expansion parameters $\langle H_l \rangle$ which turn out to be constants*:

$$\langle H_l \rangle = \left(\frac{D}{4(D-1)} \right)^{1/D} \left(\langle \frac{d}{dt}(\ln \rho^{2/D}) \rangle + \langle \frac{d\theta_l}{dt} \rangle \right)$$
$$= \left(\frac{D}{4(D-1)} \right)^{1/D} \langle \frac{d\theta_l}{dt} \rangle = \left(\frac{D}{4(D-1)} \right)^{1/D} \alpha_l, \tag{46}$$

where α_l are determined in (43) and (44).

Let us notice that Eqs.(7) and (45) together with the identity $\sum_{l=1}^{D} \theta_l \equiv 0$ imply that

$$\sum_{l=1}^{D} H_l = \frac{d}{dt}(\ln V). \tag{47}$$

Taking average on both sides of Eq.(47), we obtain the result that *the sum of the averages of the expansion parameters equals zero*, due to Eqs.(46) and (44) or alternatively from Eqs.(37) and (41). This shows that the definition (45) of the expansion parameters is consistent with the quantum stabilization of the volume of the universe, $\langle V \rangle = const$, Eq.(41).

From (45) and (46), we see that the time behavior of $\langle \ln a_l \rangle$ is given by

$$\langle \ln a_l \rangle = \left(\frac{D}{4(D-1)} \right)^{1/D} \alpha_l t + \tilde{\gamma}_l, \quad l = 1, 2, ..., D, \tag{48}$$

where $\tilde{\gamma}_l$ are arbitrary integration constants.

5   Inflation-Compactification as a Quantum Effect

We will now see that the results of Sec.4 allow to realize a dynamical explanation of the asymmetry in the sizes of extra and ordinary dimensions in the context of quantum cosmology.

At the classical level, there is no difference on whether we use a_l or $\ln a_l$ as our variables. If Eq.(48) were to hold classically, we could conclude that some dimensions exponentially expand and others exponentially contract, depending on the sign of α_l, as given by (43) and (44). In our case, the behavior of the universe is intrinsically quantum mechanical and we will refer to a *"quantum inflationary phase"* for a given dimension l if the expectation value of the expansion parameter $\langle H_l \rangle = const > 0$. Likewise we will refer to a *"quantum deflationary phase"* for a given dimension l if the expectation value of the expansion parameter $\langle H_l \rangle = const < 0$.

A case of particular interest is when the expectation values of the expansion parameters of three of the dimensions are identical and at the same time the expectation values of the expansion parameters of the remaining $D - 3$ dimensions are also identical. In such a case $\alpha_1 = \alpha_2 = \alpha_3 \equiv \alpha$ and $\alpha_4 = \alpha_5 = ... = \alpha_D \equiv \tilde{\alpha}$. Then it follows from the identity $\sum_{l=1}^{D} \alpha_l = 0$ that $\tilde{\alpha} = -3\alpha/(D - 3)$ and we get by using Eq.(43):

$$\alpha = \alpha(n) = \pm \frac{2\omega^2}{D[\mu - (2n + 1)|\omega|]} \sqrt{\frac{(D - 3)(D - 1)}{3}} |F|_n, \qquad (49)$$

$$\tilde{\alpha} = \tilde{\alpha}(n) = \mp \frac{2\omega^2}{D[\mu - (2n + 1)|\omega|]} \sqrt{\frac{3(D - 1)}{D - 3}} |F|_n, \qquad (50)$$

where $|F|_n \equiv \sqrt{F_n^2}$ and F_n^2 is determined by Eq.(26).

Invoking our definitions of quantum inflationary phase and of quantum deflationary phase, we observe that one set of dimensions is in a quantum inflationary phase and simultaneously another set of dimensions is in quantum deflationary phase. This situation is described by the following equations:

$$\langle H_i \rangle = \alpha(n) \quad for \quad i = 1, 2, 3;$$
$$\langle H_j \rangle = \tilde{\alpha}(n) \quad for \quad j = 4, ..., D. \qquad (51)$$

According to Eqs.(49) and (50), choosing the three dimensional subspace to be expanding, we get a simultaneous contraction of the extra dimensions. During this quantum "inflation-compactification" process, the expectation value of the "volume" of the universe, V, remains constant determined by Eq.(41).

6   Discussion

The minisuperspace model of quantum cosmology we discussed here demonstrates a very interesting feature which is absent, as far as we know, in all other known quantum cosmology models. Namely, a widespread belief that the cosmic time, which one uses in classical cosmology, disappears in quantum cosmology altogether, seems to be not always right. In the presented model we have seen that quantum mechanical averages of certain cosmological quantities can explicitly depend on the same cosmic time which was used in the appropriate classical cosmological model. Short explanation of the essence of the idea was given in the introduction, and for technical details see Sec.4 and Ref.[8]. Notice that the anisotropy in the evolution of the universe is an essential element which provides this unique feature of our Kaluza-Klein model.

It has been found that quantum effects stabilize the volume of the universe, so that there can be avoidance of the cosmological singularity. The stabilization of volume is consistent with a new quantum effect: existence of a quantum inflationary phase for some dimensions and simultaneous quantum deflationary phase for the remaining dimensions. This effect can be responsible for a visible asymmetry between ordinary and extra dimensions. One can show[9] that the above results also follow if instead of dust we introduce a massive scalar field whose homogeneous degrees of freedom are described quantum mechanically.

References

1. K. Kuchar, in *Proceedings of the 4th Canadian Conference on General Relativity and Relativistic Astrophysics*, ed. G. Kunstatter, D.E. Vincent and J.G. Williams (World Scientific, Singapore, 1992).
2. P.A.M. Dirac, *Lectures on Quantum Mechanics*, (Belfer Graduate School of Science, Yeshiva University, New York, 1964)
3. Lee Smolin, gr-qc/0104097.
4. J.J. Halliwell, gr-qc/0208018.
5. A. Kheyfets and W.A. Miller, *Int. J. Mod. Phys.* A **15**, 4125 (2000).
6. A.P. Gentle, N.D. George, A. Kheyfets and W.A. Miller, gr-qc/0302044.
7. K. Kuchar, *J. Math. Phys.* **24**, 2122 (1983).
8. E.I. Guendelman and A.B. Kaganovich, *Int. J. Mod. Phys.* D **2**, 221 (1993).
9. E.I. Guendelman and A.B. Kaganovich, *Mod. Phys. Lett.* A **9**, 1141 (1994).

SECTION VI
NEW PARADIGMS

String Theory, Space and Time
  C. M. Hull

Quantized Space and Time
  P. Schupp

STRING THEORY, SPACE AND TIME

C. M. HULL

Theoretical Physics Group, The Blackett Laboratory,
Imperial College London, Prince Consort Road, London SW7 2BW, United Kingdom
E-mail: c.hull@imperial.ac.uk

String theory and its implications for the nature of space and time are discussed. Duality symmetries are reviewed, with particular emphasis on the way in which string winding modes and brane wrapping modes can lead to new spatial dimensions. Brane world-volumes wrapping around Lorentzian tori can give rise to extra time dimensions and in this way dualities can change the number of time dimensions as well as the number of space dimensions. This suggests that brane wrapping modes and spacetime momenta should be on an equal footing and M–theory should not be formulated in a spacetime of definite dimension or signature.

1 Introduction

String theory is a candidate for a quantum theory of gravity that gives a unified model of all forces and matter. It has passed many rigorous theoretical tests and has revealed a rich symmetry structure and great mathematical beauty. There is as yet no direct experimental evidence for the theory, but evidence for supersymmetry would be an important sign, and this is to be sought at the LHC. Already it has had many theoretical successes, with important spin-offs leading to dramatic developments in mathematics and in supersymmetric gauge theory. The fact that it is our best candidate for a quantum theory of gravity means that it is worth asking what string theory has to say about the quantum nature of space and time.

Indeed, string theory has some dramatic implications for spacetime. It requires that spacetime have extra hidden dimensions, compactified to a size so small that they remain unseen. In general relativity, configurations related by diffeomorphisms are regarded as physically equivalent (so that spacetime is a manifold) whereas in string theory there are new symmetries, referred to as dualities, that give new relativity principles. In string theory, spacetime is not 'absolute', in the sense that spacetimes which have different geometry, topology and even dimension give rise to the same physics if they are related by string theory symmetries. Moreover, these dualities can mix geometry with matter degrees of freedom, so that the familiar distinction between geometry and matter is lost.

The extra dimensions are usually taken to be spatial dimensions, but it is interesting to ask whether extra hidden timelike dimensions are possible. This is one of the issues that will be addressed here, and it will be argued that there are duality symmetries that can change the signature of spacetime and lead to relations between geometries with different numbers of time dimensions. As a result, the signature is no longer absolute, and other regimes of the theory with different spacetime signatures are possible. Of course, the region of the universe that we live in appears to have only one time, but there could be other regions with other signatures. Some of the problems in the discussion of regimes of more than one time are intimately related to issues that arise in addressing whether quantum physics can make sense with periodic time. If geometry, topology, dimension and even signature can change, the question arises as to what is the invariant structure underlying the theory.

One of the first problems that arises is that we don't yet know what string theory really is. We don't yet know its fundamental formulation and its degrees of freedom. What we do know is how to formulate it in various limiting regimes. For example, we know that at weak coupling, it is described not by a particle theory, but by a theory of relativistic strings moving and vibrating in 10-dimensional spacetime. Each Fourier mode of a vibrating string corresponds to a particle, with mass proportional to the frequency of oscillation. The infinite set of Fourier modes then leads to an infinite tower of massive particles. These include a massless graviton, resulting in a quantum theory of gravity together with a finite set of massless or light particles and an infinite number of massive particles. However, at stronger coupling, the nature of the theory changes. In some circumstances, an extra spatial dimension opens up at strong coupling, whose size is determined by the strength of the coupling, so that the coupling becomes geometrical. The strings become membranes wrapped around the extra dimension, and there are other extended objects, known as branes, which enter into the theory on the same footing as the membranes. However, it appears that it is not a theory of membranes or of branes: the branes are not the fundamental degrees of freedom, and the picture of branes moving in a spacetime background is at best an effective low-energy description.

For Einstein, space and time have no separate meaning, but are unified in spacetime geometry. This is dynamical and describes gravity. However, general relativity breaks down at singularities such as those that arise in black holes and in big bang cosmologies, and new physics must come into play. Quantum effects should be important in such circumstances, perhaps smoothing out the singularities.

We know that the microscopic world is a quantum world, with quantum theory governing all matter and forces. However, in quantum mechanics, time plays a distinguished role, while moving to quantum field theory, which is Lorentz covariant, only works for a very restricted class of theories. In attempting to quantize Einstein's theory, one expects quantum fluctuations in the spacetime geometry, and one can calculate scattering amplitudes for gravitons. In quantum general relativity, these amplitudes are plagued by infinities and quantizing Einstein's theory does not give a well-defined theory, and so one needs a better theory if one is to find a quantum relativistic theory of gravity. In string theory, one obtains well-defined amplitudes, and a well-defined perturbation theory.

The picture of gravitons moving in a fixed background spacetime is not good in regimes in which the fluctuations can overwhelm the background, and the question arises as to what happens at the Planck scale. A number of suggestions have been made. One possibility is that spacetime remains a continuum, but at the Planck scale it has a foam-like structure with a complex topology related to the formation of virtual mini black holes. Another possibility is that spacetime has a discrete structure, perhaps a random lattice with Planck scale spacing between points. There could be a new kind of geometry emerging, such as a a non-commutative geometry. Finally, it could be that there is no spacetime at all, so that geometry is not fundamental at all but emerges from some more primitive structure, such as is found in matrix models. Perturbative string theory is formulated in terms of quanta propagating in a fixed background spacetime geometry and so suffers from the same drawbacks. Understanding the non-perturbative structure of string theory at the fundamental level requires addressing the nature of spacetime at the Planck scale.

2 String Theory, M-Theory and Duality

String theory is defined as a perturbation theory in the string coupling constant g_s, which is valid when g_s is small. The fundamental quanta are the excitations of relativistic strings moving in spacetime and comprise of a finite set of massless particles plus an infinite tower of massive particles with the scale of the mass set by the string tension $T = 1/l_s^2$, expressed in terms of a string length scale l_s. If the spacetime has some circular dimensions, or more generally has some non-contractible loops, the spectrum will also include winding modes in which a closed string winds around a non-contractible loop in spacetime. These have no analogue in local field theories and are responsible for some of the key differences between string theories and field theories. Physical quantities are calculated through a path integral over string histories, which can be calculated perturbatively in g_s using stringy Feynman rules, with string world-sheets of genus n contributing terms proportional to g_s^n. In the supersymmetric string theories, these contributions are believed to be finite at each order in g_s, giving a perturbatively finite quantum theory of gravity unified with other forces.

There are *five* distinct perturbative supersymmetric finite string theories, all in $9 + 1$ dimensions (i.e. nine space and one time), the type I, type IIA and type IIB string theories, and the two heterotic string theories with gauge groups $SO(32)$ and $E_8 \times E_8$. The massless degrees of freedom of each of these theories are governed by a 10-dimensional supergravity theory, which is the low-energy effective field theory. It was a long-standing puzzle as to why there should be five such theories of quantum gravity rather than one, and this has now been resolved. It is now understood that these are all equivalent non-perturbatively and that these distinct perturbation theories arise as different perturbative limits of a single underlying theory. [1,2] We do not have an intrinsic formulation of this underlying non-perturbative theory yet, but the relationships between the string theories has been understood through the discovery of dualities linking them. A central role in the non-perturbative theory is played by the p-branes. These are p-dimensional extended objects, so that a 0-brane is a particle, a 1-brane is a string, a 2-brane is a membrane and so on. In the perturbative superstring theories there is a 1-brane which is the fundamental string providing the perturbative states of the theory, while the other branes arise as solitons or as D-branes, [3] which are branes on which fundamental strings can end. The type II string theories have a fundamental string and a solitonic 5-brane and a set of Dp-branes, where $p = 0, 2, 4, 6, 8$ for the IIA string theory and $p = 1, 3, 5, 7, 9$ for the IIB string theory.

There are duality symmetries of string theories that relate brane degrees of freedom to fundamental quanta, so that all the branes are on the same footing. If some of the spacetime dimensions are wrapped into some compact space K, so that the spacetime is $M \times K$ for some M, then branes can wrap around homology cycles of K and these give extra massive states in the compactified theory on M. For example, a p-brane wrapping around an n-cycle with $n \leq p$ gives a $p - n$ brane in the compactified theory. These brane wrapping modes generalise the string winding modes and are related to the perturbative states by U-dualities, [1] and play an important role in the duality symmetries, as we shall see.

One of the best-understood dualities is T-duality, [4] which relates string theory on a spacetime $S^1 \times M$ with a circular dimension of radius R to a string theory on $\tilde{S}^1 \times M$ where the circular dimension is now of radius

$$\tilde{R} = \frac{l_s^2}{R} \tag{1}$$

so that the radii $R, \tilde{R}$ are inversely proportional. For bosonic and heterotic string theories, T-duality is a self-duality, so that heterotic (bosonic) string theory on a large circle is equivalent to heterotic (bosonic) string theory on a small circle, while it maps the type IIA string theory to the type IIB theory, with the result that type IIA string theory on a large circle is equivalent to type IIB string theory on a small circle. [5,6] T-duality relates perturbative states to perturbative states, as does mirror symmetry which relates a superstring theory compactified on a Calabi-Yau manifold K to a superstring theory compactified on a topologically distinct Calabi-Yau manifold, the mirror $\tilde{K}$ of K.

There are also non-perturbative dualities. For example the type IIA string theory compactified on $K3$ is equivalent to the heterotic string theory compactified on the 4-torus, [1] T^4 while the type I theory with string coupling g_s is equivalent to the $SO(32)$ heterotic string theory[2,7,8,9] with string coupling $\tilde{g}_s = 1/g_s$. This is an example of a strong-weak coupling duality relating the strong-coupling regime of one theory to the weak-coupling regime of another. Such dualities are important as they allow the description of strong-coupling physics in terms of a weakly-coupled dual theory.

M–theory arises as the strong-coupling limit of the IIA string theory. [2] The IIA string is interpreted as an 11-dimensional theory compactified on a circle of radius $R = l_s g_s$. Then at strong coupling, the extra dimension decompactifies to give a theory in 11 dimensions which has 11 dimensional supergravity as a low-energy limit. We will refer to this 10+1 dimensional theory as M–theory. Duality transformations relate this to each of the five string theories, and the string theories and M–theory can all be thought of as arising as different limits of a single underlying theory. The IIA string theory is obtained by compactifying M–theory on a circle, the IIB string is obtained from the IIA by T-duality or directly from compactifying M–theory on a 2-torus and taking the limit in which it shrinks of zero size, [10] the $E_8 \times E_8$ heterotic string is obtained by modding out M–theory on a circle by a Z_2 symmetry or equivalently from compactifying M–theory on a line interval, [11] the type I theory is obtained from the IIB string by orientifolding (modding out by world-sheet parity) [5,12] and the $SO(32)$ heterotic string is the strong coupling limit of this. [2] The type I theory and the $SO(32)$ heterotic string (as well as the type I' string) can be obtained directly from M–theory compactified on a cylinder, while the massive IIA string theory is obtained from a limit of M–theory compactified on a T^2 bundle over a circle. [13]

In D-dimensional general relativity or supergravity, a spacetime with a large circle S^1 is physically distinct from one with a small circle $\tilde{S}^1$, and a spacetime $M \times K$ is physically distinct from the mirror spacetime $M \times \tilde{K}$, but in string theory these dual pairs of spacetimes define the same string theory and so define the same physics. The heterotic string on $M \times T^4$ is equivalent to the type IIA string on $M \times K3$, even though T^4 and $K3$ are very different spaces with different properties (e.g. they have different topologies and different curvatures) and there is no invariant answer to the question: what is the spacetime manifold? In the same way that spacetimes related by diffeomorphisms are regarded as equivalent, so too must spacetimes related by dualities, and the concept of spacetime manifold should be replaced by duality equivalence classes of spacetimes (or, more generally, duality equivalence classes of string or M–theory solutions).

In the usual picture, the five superstring theories and the 11-dimensional theory arising as the strong coupling limit of the IIA string (referred to as M–theory here) are depicted as being different corners of the moduli space of the mysterious fundamental theory underpinning all of these theories (sometimes also referred to as M–theory, although we shall resist this usage here). More precisely, compactifying string theory or M–theory gives a theory depending on the moduli of metrics and antisymmetric tensor gauge fields on the compactification space. Each modulus gives rise to a scalar field in the compactified theory and the expectation value of any of the scalar fields can be used to define a coupling constant. One can then examine the perturbation theory in that constant. For some choices it will give a field theory, for others it will give a perturbative string theory and different perturbative string theories will correspond to different choices of coupling. [14] The string theories and M–theory are each linked to each other by chains of dualities and so there is only one basic theory.

Other 'corners' corresponding to particular limits of the theory have been understood to correspond to field theories without gravity. For example the IIB string theory in the background given by the product of 5-dimensional anti-de Sitter space and a 5-sphere is equivalent to $N = 4$ supersymmetric Yang-Mills theory in four dimensions, with similar results for theories in other anti-de Sitter backgrounds, [15] and certain null compactifications are equivalent to matrix models. [16]

Many dualities have now been found which can relate theories with different gauge groups, different spacetime dimensions, different spacetime geometries and topologies, different amounts of supersymmetry, and even relate theories of gravity to gauge theories. Thus many of the concepts that had been thought absolute are now understood as relative: they depend on the 'frame of reference' used, where the concept of frame of reference is generalised to include the values of the various coupling constants. For example, the description of a given system when a certain coupling is weak can be very different from the description at strong coupling, and the two regimes can have different spacetime dimension, for example. However, in all this, one thing that has remained unchanged is the number of time dimensions; all the theories considered are formulated in a Lorentzian signature with one time coordinate, although the number of spatial dimensions can change. Remarkably, it turns out that dualities can change the number of time dimensions as well, giving rise to exotic spacetime signatures. [17] The resulting picture is that there should be some underlying fundamental theory and that different spacetime signatures as well as different dimensions can arise in various limits. The new theories are different real forms of the complexification of the original M–theory and type II string theories, perhaps suggesting an underlying complex nature of spacetime.

We will now proceed to examine some of these dualities in more detail, and in particular to focus on the way in which extra spacetime dimensions can emerge from brane wrapping modes.

3   Branes and Extra Dimensions

3.1   Compactification on S^1

For a *field theory* compactified from D dimensions on a circle S_R^1 of radius R, the momentum p in the circular dimension will be quantised with $p = n/R$ for some integer n. In the limit $R \to 0$ this becomes divergent, so that finite-momentum states must move in the remaining $D - 1$ dimensions and are described by the dimensionally reduced theory in $D-1$ dimensions. For finite R, the states carrying internal momentum can be interpreted as states in $D - 1$ dimensions with mass (taking the D-dimensional field theory to be massless for simplicity)

$$M = |n|/R \tag{2}$$

The set of all such states for all n gives the 'Kaluza-Klein tower' of massive states arising from the compactification. If the original field theory includes gravity, there will be an infinite tower of massive gravitons, and if the theory is supersymmetric, then the tower fits into supersymmetry representations. In the limit $R \to 0$, the masses of all the states in these towers become infinite and they decouple, leaving the massless dimensionally reduced theory in $D - 1$ dimensions. On the other hand, taking the decompactification limit $R \to \infty$, all the states in the tower become massless and combine with the massless $D - 1$ dimensional fields to form the massless fields in D dimensions. Such a tower becoming massless is often a signal of the decompactification of an extra dimension.

For a *string theory* the situation is very different, due to the presence of string *winding modes* which become light as the circle shrinks. A string can wind m times around the circular dimension, and the corresponding state in the $D-1$ dimensional theory will have mass mRT where T is the string tension (into which a factor of 2π has been absorbed). The set of all such states for all m forms a tower of massive states and in the limit $R \to 0$ these become massless, so that there is an infinite tower of states becoming massless (and fitting into supergravity multiplets, in the case of the superstring). This signals the opening up of a new circular dimension of radius

$$\tilde{R} = 1/TR \tag{3}$$

with the string winding mode around the original circle of mass

$$M = mRT = m/\tilde{R} \tag{4}$$

reinterpreted as a momentum mode in the dual circle of radius $\tilde{R}$. Similarly, the momentum modes on the original circle ($M = n/R$) can now be interpreted as string winding modes around the dual circle ($M = nT\tilde{R}$), and the new theory in D dimensions is again a string theory. A state with momentum n/R and winding number m will have mass

$$M = \frac{n}{R} + mRT = \frac{n}{R} + \frac{m}{\tilde{R}} \tag{5}$$

and this is clearly invariant under the T-duality transformation $m \leftrightarrow n$, $R \leftrightarrow \tilde{R}$ interchanging the momentum and winding numbers and inverting the radius. Then the original string theory on $M_{D-1} \times S^1$ (with M_{D-1} some $D-1$-dimensional spacetime) is equivalent to a string theory on $M_{D-1} \times \tilde{S}^1$ where $\tilde{S}^1$ has radius $\tilde{R}$, with the momentum modes of one theory corresponding to the winding modes of the other, and this equivalence is known as a T-duality. In the limit $R \to 0, \tilde{R} \to \infty$, the decompactified T-dual theory has full D dimensional Lorentz invariance. If the first string theory is a bosonic (heterotic) string, so is the second, while if one is a type IIA string theory, the other is a type IIB string theory. Then the type IIA string theory compactified on a circle of radius R is equivalent to the type IIB string theory compactified on a circle of radius $\tilde{R}$. For 10-dimensional supergravity compactified on a circle of radius R, taking $R \to 0$ will give a 9-dimensional supergravity theory while for a string theory a new dimension opens up to replace the one of radius R that has shrunk to zero size. If R is much larger than l_s, the desciption in terms of string theory on S^1 is useful, while for $R << l_s$, the T-dual description in terms of string theory on $\tilde{S}^1$ is more appropriate, with the light states having a conventional description in terms of momentum modes on $\tilde{S}^1$ instead of winding modes on S^1.

3.2 IIA String Theory and M-Theory

The type IIA string theory in 9+1 dimensions has D0-brane states, which are particle-like non-perturbative BPS states, with quantized charge n (for integers n) and mass

$$M \sim \frac{|n|}{g_s l_s} \tag{6}$$

The state of charge n can be thought of as composed of n elementary D0-branes. In the strong coupling limit $g_s \to \infty$, these states all become massless. Moreover, the D0-brane states for a given n fit into a short massive supergravity multiplet with spins ranging from zero to two and so at strong coupling there is an infinite number of gravitons becoming massless. It was proposed by Witten [2] that this tower of massless states should be interpreted as a Kaluza-Klein tower for an extra circular dimension of radius

$$R_M = g_s l_s \tag{7}$$

Then the strong coupling limit of the IIA string theory is interpreted as the limit in which $R_M \to \infty$ so that the extra dimension decompactifies to give a theory in 10+1 dimensions, and this is M–theory. Moreover, for the IIA string theory in $D = 10$ Minkowski space, the strong coupling limit is invariant under the full 11-dimensional Lorentz group and the effective field theory describing the massless degrees of freedom of M–theory is 11-dimensional supergravity. The radius can be rewritten in terms of the 11-dimensional Planck length l_p as

$$R_M = g_s^{2/3} l_p \tag{8}$$

The IIA string theory is really only defined perturbatively for very small coupling g_s. It can now be 'defined' at finite coupling g_s as M–theory compactified on a circle of radius R_M, so that the problem is transferred to the one of defining M–theory. However, at low energies we see that the non-perturbative IIA theory is described by 11-dimensional supergravity compactified on a circle, and this leads to important non-perturbative predictions, so that this viewpoint can be useful even though we still know rather little about M–theory.

The IIA string has Dp-branes for all even p, while M–theory has a 2-brane or membrane and a 5-brane. All the branes of the IIA string theory have an M–theory origin. For example, an M–theory membrane will give the fundamental string of the IIA theory if it wraps around the circular dimension and the D2-brane if it does not.

3.3 Compactification on T^2

For a D dimensional field theory compactified on a 2-torus there will be momentum modes with masses

$$M \sim \sqrt{\frac{p^2}{R_1^2} + \frac{q^2}{R_2^2}} \tag{9}$$

where R_1, R_2 are the radii of the circular dimensions, p, q are integers and for simplicity we take the torus to be rectangular. These will decouple in the limit $R_1, R_2 \to 0$ leaving a theory in $D - 2$ dimensions. For example, for 11-dimensional supergravity, this limit will give the dimensionally reduced 9-dimensional maximal supergravity theory.

We now compare this with M–theory compactified on T^2. Consider first the circle of radius R_2, say. M–theory compactified on this circle is equivalent to the IIA string theory with coupling constant $g_s = (R_2/l_p)^{3/2}$, and so the limit $R_2 \to 0$ is the weak coupling limit of this IIA string theory. We now have the IIA string theory compactified on a circle of radius R_1, and by T-duality this is equivalent to the IIB string theory compactified on a circle of radius $\tilde{R}_1 = 1/T R_1$. Taking the limit $R_1 \to 0$ is then the limit in which $\tilde{R}_1 \to \infty$ and an extra circle opens up to give the IIB string theory in 9+1 dimensions. The IIA string winding modes provide the tower of states that become massless in the limit and which are re-interpreted as momentum modes on the circle of radius $\tilde{R}_1$. Moreover, these IIA string winding modes are M–theory membranes wrapped around the 2-torus. These membrane wrapping modes have mass

$$M \sim |n| T_2 R_1 R_2 \tag{10}$$

where the membrane tension is $T_2 = 1/l_p^3$.

Then we have the result [10] M–theory compactified on a general 2-torus of area A and modulus τ is equivalent to the IIB string theory compactified on a circle of radius

$$R_B = \frac{l_p^3}{A} \tag{11}$$

with string coupling g_s and axionic coupling θ (defined as the expectation value of the scalar field in the Ramond-Ramond sector) given by

$$\tau = \theta + i\frac{1}{g_s} \tag{12}$$

The states of the IIB string carrying momentum in the circular dimension arise from membranes wrapping the 2-torus while the (p,q) string of the IIB theory winding round the circular dimension (with fundamental string charge p and D-string charge q) arises from M–theory states carrying momentum p/R_1 and q/R_2 in the compact dimensions. Then in the limit $A \to 0$, we lose two of the dimensions, as in the field theory, leaving a theory in 8+1 dimensions, but a new spatial dimension opens up to give a theory in 9+1 dimensions.

3.4   Compactification on T^3

For 11-dimensional supergravity compactified on a 3-torus, the limit in which the radii R_1, R_2, R_3 all tend to zero gives the dimensional reduction to the maximal supergravity in 8 dimensions. For M–theory on T^3, membranes can wrap any of the three 2-cycles of T^3. From the last section, we know that if R_1 and R_2 both shrink to zero while R_3 stays fixed, an extra dimension opens up with radius $\tilde{R}_3 = l_p^3/R_1 R_2$ and the tower of membrane wrapping states is reinterpreted as a Kaluza-Klein tower for this extra dimension. The same picture applies to each of the three 2-cycles, and so if all 3 radii shrink, there are three extra dimensions opening up, with radii $\tilde{R}_i$ given by

$$\tilde{R}_i = \frac{l_p^3}{R_j R_k}, \qquad i \neq j \neq k \tag{13}$$

Then when the original three torus shrinks to zero size, three dimensions are lost but three new ones emerge, so we are again back in 11 dimensions and the 11-dimensional theory is again M–theory. [18] Thus M–theory compactified on a dual T^3 with radii R_1, R_2, R_3 is equivalent to M–theory compactified on the dual T^3 with radii $\tilde{R}_1, \tilde{R}_2, \tilde{R}_3$ given by (13).

3.5   Compactification on T^4

It is tempting to apply these arguments to higher tori. For example, T^4 has six 2-cycles, and membranes can wrap any of them. Compactifying $D = 11$ supergravity on a T^4 and taking all four radii $R_i \to 0$ gives a $D = 7$ field theory, but for M–theory there are six towers of states becoming massless in the limit arising from membranes wrapping each of the six shrinking 2-cycles. If each of these towers is interpreted as a Kaluza-Klein tower, this would give 6 extra dimensions in addition to the 7 original dimensions remaining, giving a total of 13 dimensions. However, there is no conventional supersymmetric theory in 13 dimensions, so it is difficult to see how such a theory could emerge. In fact the situation here is more complicated, and the 6 towers have a different interpretation here. The difference here is that there is also a string in the compactified theory arising from the M–theory 5-brane wrapped around the T^4 which becomes 'light' at the same time as the 6 towers of membrane wrapping modes. [14] It turns out that M–theory on T^4 is dual to IIB string theory on T^3. The M–theory 5-brane wrapped around the T^4 gives the fundamental string of the IIB theory moving in 7 dimensions. [14] Compactifying the IIB string on T^3 gives in addition three momentum modes and three winding modes, fitting into a **6** of the T-duality group $SO(3,3)$, and these correspond to the 6 towers of membrane wrapping modes, which themselves transform as a **6** of the torus group $SL(4) \sim SO(3,3)$. Thus only 3 of the 6 towers can be interpreted as momentum modes for an extra dimension, the other three being interpreted as string winding modes, and the spacetime dimension of the dual theory is 10, not 13. Note that there is no invariant way of choosing which three of the six correspond to spacetime dimensions, as T-duality transformations will relate momentum and winding modes and change one subset of three to another. In this case, taking the limit in which the T^4 on which the M–theory is compactified shrinks to zero size does not correspond to a decompactification limit of the dual theory, but to the weak coupling limit in which the coupling constant g_s of the compactified IIB string theory tends to zero. [14]

4   Branes and Space and Time

We have seen that wrapped branes are associated with towers of massive states and that in some cases these can be interpreted as Kaluza-Klein towers for extra dimensions. In a limit in which such a tower becomes massless (e.g. $R_i \to 0$ for toroidal compactifications, or $g_s \to \infty$ for the IIA D0-branes), the corresponding dimension decompactifies and new dimensions unfold. The presence of an enlarged Lorentz symmetry puts the new braney dimensions on an equal footing with the other dimensions, and the full theory includes gravity in the enlarged space. The number of dimensions lost in the limit is not always the same as the number of extra dimensions, so that the total number of spacetime dimensions can change (as in the relations between 11-dimensional M–theory and 10-dimensional type II string theories considered in sections 3.2 and 3.3). We have also seen that, as in the case of M–theory on T^4, the towers of wrapped brane states cannot always be interpreted in terms of extra dimensions, and it is necessary to perform a more complete analysis to see what is going on.

In all of the above cases, branes were wrapped around spacelike cycles and the extra dimensions that arose were all spacelike. A brane world-volume can also wrap around timelike cycles, and we will see that in such cases the extra dimensions can be timelike, so that the signature of spacetime can change.

It is natural to ask whether it makes sense to consider compact time. There are many classical solutions of gravity, supergravity, string and M–theories with compact time and it is of interest to investigate their properties. Compact time does not appear to be a feature of our universe, but almost all spacetimes that are studied are also unrealistic.The presence of closed timelike loops means that the physics in such spaces is unusual, but it has often been fruitful in the past to study solutions that have little in common with the real world. An important issue with these solutions (as with many others) is whether a consistent quantum theory can be formulated in such backgrounds. If time were compact but with a huge period, it is not clear how that would manifest itself.

With a compact time, it is straightforward to solve classical field equations, imposing periodic boundary conditions in time instead of developing Cauchy data. Can quantum theory make sense with compact time? There is no problem in solving Schrödinger or wave equations with periodic boundary conditions, but it is difficult to formulate any concept of measurement or collapse of a wave-function, as these would be inconsistent with periodic time: if a superposition of states collapsed to an eigenstate of an observable in some measurement process, it must already have been in that eigenstate from the last time it was measured. In string theory, it is straightforward to study the solutions of the physical state conditions, but there are new issues that arise from string world-sheets (and brane world-volumes) wrapping around the compact time. It has proved very fruitful to consider such compactifications in string theory. For example, the compactification of all 25+1 dimensions in bosonic string theory on a special Lorentzian torus played a central role in the work of Borcherds on the construction of vertex algebras and their application to the monster group. [19]

5 Compactification on Lorentzian Tori and Signature Change

5.1 Compactification on a Timelike Circle

Consider spacetimes of the form $M_{D-1} \times S^1$ where S^1 is a timelike circle of radius R and M_{D-1} is a Riemannian space. The time component of momentum is quantized

$$p^0 = \frac{n}{R} \tag{14}$$

and in the limit $R \to 0$, only the states with $p^0 = 0$ survive. For a field theory, the result is a dimensional reduction to a Euclidean field theory in $D - 1$ dimensions, on M_{D-1}. For example, dimensionally reducing $D = 11$ supergravity on a timelike circle gives a supergravity theory in 10 Euclidean dimensions, denoted the IIA_E supergravity theory. [17] Timelike reductions of supergravity theories have been considered by a number of authors. [20,21,22] The field theory resulting from such a timelike reduction will in general have fields whose kinetic terms have the wrong sign. For example, the D dimensional graviton will give a graviton, a scalar and a vector field in $D - 1$ dimensions on reducing on a circle, and if the circle is timelike, then the vector field will have a kinetic term of the wrong sign. Then the action for the physical matter fields of the reduced theory in $D - 1$ Euclidean dimensions will not be positive. This apparent problem is the result of the truncation to $p^0 = 0$ states. If this truncation is not made and if the full Kaluza-Klein towers of states with $p^0 = n/R$ for all n are kept, then the theory is the full unitary D dimensional theory on a particular background, and the D dimensional gauge invariance can be used to choose a physical gauge locally with a positive action and states with positive norm. In general such a gauge choice cannot be made globally and there will be zero-mode states for which the action will not be positive. For example, the states with $p^0 = 0$ are governed by the non-positive dimensionally reduced action in $D-1$ dimensions. For a Yang-Mills theory reduced on a timelike circle, the time component of the vector potential A_0 gives a scalar field in $D-1$ dimensions with a kinetic term of the wrong sign. For the full D dimensional theory compactified on the timelike circle, the negative-norm A_0 can be brought to a constant by D-dimensional gauge transformations, but one cannot gauge away the degrees of freedom associated with Wilson lines winding around the compact time dimension. (The fields with kinetic terms of the wrong sign can be handled in the path integral in the same way as the negative-action gravitational conformal mode is sometimes dealt with, namely by analytic continuation so that the offending field becomes imaginary. [24]

In a string theory, however, there will be winding modes in which the 1+1 dimensional string world-sheet winds around the compact time dimension, giving a spacelike 'world-line' in the compactified theory in $D-1$ dimensions. As in the spacelike case, as $R \to 0$, a dual circle opens up with radius $\tilde{R} = 1/TR$, and the new circle is again timelike. The winding number becomes the p^0 of the dual theory, and in this way a superstring theory in 9+1 dimensions compactified on a timelike circle of radius R is T-dual to a superstring theory in 9+1 dimensions compactified on a timelike circle of radius $\tilde{R}$. Such timelike T-dualities were considered for the bosonic and heterotic strings by Moore [23], and they take the bosonic string theory to the bosonic string theory and the heterotic string theory to the heterotic string theory. However, for type II theories there is a surprise. It is straightforward to see that timelike T-duality cannot take the IIA string theory to either the IIB string or the IIA string, [24] but must take it to a 'new' theory, denoted the IIB^* string theory. [24] Similarly, timelike T-duality takes the IIB string to a IIA^* string theory. [24]

The IIA^* and IIB^* strings are taken into each other by T-duality on a spacelike circle, and the IIA^* (IIB^*) theory is obtained from the IIA (IIB) string theory by acting with $(i)^{F_L}$ [24] (where F_L is the left-handed fermion number). The supergravity limits of the IIA^* and IIB^* have non-positive actions for the matter fields (the kinetic terms for the fields in the R-NS and R-R sectors have the wrong sign) so that the low-energy field theories are non-unitary, but the IIA^* and IIB^* string theories compactified on a timelike circle are equivalent to the IIA and IIB string theories on the dual timelike circle. Then, at least when on a timelike circle, the IIA^* and IIB^* string theories are precisely the timelike compactifications of the usual IIA and IIB string theories, albeit written in dual variables. The supergravity limit for the IIA or IIB variables is the conventional one, while the supergravity limit for the dual variables is non-unitary. A physical gauge can then be chosen locally for the IIA^* and IIB^* string theories on a timelike circle, and any lack of unitarity or positivity is due to zero-modes.

If time is compact and the physics is periodic in time, the requirements for a sensible theory are not the same as in Minkowski space. A theory that is unstable in Minkowski space (perhaps due to negative energy configurations) need not be pathological if time is compact: the periodic boundary conditions forbid any runaway solutions and the system will always return to its starting point after a period. A nonunitary theory in Minkowski space will not conserve probability, but with periodic time, any probability that is lost will always come back, as the solutions of the wave equations are required to be periodic. This suggests that the timelike compactifications of the IIA^* and IIB^* string theories should be consistent, although the question remains as to the status of the decompactification limit in which the radius of the timelike circle becomes infinite. Similar considerations will apply to the other new theories [17] described in this section.

5.2 Compactification on $T^{1,1}$

Consider now compactification on the Lorentzian torus $T^{1,1}$ with one spacelike circle and one timelike one. (We will use the notation $T^{s,t}$ for a torus with s spacelike circles and t timelike ones.) For 11-dimensional supergravity, the limit $R_s, R_t \to 0$ gives a 9-dimensional Euclidean supergravity theory. For M–theory on a Euclidean torus T^2, we saw in section 2.3 that in the limit in which the torus shrank to zero size, one new spacelike dimension opened up to give the IIB string theory in 9+1 dimensions. Here we expect something similar to happen. Considering first the compactification on the spacelike circle of radius R_s, when R_s is small we obtain the IIA string theory with coupling constant $g_s = (R/l_p)^{3/2}$. The compactification of this on a timelike circle of radius R_t is T-dual to the IIB^* string theory compactified on a timelike circle of radius

$$\tilde{R}_t = \frac{1}{TR_t} \tag{15}$$

Then taking the limit $R_t \to 0$, we obtain a theory in the expected 9 spacelike dimensions together with a new time dimension which opens up, the T-dual of the original timelike dimension. The membranes wrapping around $T^{1,1}$ have become the modes carrying the time component of momentum p^0 of the dual IIB^* theory, and so wee obtain the result [17] that M–theory compactified on $T^{1,1}$ with radii R_s, R_t is dual to the IIB^* string theory compactified on a timelike circle of radius $l_p^3/R_s R_t$.

5.3 Compactification on $T^{2,1}$

We have seen in section 2.3 that M–theory compactified on a Euclidean 2-torus T^2 gains a new spatial dimension in the limit in which the 2-torus shrinks to zero size, replacing the two which have disappeared, so that the original theory in (10,1) dimensions becomes a theory in (9,1) dimensions: $(9,1) = (10,1) - (2,0) + (1,0)$. Similarly, we have seen in section 4.1 that M–theory compactified on a Lorentzian 2-torus $T^{1,1}$ gains a new time dimension in the limit in which the 2-torus shrinks to zero size, replacing the $(1,1)$ dimensions which have disappeared so that the original theory in (10,1) dimensions again becomes a theory in (9,1) dimensions: $(9,1) = (10,1) - (1,1) + (0,1)$. Thus a shrinking T^2 is associated with an extra space dimension while a shrinking $T^{1,1}$ is associated with an extra time dimension.

For M–theory on a shrinking Euclidean T^3, an extra space dimension emerges for each of the three shrinking 2-cycles, so that the three toroidal dimensions which are lost are replaced by three new spatial dimensions, and we end up back in M–theory in (10,1) dimensions: $(10,1) = (10,1) - 3 \times (1,0) + 3 \times (1,0)$.

Consider now the compactification on a Lorentzian 3-torus $T^{2,1}$ with two space-like and one timelike circles. In the limit in which the torus shrinks to zero size, 2+1 dimensions are lost leaving 8 Euclidean dimensions and reducing 11-dimensional supergravity on $T^{2,1}$ indeed gives a supergravity in (8,0) dimensions. In M–theory, if the discussion above applies here, we expect an extra space dimension for every shrinking T^2 and an extra time dimension for every shrinking $T^{1,1}$. The torus $T^{2,1}$ has two Lorentzian 2-cycles and one Euclidean one, so that this suggests there should be an extra two time dimensions and one space dimension that open up in this limit, giving a theory in 11 dimensions with two-timing signature $(9,2) = (8,0) + (1,0) + 2 \times (0,1)$. If all the towers of wrapped membranes give extra dimensions, this must be the result, but we have seen that in some cases towers of wrapped brane states can have other meanings. A more careful analysis shows that this interpretation is indeed correct and taking M–theory on a shrinking $T^{2,1}$ gives a new theory in 9+2 dimensions. [17]

Then dualities can change the number of time dimensions as well as the number of space dimensions. This new theory in 9+2 dimensionsis the M^* theory, [17] and it has an effective field theory which is a new supergravity theory in 9+2 dimensions. M–theory compactified on $T^{2,1}$ is equivalent to M^* theory compactified on a two-time torus $T^{1,2}$, with the sizes of the circles related by a formula similar to (13).

5.4   Compactifications of M^* Theory

We can now investigate the compactifications of M^* theory on various tori. [17] Compactifying the M^* theory on a timelike circle gives the IIA^* string theory in 9+1 dimensions, while compactifying on a spacelike circle gives a new IIA-like string theory in 8+2 dimensions. Next consider the compactification on 2-tori in the limit in which they shrink to zero size. For $T^{0,2}$ this gives the IIB string (compactification on the first circle gives the IIA^* theory and the second then gives its T-dual on a timelike circle), for $T^{1,1}$ it gives the IIB^* theory and for $T^{2,0}$ it gives a new IIB-like theory in 7+3 dimensions. Thus a shrinking $T^{0,2}$ gives an extra time dimension, a shrinking $T^{1,1}$ gives an extra space dimension and a shrinking $T^{2,0}$ gives an extra time dimension. This can now be used to find the results of compactification on a shrinking three-torus. For $T^{1,2}$ there are two $T^{1,1}$ cycles and one Euclidean T^2 cycle giving a theory in $(9,2) - (1,2) + 2 \times (1,0) + (0,1) = (10,1)$ dimensions and we are back in M–theory, for $T^{2,1}$ there are two T^2 cycles and one $T^{1,1}$ cycle giving a theory in $(9,2) - (1,2) + 2 \times (0,1) + (1,0) = (9,2)$ dimensions and we are back in M^* theory, while for $T^{3,0}$ there are three Euclidean T^2 cycles giving a theory in $(9,2) - (3,0) + 3 \times (0,1) = (6,5)$ dimensions, giving a new theory in 6+5 dimensions. This theory is the M' theory, [17] and M^* theory compactified on $T^{3,0}$ is equivalent to the M' theory compactified on a dual $T^{0,3}$.

The above analysis can then be repeated for this new M' theory, and it turns out that only 11-dimensional theories that arise are the M, M^* and M' theories, with signatures (10,1), (9,2) and (6,5), together with the mirror theories in signatures (1,10), (2,9) and (5,6). Reduction on circles gives IIA-like theories in signatures

10+0, 9+1, 8+2, 6+4 and 5+5 while reducing on 2-tori gives IIB-like theories in signatures 9+1, 7+3, and 5+5. (There are of course also mirror string theories in the signatures 1+9, 2+8 etc with space and time interchanged.)

In each of these 10 and 11 dimensional cases there is a corresponding supergravity limit and it is a non-trivial result that these supergravities exist, and it is unlikely that there are maximal supergravities in signatures outside this list. These theories are linked to each other by an intricate web of dualities, [17] some of which have been outlined above, and in particular all are linked by dualities to M–theory.

Each of these theories has a set of branes of various world-volume signatures. [25,26] For the M–type theories, M–theory has branes of world-volume signature 2+1 and 5+1 (the usual M2 and M5 branes), M^* theory has branes of world-volume signature 3+0,1+2 and 5+1 while M' theory has branes of world-volume signature 2+1,0+3, 5+1, 3+3 and 1+5.

6 Discussion

In a field theory, compactification and then shrinking the internal space K to zero size gives a dimensionally reduced field theory in lower dimensions. In compactified string theory or M–theory, however, new dimensions can emerge when the internal space shrinks, with the Kaluza-Klein towers for the new dimensions corresponding to the brane wrapping modes in which branes wrap around cycles of K. In some cases (e.g. toroidal compactifications of string theory or M–theory on T^3) the number of new dimensions equals the number that are lost and one regains the original spacetime dimension, while in others (such as M–theory compactified on T^2) the number of new dimensions is different from the number that are lost and so the dimension of spacetime changes (for M–theory on T^2 it changes from 11 to 10).

Clearly, the notion of what is a spacetime dimension is not an invariant concept, but depends on the 'frame of reference', in the sense that it will depend on the values of various moduli. A given tower of BPS states could have a natural interpretation as a Kaluza-Klein tower associated with momentum in a particular compact spacetime dimension for one set of parameters, but could have an interpretation as a tower of brane wrapping modes for other values, and we have seen many examples of this in the preceding sections. We are used to considering field theories in spacetimes of given dimension and signature, but any attempt to formulate M–theory or string theory as a theory in a given spacetime dimension or signature will be misleading. In particular, the theory underpinning all these theories has a limit which behaves like a theory in 10+1 dimensions with a supergravity limit and systematic corrections, but cannot at the fundamental level be a theory in 10+1 dimensions, as it has some limits which live 9+1 dimensions and others that live in 9+2 or 6+5 dimensions.

The supersymmetry algebra in 10+1 dimensions is

$$\{Q,Q\} = C(\Gamma^M P_M - \frac{1}{2!}\Gamma^{M_1 M_2} Z_{M_1 M_2} - \frac{1}{5!}\Gamma^{M_1 \ldots M_5} Z_{M_1 \ldots M_5}) , \qquad (16)$$

where C is the charge conjugation matrix, P_M is the energy-momentum 11-vector and $Z_{M_1 M_2}$ and $Z_{M_1 \ldots M_5}$ are 2-form and 5-form charges, associated with brane charges. [28] There are 11+55+462=528 charges on the right-hand-side, which can be assembled into a symmetric bi-spinor $X_{\alpha\beta}$. Compactifying and then dualising, one finds that some of the brane charges become momenta of the dual theory and some of the momenta become brane charges of the dual theory, so that the split of the bi-spinor X into an 11-momentum and brane charges changes under duality.

This suggests that rather than trying to formulate the theory in 10+1 dimensions, all 528 charges should be treated in the same way. There seem to be at least two ways in which this might be done. The first would be a geometrical one in which all 528 charges were treated as momenta and there is an underlying spacetime of perhaps 528 dimensions. The duality symmetries could then act geometrically, and there would be perhaps some dynamical way of choosing 11 of the dimensions as the preferred ones, e.g. through the 'world' being an 11-dimensional surface in this space. For example, in considering T-duality between a string theory on a space $M \times S^1$ and one in the dual space $M \times \tilde{S}^1$, it is sometimes useful to consider models on $M \times S^1 \times \tilde{S}^1$ in which both the circle of radius R and the dual circle of radius $\tilde{R}$ are present, with different projections giving the two T-dual models. [27]

We have seen that different spacetimes related by dualities can define the same physics, so that the notion of spacetime geometry cannot be fundamental. This suggests that different degrees of freedom should be used, with spacetime emerging as a derived concept. An alternative 'anti-geometrical' formulation would be one in which none of the charges were geometrical, but instead an algebraic approach similar to that of matrix theory was used. For example, M–theory could be compactified to 0,1 or 2 dimensions to give a theory that would be expected to have duality symmetry [29] E_{11}, E_{10} or E_9 where E_9 is an affine E_8, E_{10} is a hyperbolic algebra discussed, for example, in [30] and E_{11} might be some huge algebraic structure associated with the E_{11} Dynkin diagram. In one dimension the theory might be some matrix quantum mechanics associated with E_{10} while in zero dimensions it would be some form of non-dynamical matrix theory. At special points in the moduli space, some of the charges would be associated with extra dimensions that are decompactifying. At different points, different numbers of space and time dimensions emerge, and the duality structure discussed above can be understood as a consequence of the E_{10} or E_{11} symmetries of the compactified theory.[31][32] It has been suggested that E_{10} or E_{11} could be symmetries of the full M-theory, [33][34][35] and if this is the case, the signature-changing dualities would be an important part of this structure.

The five superstring theories and M–theory are different corners of the moduli space of some as yet unknown fundamental theory and the dualities linking them all involve compactification on Riemannian spaces. If this is extended to include compactification on spaces with Lorentzian signature a richer structure emerges. The strong coupling limit of the type IIA superstring is M–theory in 10+1 dimensions whose low energy limit is 11-dimensional supergravity theory. The type I, type II and heterotic superstring theories and certain supersymmetric gauge theories emerge as different limits of M–theory. The M–theory in 10+1 dimensions is linked via dualities to M^* theory in 9+2 dimensions and M'-theory in 6+5 dimensions. Various limits of these give rise to IIA-like string theories in 10+0, 9+1,8+2,6+4 and 5+5 dimensions, and to IIB-like string theories in 9+1,7+3, and 5+5 dimensions. The field theory limits are supergravity theories with 32 supersymmetries in 10 and 11 dimensions with these signatures, many of which are new. Further dualities similar to those of Maldacena [15] relate these to supersymmetric gauge theories in various signatures and dimensions, such as 2+2, 3+1 and 4+0. These new string theories and M–type theories in various spacetime signatures can all be thought of as providing extra corners of the moduli space. Some corners are stranger than others, but in any case we can only live in one corner (perhaps M–theory compactified on the product of a line interval and a Calabi-Yau 3-fold) and there is no reason why other corners might not have quite unfamiliar properties.

Theories in non-Lorentzian signatures usually have many problems, such as lack of unitarity and instability. However, the theories considered here are related to M–theory via dualities and so are just the usual theory expressed in terms of unusual variables. For example, the M^* theory in 9+2 dimensions compactified on $T^{1,2}$ is equivalent to M–theory compactified on $T^{2,1}$, and so the compactified M^* theory will make sense provided M–theory compactified on a Lorentzian torus is a consistent theory. Then the problems with formulating a theory in 9+2 dimensions are in this case only apparent, as the theory can be rewritten as a theory in 10+1 dimensions using different variables, so that the extra time dimension is replaced by the degrees of freedom associated with branes wrapped around time.

There are several possible generalisations of the notion of a particle to general signatures. A physical particle or an observer in Lorentzian spacetime with signature $(S, 1)$ follows a timelike (or null) world-line while a tachyon would follow a spacelike one. In a space of signature (S, T), one can again consider worldlines of signature $(0, 1)$, but other generalisations of particle might include branes with worldvolumes ('time-sheets') of signature $(0, t)$ with $t \leq T$, sweeping out some or all of the times. In a general signature (S, T), it is natural to consider branes of arbitrary signature (s, t) with $s \leq S$ and $t \leq T$, and the conditions on (s, t) for these to be supersymmetric are straightforward to find. [25]

In conclusion, we have reviewed part of the intricate web of duality symmetries linking many apparently different theories. As the theories are all related in this way, they should all be regarded as corners of a single underlying theory. In particular, two dual theories can be formulated in spacetimes of different geometry, topology and even signature and dimension, and so all of these concepts must be relative rather than absolute, depending on the values of certain parameters or couplings, and such a relativity principle should be a feature of the fundamental theory that underlies all this.

References

1. C.M. Hull and P.K. Townsend, Nucl. Phys. **B438** (1995) 109 hep-th/9410167.
2. E. Witten, Nucl. Phys. **B443** (1995) 85, hep-th/9503124
3. J. Polchinski, hep-th/9611050
4. A. Giveon, M. Porrati and E. Rabinovici, Phys. Rep. **244** (1994) 77.
5. J. Dai, R.G. Leigh and J. Polchinski, Mod. Phys. Lett. **A4** (1989) 2073;
6. M. Dine, P. Huet and N. Seiberg, Nucl. Phys. **B322** (1989) 301.
7. C.M. Hull, Phys. Lett. **B357** (1995) 545, hep-th/9506194.

8. A. Dabholkar, Phys. Lett. **B357** (1995) 307, hep-th/9506160.

9. J. Polchinski and E. Witten, Nucl.Phys. **B460** (1996) 525, hep-th/9510169.

10. P. Aspinwall, Nucl. Phys. Proc. Suppl. **46** (1996) 30, hep-th/9508154 ;
J.H. Schwarz, Phys.Lett. **B360** (1995) 13; Erratum-ibid. **B364** (1995) 252,
hep-th/9508143 .

11. P. Hořava and E. Witten, Nucl. Phys. **B460** (1996) 506, hep-th/9510209.

12. G. Pradisi and A. Sagnotti,Phys. Lett. **B216** (1989) 59;
M. Bianchi and A. Sagnotti, Phys. Lett. **B247** (1990) 517; Nucl. Phys. **B361**
(1990) 519;
P. Hořava, Nucl. Phys. **B327** (1989) 461; Phys. Lett. **B231** (1989) 251.

13. C.M. Hull, JHEP **9811** (1998) 27, hep-th/9811021 .

14. C.M. Hull, Nucl. Phys. **B468** (1996) 113 hep-th/9512181.

15. J. Maldacena, hep-th/9711200.

16. T. Banks, W. Fischler, S. H. Shenker and L. Susskind, Phys. Rev. D **55**, 5112
(1997) [arXiv:hep-th/9610043].

17. C.M. Hull, JHEP **11** (1998) 017 hep-th/9807127.

18. W. Fischler, E. Halyo, A. Rajaraman and L. Susskind, Nucl.Phys. **B501**
(1997) 409, hep-th/9703102.

19. P. Goddard, 'The Work of R.E. Borcherds', to appear in *Proceedings of the
International Congress of Mathematicians*, and references therein.

20. C. M. Hull and B. Julia, Nucl. Phys. **B534** (1998) 250, hep-th/9803239.

21. E. Cremmer, I.V. Lavrinenko, H. Lu, C.N. Pope, K.S. Stelle and T.A. Tran,
Nucl.Phys. **B534** (1998) 250, hep-th/9803259.

22. K. S. Stelle, hep-th/9803116.

23. G. Moore, hep-th/9305139,9308052.

24. C.M. Hull, JHEP **07** (1998) 021, hep-th/9806146.

25. C.M. Hull and R.R. Khuri, Nucl. Phys. **B536** (1998) 219 hep-th/9808069.

26. C. M. Hull and R. R. Khuri, Nucl. Phys. B **575** (2000) 231 [arXiv:hep-
th/9911082]; Nucl. Phys. B **536** (1998) 219 [arXiv:hep-th/9808069].

27. C. M. Hull, arXiv:hep-th/0406102.

28. C.M. Hull, Nucl. Phys. **B509** (1998) 216, hep-th/9705162.

29. B. Julia, in *Superspace and Supergravity*, eds. S.W. Hawking and M. Rocek
(Cambridge University Press, 1981)

30. B. Julia, in *Lectures in Applied Mathematics*, Vol. **21** (1985) 355;
H. Nicolai, *Phys. Lett.* **B276** (1992) 333

31. A. Keurentjes, arXiv:hep-th/0404174.

32. A. Keurentjes, Nucl. Phys. B **697**, 302 (2004) [arXiv:hep-th/0402090].

33. T. Damour, M. Henneaux and H. Nicolai, Phys. Rev. Lett. **89**, 221601 (2002)
[arXiv:hep-th/0207267]; Class. Quant. Grav. **20**, R145 (2003) [arXiv:hep-
th/0212256].

34. P. C. West, Class. Quant. Grav. **18**, 4443 (2001) [arXiv:hep-th/0104081]; Class.
Quant. Grav. **20**, 2393 (2003) [arXiv:hep-th/0212291]; Phys. Lett. B **575**, 333
(2003) [arXiv:hep-th/0307098].

35. F. Englert and L. Houart, JHEP **0401**, 002 (2004) [arXiv:hep-th/0311255].

QUANTIZED SPACE AND TIME

PETER SCHUPP

International University Bremen, Campus Ring 8, 28759 Bremen, Germany
E-mail: p.schupp@iu-bremen.de

This is a concise introduction to noncommutative space-time structures, noncommutative gauge theory, and the construction of the noncommutative standard model in the perturbative approach based on star products and the Seiberg-Witten map.

1 Introduction

The concept of space-time as a smooth manifold cannot be extended to arbitrarily short distances. High precision localization requires large energy-densities that distort space-time. This points to the existence of a fundamental uncertainty in the position of space-time events, which would in fact be natural in a theory where space-time coordinates are promoted to noncommutative operators[1]. The mathematical framework that can take such a microscopic structure of space-time into account is noncommutative geometry.

In string theory noncommutativity arises quite generically in the presence of background fields – on equal footing with the open string metric and without direct reference to the Planck scale. Noncommutative effects may thus appear at much lower energies. Current conservative limits on the noncommutativity scale from accelerator experiments are of the order of a few TeV; much more stringent limits exist for low-energy physics[2].

In this note we shall focus on an effective, perturbative formulation of particle physics on noncommutative space-time with realistic gauge group and particle content. We want to describe the scattering of particles taking effects due to space-time noncommutativity only in the interaction region into account. We shall therefore assume that space-time has an asymptotically commutative description at large distances. This is in accordance with everyday observations, experimental facts and the expectation that the quantum structure of space-time should be only relevant at very short distances. There may even be a phase transition from a noncommutative to an exactly commutative regime at low energies.

1.1 Models of noncommutative space-time

The basic idea of noncommutative geometry[3] is to focus on the algebra of functions on a space-time manifold rather than on a set of points. In the noncommutative realm this algebra is promoted to an arbitrary associative algebra while the original notion of space-time "points" looses its meaning. Let us consider some concrete examples of noncommutative space-time structures[4].

In the *canonical structure* the usual coordinate functions x^μ are promoted to operators $\hat{x}^\mu$ with commutation relations

$$[\hat{x}^\mu, \hat{x}^\nu] = i\theta^{\mu\nu}, \quad [\hat{x}^\xi, \theta^{\mu\nu}] = 0, \tag{1}$$

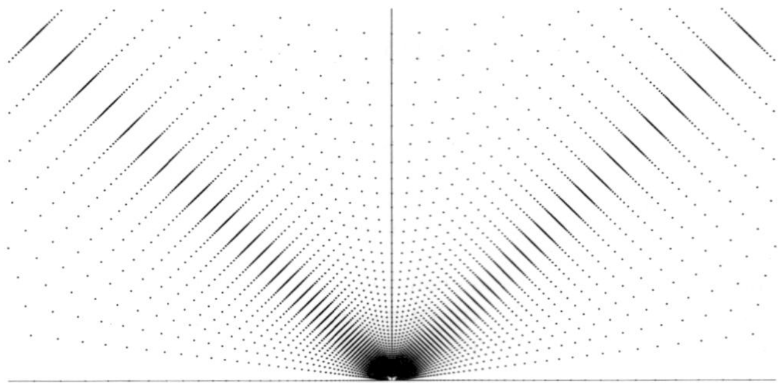

Figure 1. Space-time lattice structure of quantum Minkowski space: Shown are eigenvalues of position (distance from origin) versus time in spacelike and timelike regions[6].

where $\theta^{\mu\nu}$ is an antisymmetric matrix with real constant entries.

The *Lie structure* has commutation relations that close linearly in the $\hat{x}^\xi$,

$$[\hat{x}^\mu, \hat{x}^\nu] = iC^{\mu\nu}{}_\xi \hat{x}^\xi, \qquad (2)$$

where the structure constants $C^{\mu\nu}{}_\xi$ satisfy the Jacobi identity. Such a noncommutative structure defines fuzzy spheres[5], where one considers all representations of $SU(2)$ up to a certain maximum spin cutoff. It can also be used to define noncommutative versions of $\mathbb{R}^n$. An interesting special case of the Lie structure is $[\hat{x}^\mu, \hat{x}^\nu] = i(v^\mu \hat{x}^\nu - v^\nu \hat{x}^\mu)$ with a constant vector v^μ. For a time-like vector this is known as κ-Minkowski space.

Quantum space structures can be written in terms of generators modulo relations that are typically quadratic in the generators with a deformation parameter $q \geq 1$. Quantum Minkowski space

$$\epsilon_q{}^i{}_{kj}\hat{x}^j \hat{x}^k = (1 - q^2)\hat{x}^0 \hat{x}^i, \quad \hat{x}^0 \hat{x}^i = \hat{x}^i \hat{x}^0, \qquad (3)$$

is covariant under generalizations of the usual Lorentz and Poincaré symmetries called quantum groups. An interesting aspect of quantum spaces is the existence of lattice-like space-time structures, see figure 1. The lattice points correspond to space-time eigenvalues in Hilbert space representations of the deformed Heisenberg algebra over a quantum space[6]. A very interesting aspect of this model is that the time-eigenvalues are discrete even though the time-coordinate commutes with the other operators, see (3), and one would thus expect a continuous spectrum for it. The discrete spectrum of time arises in the irreducible representations of the Heisenberg algebra. This is possible because inequivalent representations are labelled by a continuos parameter $1 \leq \tau < q$ and a particular fixed value of τ has been chosen in figure 1. A speculative physical interpretation of this phenomenon could be a phase-transition from contiuous time and space at low energies to a lattice-like space-time structure with fixed τ at high energies.

How can we detect noncommutativity experimentally? Are there particle physics signatures for space-time noncommutativity? To answer these questions we need to formulate quantum field theory on noncommutative space-time. This is a hard task and the appropriate method depends on the particlular problem that one tries to address. Our strategy shall be to treat noncommutativity perturbatively.

1.2 Star products

The noncommutative algebras that we have introduced can be written as star products on the space of functions over a suitable manifold. The canonical structure (1) is, e.g., realized by the Moyal-Weyl star product

$$f \star g = \sum_{n=0}^{\infty} \frac{1}{n!} \left(\frac{i\hbar}{2} \right)^n \theta^{\mu_1 \nu_1} \dots \theta^{\mu_n \nu_n} \partial_{\mu_1} \dots \partial_{\mu_n} f \cdot \partial_{\nu_1} \dots \partial_{\nu_n} g, \tag{4}$$

where f and g are ordinary C^∞ functions on $\mathbb{R}^n$ (flat space-time) and $\hbar$ is the (formal) deformation parameter.[a] We find $x^\mu \star x^\nu - x^\nu \star x^\mu = i\hbar\theta^{\mu\nu}$ for coordinate functions on $\mathbb{R}^n$. More generally a local star product $\star$ is an associative $\mathbb{C}[[\hbar]]$-bilinear product of C^∞ functions on a suitable manifold M, written as a formal power series in bidifferential operators starting with the commutative pointwise product of functions. The star product is a quantization of a given Poisson structure $\{f, g\} = \theta^{ij} \partial_i f \partial_j g$:

$$[f \overset{\star}{,} g] = i\hbar\theta^{ij} \partial_i f \partial_j g + \mathcal{O}(\hbar^2). \tag{5}$$

Two star products $\star$, $\star'$ are called equivalent if they are related by a formal linear differential operator $\mathcal{D}$ starting with the identity, such that $\mathcal{D}(f \star' g) = \mathcal{D}f \star \mathcal{D}g$. $\mathcal{D}$ is formally invertible and associativity of $\star'$ follows from associativity of $\star$.

Star products provide a convenient language to formulate noncommutative field theory. The semi-classical Poisson limit and the classical "commutative" limit $\hbar \to 0$ are directly build into the definition of a star product. This facilitates the perturbative computation of the scattering of particles that enter from an asymptotically commutative region into a scattering region, where noncommutative effects are relevant.

2 Gauge theory on noncommutative space-time

The construction of a gauge theory on a given non-commutative space can be based on a few fundamental concepts: covariant coordinates (and functions), locality, and gauge equivalence and consistency conditions[4,7,8,9].

2.1 Covariant coordinates

Let us consider an infinitesimal non-commutative local gauge transformation $\hat{\delta}$ of a fundamental matter field that carries a representation ρ_Ψ

$$\hat{\delta}\widehat{\Psi} = i\rho_\Psi(\widehat{\Lambda}) \star \widehat{\Psi}. \tag{6}$$

In the non-Abelian case $\widehat{\Psi}$ is a vector, $\rho_\Psi(\widehat{\Lambda})$ is a matrix and $\star$ includes matrix multiplication. The product of a field and a coordinate, $\widehat{\Psi} \star x^\mu$, yields a new field that transforms just like $\widehat{\Psi}$. The opposite product $x^\mu \star \widehat{\Psi}$, however, is not a covariant

---

[a] In the following sections we will not always write $\hbar$ explicitly.

object because the gauge parameter does not commute with x^μ. In analogy to the covariant derivatives of ordinary gauge theory we thus need to introduce covariant coordinates[4] $X^\mu = x^\mu + \theta^{\mu\nu}\widehat{A}_\nu$, where $\widehat{A}_\nu$ is a non-commutative analog of the gauge potential with the following transformation property:[b]

$$\hat{\delta}\widehat{A}_\mu = \partial_\mu\widehat{\Lambda} + i[\widehat{\Lambda}\stackrel{\star}{,}\widehat{A}_\mu].\tag{7}$$

More generally, covariant functions $\mathcal{D}(f)$[7], where f is an ordinary function (or a matrix-valued function) and $\mathcal{D}$ is an invertible linear differential operator, transform covariantly under gauge transformations like $\delta\mathcal{D}(f) = i[\widehat{\Lambda}\stackrel{\star}{,}\mathcal{D}(f)]$. The product of a covariant function and a field $\widehat{\Psi}$ *is* a covariant object that transforms like $\widehat{\Psi}$. The covariantizing map $\mathcal{D}$ turns $\star$ into an equivalent star product $\star'$. The space of fields (sections) is naturally a left $\star'$-module and a right $\star$-module. From the covariant coordinates one can construct further covariant objects: The covariant derivative

$$\widehat{D}_\mu\widehat{\Psi} = \partial_\mu\widehat{\Psi} - i\rho_\Psi(\widehat{A}_\mu)\star\widehat{\Psi},\tag{8}$$

is related to the covariant expression $\rho_\Psi(X^\mu)\star\widehat{\Psi} - \widehat{\Psi}\star x^\mu$ which is simply the difference between the left and right actions of x^μ on $\widehat{\Psi}$. The corresponding non-commutative field strength

$$\widehat{F}_{\mu\nu} = \partial_\mu\widehat{A}_\nu - \partial_\nu\widehat{A}_\mu - i[\widehat{A}_\mu\stackrel{\star}{,}\widehat{A}_\nu], \qquad \hat{\delta}\widehat{F}_{\mu\nu} = i[\widehat{\Lambda}\stackrel{\star}{,}\widehat{F}_{\mu\nu}],\tag{9}$$

is related to the commutator of covariant coordinates X^μ. (In the following we shall often omit the symbol ρ_Ψ, when its presence is obvious.)

We can now write a noncommutative generalization of the Yang-Mills action:

$$\widehat{S} = \int d^4x\, \frac{-1}{2g^2}\, Tr(\widehat{F}_{\mu\nu}\star\widehat{F}^{\mu\nu}) + \overline{\widehat{\Psi}}\star i\,\widehat{\slashed{D}}\widehat{\Psi}.\tag{10}$$

This action is invariant under noncommutative gauge transformations, because of the trace property of the integral with respect to the Moyal-Weyl star product.

$$\int d^4x\, f\star g = \int d^4x\, g\star f = \int d^4x\, f\cdot g.\tag{11}$$

Applying the usual QFT techniques directly to this action we find some intriguing properties and problems:

- The choice of gauge groups appears to be restricted to $U(N)$ in the fundamental representation. Related to this the Abelian charge can take on only the discrete values $Q/e = -1, 0, +1$.

- There appears to be no natural definition of tensor products. It is not clear a priori how to write covariant Yukawa terms.

- The meaning of noncommutative "in" and "out"-states is opaque.

- The β-function of non-commutative $U(1)$ gauge theory behaves like that of a non-abelian theory. There is UV/IR mixing.

---

[b]Here and in the following we use $\theta^{\mu\nu}$ to lower indices, yielding expressions that resemble ordinary Yang-Mills theory. This is only possible in the case of constant $\theta^{\mu\nu}$. Otherwise one should work with $\tilde{A}^\mu$, where $X^\mu = \mathcal{D}(x^\mu) = x^\mu + \tilde{A}^\mu$[7].

In the following we shall discuss how to overcome these obstacles and construct a noncommutative generalization of the standard model.

2.2 *Locality, gauge equivalence and consistency conditions*

So far it has not been important that the noncommutative structures that we considered can be realized as star products. We shall now introduce another important concept that will help us to solve the apparent problems mentioned at the end of the previous section. That concept is the principle of locality. A star product is a formal power series starting with the ordinary product plus higher order derivative terms that are chosen so as to yield an associative product. The star product can be pictured as a tower build upon the leading Poisson tensor $\theta^{\mu\nu}$. It is a natural question to ask whether it is possible to express also the non-commutative fields $\hat{A}$, $\hat{\Psi}$ and the non-commutative gauge parameter $\hat{\Lambda}$ in a similar fashion as towers build upon the corresponding ordinary fields A, Ψ and ordinary gauge parameter Λ. This is indeed the case; the non-commutative fields and parameter can be expressed as local functions of their commutative counterparts

$$\hat{A}_\mu[A] = A_\mu + \frac{1}{4}\theta^{\xi\nu}\left\{A_\nu, \partial_\xi A_\mu + F_{\xi\mu}\right\} + \ldots , \tag{12}$$

$$\hat{\Psi}[\Psi, A] = \Psi + \frac{1}{2}\theta^{\mu\nu}\rho_\Psi(A_\nu)\partial_\mu\Psi + \frac{1}{4}\theta^{\mu\nu}\rho_\Psi(\partial_\mu A_\nu)\Psi + \ldots , \tag{13}$$

$$\hat{\Lambda}_\Lambda[A] = \Lambda + \frac{1}{4}\theta^{\mu\nu}\{A_\nu, \partial_\mu\Lambda\} + \ldots , \tag{14}$$

where $F_{\mu\nu} = \partial_\mu A_\nu - \partial_\nu A_\mu - i[A_\mu, A_\nu]$ is the ordinary field strength. Here we understand a local function of a field to be a formal series that *at each order* in θ depends on the field and a *finite* number of derivatives of the field. (Summing all orders in θ, e.g. by replacing the formal parameter $\hbar$ by some fixed number, we of course obtain expressions with arbitrarily high derivatives, i.e., a nonlocal object.) The expressions (12)–(14) have the remarkable property that ordinary gauge transformations $\delta_\Lambda A_\mu = \partial_\mu\Lambda + i[\Lambda, A_\mu]$ and $\delta_\Lambda \Psi = i\Lambda \cdot \Psi$ induce noncommutative gauge transformations (6), (7) of $\hat{A}$ and $\hat{\Psi}$ with parameter $\hat{\Lambda}$: The expansions of the noncommutative fields satisfy the gauge equivalence conditions

$$\delta_\Lambda \hat{A}_\mu[A] = \partial_\mu\hat{\Lambda}_\Lambda[A] + i[\hat{\Lambda}_\Lambda[A] \stackrel{\star}{,} \hat{A}_\mu[A]] , \tag{15}$$

$$\delta_\Lambda \hat{\Psi}[\Psi, A] = i\hat{\Lambda}_\Lambda[A] \star \hat{\Psi}[\Psi, A]. \tag{16}$$

Here is an example how the relevant series are resummed to verify the gauge equivalence condition in the abelian case for a matter field ψ:

$$\delta_\lambda\hat{\psi} = i\lambda\psi + \tfrac{1}{2}\theta^{ij}(\partial_j\lambda)\partial_i\psi + \tfrac{1}{2}\theta^{ij}a_j\partial_i(i\lambda\psi) + \ldots$$

$$= i(\lambda + \tfrac{1}{2}\theta^{ij}a_j\partial_i\lambda + \ldots) \star (\psi + \tfrac{1}{2}\theta^{ij}a_j\partial_i\psi + \ldots)$$

$$= i\hat{\lambda} \star \hat{\psi} .$$

The expressions (12), (14) for the noncommutative field strength and noncommutative gauge parameter in terms of their classical counterparts are known as Seiberg-Witten (SW) maps. They were introduced by Seiberg and Witten as field redefinitions in the context of string theory[10].

Any pair of non-commutative gauge parameters $\hat{\Lambda}_\alpha[a]$, $\hat{\Lambda}_\beta[a]$ has to satisfies the following consistency condition (cocycle condition)[9]

$$[\hat{\Lambda}_\alpha[A] \overset{\star}{,} \hat{\Lambda}_\beta[A]] + i\delta_\alpha\hat{\Lambda}_\beta[A] - i\delta_\beta\hat{\Lambda}_\alpha[A] = \hat{\Lambda}_{[\alpha,\beta]}[A], \tag{17}$$

which follows from $[\delta_\alpha, \delta_\beta]\widehat{\Psi}[\Psi, A] = \delta_{-i[\alpha,\beta]}\widehat{\Psi}[\Psi, A]$. The consistency condition (17) is important for the practical calculation of SW maps, since it involves only the gauge parameter and because it reduces the task to a cohomological problem[9,11]. The gauge equivalence and consistency conditions do not uniquely determine SW maps. To first order in θ we have the freedom of classical field redefinitions and noncommutative gauge transformations. We have used that freedom to choose maps with Hermitean $\hat{A}_\mu$ and $\hat{\Lambda}$. The constants that parametrize the freedom in the SW map receive quantum corrections in noncommutative gauge theory[12]. The freedom in the Seiberg-Witten map is also important in the context of tensor products of fields and gauge groups.

Infinitesimal gauge transformations can be exponentiated to yield finite gauge transformations with "NC group elements" $\widehat{G}_g[A]$ corresponding to $g = \exp(i\Lambda)$:

$$\widehat{G}_g[A] \star \widehat{\Psi} = \exp(\delta_\Lambda)\widehat{\Psi} = \left(1 + i\widehat{\Lambda} + \frac{i}{2}(\delta_\Lambda\widehat{\Lambda}) - \frac{1}{2}\widehat{\Lambda} \star \widehat{\Lambda} + \dots\right) \star \widehat{\Psi}. \tag{18}$$

(Care has to be taken, because $\widehat{\Lambda} = \widehat{\Lambda}_\Lambda[A]$ depends on the gauge potential.) In terms of the "NC group element" $\widehat{G}_g[A]$ the consistency condition becomes

$$\widehat{G}_{g_1}[A_{g_2}] \star \widehat{G}_{g_2}[A] = \widehat{G}_{g_1 \cdot g_2}[A], \tag{19}$$

where A_{g_2} is the gauge transformed ("shifted") gauge potential

$$A_{g_2} \equiv g_2 \cdot A \cdot g_2^{-1} + ig_2 dg_2^{-1}. \tag{20}$$

The "NC group law" (19) is the starting point for noncommutative vector bundles with nontrivial gauge fields and for the construction of noncommutative gerbes[13].

2.3 θ-expanded noncommutative Yang-Mills action

Using the SW maps (12) and (13) to expand the fields $\widehat{A}$ and $\widehat{\Psi}$ in the action (10) yields an action that is written in ordinary fields. Expanding to first order in θ and integrating by parts we find:

$$S = \int d^4x \left(-\frac{1}{2g^2}\mathrm{tr}F_{ij} \star F^{ij} + \frac{1}{4g^2}\theta^{ij}\mathrm{tr}F_{ij}F_{kl}F^{kl} - \frac{1}{g^2}\theta^{ij}\mathrm{tr}F_{ik}F_{jl}F^{jl} \right.$$
$$\left. + \overline{\Psi}(i\not{D} - m)\Psi - \frac{1}{4}\theta^{ij}\overline{\Psi}F_{ij}(i\not{D} - m)\Psi - \frac{i}{2}\theta^{ij}\overline{\Psi}\gamma^k F_{ki}D_j\Psi\right)$$

with $D_i\Psi \equiv \partial_i\Psi - iA_i\Psi$ and $F_{ij} = \partial_i A_j - \partial_j A_i - i[A_i, A_j]$. This formulation of noncommutative gauge theory[4,8,7,14,9] has the following advantages:

- We can freely choose the gauge group, and its representations.

- Related to this, there is no problem with charges.

- Tensor products of fields and of gauge groups exist.

- Nontrivial gauge fields (and vector bundles) are possible.

We shall now discuss some of these points in more detail.

3 Noncommutative gauge fields are valued in the enveloping algebra

With the SW maps we can freely choose a structure group G (gauge group). The fields A, F and the gauge parameter Λ are valued in the Lie algebra of G while their noncommutative counterparts $\widehat{A}$, $\widehat{F}$, $\hat{\Lambda}$ are valued in the enveloping algebra of the gauge group, i.e., they contain polynomials in the Lie algebra generators

$$\hat{\Lambda} = \Lambda_a(x)T^a + \Lambda^1_{ab}(x) : T^aT^b : +\Lambda^2_{abc}(x) : T^aT^bT^c : +\ldots, \tag{21}$$

where $: :$ denotes symmetric ordering[8]. This is a general feature of space-time noncommutativity as can be seen by computing the commutator of two noncommutative gauge transformations with Lie algebra-valued gauge parameters $\hat{\Lambda} = \Lambda_a(x)T^a$ and $\hat{\Lambda}' = \Lambda'_b(x)T^b$:

$$[\hat{\Lambda} \overset{\star}{,} \hat{\Lambda}'] = \frac{1}{2}\{\Lambda_a(x) \overset{\star}{,} \Lambda'_b(x)\}[T^a, T^b] + \frac{1}{2}[\Lambda_a(x) \overset{\star}{,} \Lambda'_b(x)]\{T^a, T^b\}. \tag{22}$$

The result is valued in the enveloping algebra, because the coefficient of $\{T^a, T^b\}$ is non-zero: The functions $\Lambda_a(x)$, $\Lambda'_b(x)$ do not commute. This argument can be repeated to show that polynominals to any order in the generators of the Lie algebra will appear. Since the gauge transformation of a noncommutative gauge field involves $\hat{\Lambda}$, it must also be valued in the envoloping algebra. If the coefficients of all the terms in the resulting expansion of the gauge field were independent, we would, however, obtain a theory with an infinite number of degrees of freedom which does not make sense physically. We could of course consider a matrix representation of the Lie algebra; then $\widehat{A}$, $\widehat{F}$ and $\hat{\Lambda}$ will also be matrix-valued. We will no longer have an infinite number of degrees of freedom. For $U(N)$ in the fundamental representation this solves the problem and we do get the same number of degrees of freedom as in the commutative limit, but for $SU(N)$ A, F and Λ are traceless while $\widehat{A}$, $\widehat{F}$, $\hat{\Lambda}$ are not[4]. Considering the expansion of the noncommutative fields as functions of the classical fields via the SW map, we do not have these problems: It is still true that the noncommutative fields are valued in the enveloping algebra of the gauge group, but the coefficients in the expansion are not independent. They are fixed by the SW map and we get the same number of degrees of freedom as classically. That is one of the essential reasons for the use of the SW map in the formulation of noncommutative gauge theory.

3.1 Charge quantization

Noncommutative QED has the interesting property that the theory can apparently accommodate only charges $\pm q$ or zero for one fixed q [15]. That is so because the only couplings of the non-commutative gauge boson $\widehat{A}_\mu$ to a matter field $\widehat{\Psi}$ compatible with the non-commutative gauge transformation (7) are

$$+i\widehat{\Psi} \star \widehat{A}_\mu, \quad -i[\widehat{A}_\mu \overset{\star}{,} \widehat{\Psi}], \quad \text{and} \quad -i\widehat{A}_\mu \star \widehat{\Psi}. \tag{23}$$

Table 1. Charges of the Standard Model fields (1st generation)

	$SU(3)_C$	$SU(2)_L$	$U(1)_Y$	$U(1)_Q$
e_R	**1**	**1**	-1	-1
$\begin{pmatrix} \nu_L \\ e_L \end{pmatrix}$	**1**	**2**	$-1/2$	$\begin{pmatrix} 0 \\ -1 \end{pmatrix}$
u_R	**3**	**1**	$2/3$	$2/3$
d_R	**3**	**1**	$-1/3$	$-1/3$
$\begin{pmatrix} u_L \\ d_L \end{pmatrix}$	**3**	**2**	$1/6$	$\begin{pmatrix} 2/3 \\ -1/3 \end{pmatrix}$
$\begin{pmatrix} \phi^+ \\ \phi^0 \end{pmatrix}$	**1**	**2**	$1/2$	$\begin{pmatrix} 1 \\ 0 \end{pmatrix}$

Similar to the case of ordinary Yang-Mills theory, there are no additional factors allowed in front of these terms. For the Standard Model on NC space-time this is a major problem because of the existence of fractional charges and hypercharges, see table 1.

The problem can be traced back to the fact that the gauge fields are enveloping algebra valued. The charge quantization problem should really be seen as a problem with the degrees of freedom (more precisely the number of fundamental fields). In the θ-expanded approach based on Seiberg-Witten maps the fundamental field $a_\mu(x)$ enters as $A_\mu = eQa_\mu(x)$, where e is the coupling constant and Q is the charge operator. The resulting $\widehat{A}_\mu$ is a highly nonlinear function of Q, so that for differently charged particles we will also get different noncommutative gauge fields $\widehat{A}^q_\mu = eq\widehat{a}^q_\mu(x)$. This appears to force us to introduce a new type of photon for every charge appearing in the Standard Model. The important point is to realize that the fields $\widehat{A}^q_\mu$ for different charges q are not independent. They are all functions of the physical field $a_\mu(x)$ via the SW map. Consequently there is only one type of photon (and gluon etc.) and no restriction on the allowed charges in θ-expanded NC QED[16,17].

There is an alternative approach to the construction of noncommutative generalizations of the Standard Model that circumvents the problem of charge quantization at the expense of additional $U(1)$ factors that are then removed with the help of extra Higgs fields. In this approach the charges are still quantized but to the correct values of the usual quarks and leptons[18]. In principle the two approaches can be combined.

4 Products of fields and tensor products of gauge groups

The gauge parameter for a product $G \times G'$ of two groups is $\Lambda = \lambda + \lambda'$, where λ and λ' are valued in the respective Lie algebras of the groups G and G'. Similarly the gauge potential is $A_\mu = a_\mu + a'_\mu$. In the noncommutative case the situation is more complicated, nevertheless the noncommutative gauge parameter $\hat{\Lambda}$ depends on λ, λ', a, a' and can also be written as the sum of two terms:

$$\hat{\Lambda}_{(\lambda,\lambda')}[a,a'] = \hat{\Lambda}_\lambda[a,a'] + \hat{\Lambda}'_{\lambda'}[a,a']. \tag{24}$$

The consistency relation (17) for $\hat{\Lambda}_{(\lambda,\lambda')}[a,a']$ implies a total of three relations for $\hat{\Lambda}_\lambda[a,a']$ and $\hat{\Lambda}'_{\lambda'}[a,a']$: $\hat{\Lambda}_\lambda[a,a']$ and $\hat{\Lambda}'_{\lambda'}[a,a']$ satisfy CR's separately

$$[\hat{\Lambda}_\alpha \overset{\star}{,} \hat{\Lambda}_\beta] + i\delta_\alpha\hat{\Lambda}_\beta - i\delta_\beta\hat{\Lambda}_\alpha = \hat{\Lambda}_{[\alpha,\beta]}, \tag{25}$$

$$[\hat{\Lambda}'_{\alpha'} \overset{\star}{,} \hat{\Lambda}'_{\beta'}] + i\delta_{\alpha'}\hat{\Lambda}'_{\beta'} - i\delta_{\beta'}\hat{\Lambda}'_{\alpha'} = \hat{\Lambda}'_{[\alpha',\beta']} \tag{26}$$

and there is a mixed consistency relation

$$[\hat{\Lambda}_\alpha \overset{\star}{,} \hat{\Lambda}'_{\beta'}] + i\delta_\alpha\hat{\Lambda}'_{\beta'} - i\delta_{\beta'}\hat{\Lambda}_\alpha = 0. \tag{27}$$

Note that there is no inhomogeneous term on the RHS because $[\alpha,\beta'] = 0$.

4.1 Hybrid Seiberg-Witten map

For the product of fields $\psi\,\psi'$, where ψ transforms under G and ψ' transforms under G' we could try to construct a SW map $\hat{\boldsymbol{\Psi}}[\psi,\psi',a,a']$ directly.[c] It is, however, more convenient to follow a different strategy. For this we need a hybrid SW map[16,17] that interpolates between the SW map for the gauge field (12) and the SW map for the matter field (13): Consider a field Φ that transforms on the left and on the right under two arbitrary gauge groups G_L and G_R, respectively: $\delta\Phi = i\Lambda \cdot \Phi - i\Phi \cdot \Lambda'$, where Λ is valued in (a representation of) $\mathrm{Lie}(G_L)$ and Λ is valued in (a representation of) $\mathrm{Lie}(G_R)$. We shall also assume that there are gauge fields A and A' corresponding to the respective gauge groups, that transform as $\delta A_\nu = \partial_\nu\Lambda + i[\Lambda, A_\nu]$ and $\delta A'_\nu = \partial_\nu\Lambda' + i[\Lambda', A'_\nu]$. The following hybrid Seiberg-Witten map (given to order θ)

$$\hat{\Phi}^H[\Phi,A,A'] = \Phi + \frac{1}{2}\theta^{\mu\nu}\left(A_\nu\partial_\mu\Phi + \frac{1}{2}\partial_\mu A_\nu\Phi + \partial_\mu\Phi A'_\nu + \frac{1}{2}\Phi\partial_\mu A'_\nu + iA_\nu\Phi A'_\mu\right) \tag{28}$$

has the property that the transformations of Φ, A and A' that were given above induce the following transformation of $\hat{\Phi}^H$:[d]

$$\delta\hat{\Phi}^H = i\hat{\Lambda} \star \hat{\Phi}^H - i\hat{\Phi}^H \star \hat{\Lambda}'. \tag{29}$$

Here $\hat{\Lambda} = \hat{\Lambda}_\Lambda[A]$ and $\hat{\Lambda}' = \hat{\Lambda}'_{\Lambda'}[A']$ are computed according to the usual SW map (14). There is also a corresponding formula for the covariant derivative of $\delta\hat{\Phi}^H$:

$$\hat{D}_\mu\hat{\Phi}^H = \partial_\mu\hat{\Phi}^H - i\hat{A}_\mu \star \hat{\Phi}^H + i\hat{\Phi}^H \star \hat{A'_\mu}. \tag{30}$$

---

[c]The naive choice $\hat{\psi} \star \hat{\psi}'$ does of course not work, because the gauge parameter $\hat{\lambda}'$ does not $\star$-commute with $\hat{\psi}$ in the second term of $\delta\hat{\psi} \star \hat{\psi}' = i\hat{\lambda} \star \hat{\psi} \star \hat{\psi}' + i\hat{\psi} \star \hat{\lambda}' \star \hat{\psi}'$.

[d]A similar formula with a $+$ sign in front of the second term would require the use of the opposite SW map (with $\theta \to -\theta$) for Λ'.

312

The SW map for the product of fields ψ and ψ' can now be written as

$$\widehat{\Psi}[\psi, \psi', a, a'] = \widehat{\psi}^H[\psi, a + a', a'] \star \widehat{\psi}'[\psi', a']. \tag{31}$$

The gauge transformation $\delta\psi = i\lambda\psi$, $\delta\psi' = i\lambda'\psi'$, $\delta a_\mu = \partial_\mu\lambda + i[\lambda, a_\mu]$, $\delta a'_\mu = \partial_\mu\lambda' + i[\lambda', a'_\mu]$ induces the desired transformation

$$\delta\widehat{\Psi}[\psi, \psi', a, a'] = i\widehat{\Lambda}_{(\lambda+\lambda')}[a + a'] \star \widehat{\Psi}[\psi, \psi', a, a'], \tag{32}$$

where we have used that ψ and λ' commute. We see that the given version of the hybrid SW map corresponds to a solution for the noncommutative gauge parameter that is symmetric in λ and λ'.

5   The Noncommutative Standard Model (NCSM)

The structure group of the Standard Model is $G_{SM} = SU(3)_C \times SU(2)_L \times U(1)_Y$. The gauge potential A_μ and gauge parameter Λ are valued in $\mathrm{Lie}(G_{SM})$:

$$A_\nu = g'\mathcal{A}_\nu(x)Y + g\sum_{a=1}^{3} B_{\nu a}(x)T_L^a + g_S\sum_{b=1}^{8} G_{\nu b}(x)T_S^b \tag{33}$$

$$\Lambda = g'\alpha(x)Y + g\sum_{a=1}^{3} \alpha_a^L(x)T_L^a + g_S\sum_{b=1}^{8} \alpha_b^S(x)T_S^b, \tag{34}$$

where Y, T_L^a, T_S^b are the generators of $u(1)_Y$, $su(2)_L$ and $su(3)_C$ respectively. In addition to the gauge bosons we have three families of left- and right-handed fermions and a Higgs doublet

$$\Psi_L^{(i)} = \begin{pmatrix} L_L^{(i)} \\ Q_L^{(i)} \end{pmatrix}, \qquad \Psi_R^{(i)} = \begin{pmatrix} e_R^{(i)} \\ u_R^{(i)} \\ d_R^{(i)} \end{pmatrix}, \qquad \Phi = \begin{pmatrix} \phi^+ \\ \phi^0 \end{pmatrix} \tag{35}$$

where i = 1,2,3 is the generation index and ϕ^+, ϕ^0 are complex scalar fields. We shall now apply the appropriate SW maps to the fields A_μ, $\Psi^{(i)}$, Φ, expand to first order in θ and write the corresponding NC Yang-Mills action[17]. Special care must be taken in the definition of the trace in the gauge kinetic terms and in the construction of covariant Yukawa terms.

5.1   Noncommutative Yukawa terms

The classical Higgs field $\Phi(x)$ commutes with the generators of the $U(1)$ and $SU(3)$ gauge transformations. It also commutes with the corresponding gauge parameters. The latter is no longer true in the noncommutative setting: The coefficients $\alpha(x)$ and $\alpha_b^S(x)$ of the $U(1)$ and $SU(3)$ generators in the gauge parameter are functions and therefore do not $\star$-commute with the Higgs field. This makes it hard to write down covariant Yukawa terms. The solution to the problem is the hybrid SW map (28). By choosing appropriate representations it allows us to assign separate left and right charges to the noncommutative Higgs field $\widehat{\Phi}^H$ that add up to its usual charge[17]. Here are two examples:

$$Y = \quad \begin{array}{ccccccc} \overline{\widehat{L}}_L & \star & \rho_L(\widehat{\Phi}) & \star & \widehat{e}_R & \quad & \overline{\widehat{Q}}_L & \star & \rho_Q(\widehat{\Phi}) & \star & \widehat{d}_R \\ 1/2 & & \underbrace{-1/2+1}_{1/2} & & -1 & & -1/6 & & \underbrace{1/6+1/3}_{1/2} & & -1/3 \end{array} \tag{36}$$

We see here two instances of a general rule: The gauge fields in the SW maps and in the covariant derivatives inherit their representation (charge for Y, trivial or fundamental representation for T_L^a, T_S^b) from the fermion fields $\Psi^{(i)}$ to their left and to their right.

5.2 The Minimal Noncommutative Standard Model

The trace in the kinetic terms for the gauge bosons is not unique, it depends on the choice of representation. This would not matter if the gauge fields were Lie algebra valued, but in the noncommutative case they live in the enveloping algebra. The simplest choice is a sum of three traces over the $U(1)$, $SU(2)$, $SU(3)$ sectors with $Y = \frac{1}{2}\begin{pmatrix} 1 & 0 \\ 0 & -1 \end{pmatrix}$ in the definition of $\mathbf{tr_1}$ and the fundamental representation for $\mathbf{tr_2}$ and $\mathbf{tr_3}$. This leads to the following gauge kinetic terms

$$S_{\text{gauge}} = -\frac{1}{4}\int d^4x\, f_{\mu\nu}f^{\mu\nu} - \frac{1}{2}\operatorname{Tr}\int d^4x\, F_{\mu\nu}^L F^{L\mu\nu}$$
$$-\frac{1}{2}\operatorname{Tr}\int d^4x\, F_{\mu\nu}^S F^{S\mu\nu} + \frac{1}{4}g_S\,\theta^{\mu\nu}\operatorname{Tr}\int d^4x\, F_{\mu\nu}^S F_{\rho\sigma}^S F^{S\rho\sigma}$$
$$-g_S\,\theta^{\mu\nu}\operatorname{Tr}\int d^4x\, F_{\mu\rho}^S F_{\nu\sigma}^S F^{S\rho\sigma} + \mathcal{O}(\theta^2)\,. \tag{37}$$

Note, that there are no new triple f or triple F^L-terms: Contrary to common believe triple photon couplings are not a prediction of noncommutative gauge theories; in the minimal NCSM they do not appear. This model is minimal in the sense that it deviates as little as possible from the Standard Model on commutative space-time. The full action of the Minimal Noncommutative Standard Model is[17]:

$$S_{NCSM} = \int d^4x \sum_{i=1}^{3} \overline{\widehat{\Psi}}_L^{(i)} \star i\widehat{\slashed{D}}\widehat{\Psi}_L^{(i)} + \int d^4x \sum_{i=1}^{3} \overline{\widehat{\Psi}}_R^{(i)} \star i\widehat{\slashed{D}}\widehat{\Psi}_R^{(i)}$$
$$-\int d^4x \frac{1}{2g'}\mathbf{tr_1}\widehat{F}_{\mu\nu} \star \widehat{F}^{\mu\nu} - \int d^4x \frac{1}{2g}\mathbf{tr_2}\widehat{F}_{\mu\nu} \star \widehat{F}^{\mu\nu}$$
$$-\int d^4x \frac{1}{2g_S}\mathbf{tr_3}\widehat{F}_{\mu\nu} \star \widehat{F}^{\mu\nu} + \int d^4x\left(\rho_0(\widehat{D}_\mu\widehat{\Phi})^\dagger \star \rho_0(\widehat{D}^\mu\widehat{\Phi}) \right.$$
$$\left. -\mu^2\rho_0(\widehat{\Phi})^\dagger \star \rho_0(\widehat{\Phi}) - \lambda\rho_0(\widehat{\Phi})^\dagger \star \rho_0(\widehat{\Phi}) \star \rho_0(\widehat{\Phi})^\dagger \star \rho_0(\widehat{\Phi}) \right)$$
$$-\int d^4x\left(\sum_{i,j=1}^{3} W^{ij}\left((\overline{\widehat{L}}_L^{(i)} \star \rho_L(\widehat{\Phi})) \star \widehat{e}_R^{(j)} + \overline{\widehat{e}}_R^{(i)} \star (\rho_L(\widehat{\Phi})^\dagger \star \widehat{L}_L^{(j)}) \right) \right)$$

$$+ \sum_{i,j=1}^{3} G_u^{ij} \left((\bar{\widehat{Q}}_L^{(i)} \star \rho_{\bar{Q}}(\widehat{\bar{\Phi}})) \star \widehat{u}_R^{(j)} + \bar{\widehat{u}}_R^{(i)} \star (\rho_{\bar{Q}}(\widehat{\bar{\Phi}})^\dagger \star \widehat{Q}_L^{(j)}) \right)$$

$$+ \sum_{i,j=1}^{3} G_d^{ij} \left((\bar{\widehat{Q}}_L^{(i)} \star \rho_Q(\widehat{\Phi})) \star \widehat{d}_R^{(j)} + \bar{\widehat{d}}_R^{(i)} \star (\rho_Q(\widehat{\Phi})^\dagger \star \widehat{Q}_L^{(j)}) \right) \Bigg) \qquad (38)$$

where W^{ij}, G_u^{ij}, G_d^{ij} are Yukawa couplings, $\bar{\Phi} = i\tau_2 \Phi^*$ and we have omitted the superscript ^{H} in the hybrid SW map of the Higgs.

5.3 Non-minimal versions of the NCSM

We can use the freedom in the choice of traces in kinetic terms for the gauge fields to construct non-minimal versions of the NCSM. The general form of the gauge kinetic terms is[17,19]

$$S_{\text{gauge}} = -\frac{1}{2} \int d^4x \sum_{\rho} \kappa_\rho \text{Tr}\left(\rho(\widehat{F}_{\mu\nu}) \star \rho(\widehat{F}^{\mu\nu}) \right), \qquad (39)$$

where the sum is over all unitary irreducible inequivalent representations ρ of the gauge group G. The freedom in the kinetic terms is parametrized by real coefficients κ_ρ that are subject to the constraints

$$\frac{1}{g_I^2} = \sum_{\rho} \kappa_\rho \text{Tr}\left(\rho(T_I^a)\rho(T_I^a) \right), \qquad (40)$$

where g_I and T_I^a are the usual "commutative" coupling constants and generators of $U(1)_Y$, $SU(2)_L$, $SU(3)_C$, respectively. Both formulas can also be written more compactly as

$$S_{\text{gauge}} = -\frac{1}{2} \int d^4x \, \mathbf{Tr} \frac{1}{\mathbf{G}^2} \widehat{F}_{\mu\nu} \star \widehat{F}^{\mu\nu}, \qquad \frac{1}{g_I^2} = \mathbf{Tr} \frac{1}{\mathbf{G}^2} T_I^a T_I^a, \qquad (41)$$

where the trace $\mathbf{Tr}$ is again over all representations and $\mathbf{G}$ is an operator that commutes with all generators T_I^a and encodes the coupling constants. The possibility of new parameters in gauge theories on noncommutative space-time is a consequence of the fact that the gauge fields are in general valued in the enveloping algebra of the gauge group.

The expansion in θ is at the same time an expansion in the momenta. The θ-expanded action can thus be interpreted as a low energy effective action. In such an effective low energy description it is natural to expect that all representations that appear in the commutative theory (matter multiplets and adjoint representation) are important. We should therefore consider the non-minimal version of the NCSM with non-zero coefficients κ_ρ at least for these representations. The number of new parameters in the non-minimal NCSM can be restricted by considering GUTs on noncommutative space-time[19].

6 Conclusion

A general feature of gauge theories on noncommutative space-time is the appearance of many new interactions including Standard Model-forbidden processes. The origin of these new interactions is two-fold: One source are the star products that let abelian gauge theory on NC space-time resemble Yang-Mills theory with the possibility of triple and quadruple gauge boson vertices. The other source are the gauge fields in the Seiberg-Witten maps for the gauge and matter fields. These can be pictured as a cloud of gauge bosons that dress the original 'commutative' fields and that have their origin in the interaction between gauge fields and the NC structure of space-time. One of the perhaps most striking effects and a possible signature of space-time noncommutativity is the spontaneous breaking of continuous and discrete space-time symmetries: The actions that we have written are invariant under the usual space-time symmetries if we also transform $\theta^{\mu\nu}$ as a tensor. If we however consider $\theta^{\mu\nu}$ as a spectator or as the vacuum expectation of some background field, then we do find processes that violate certain space-time symmetries and the corresponding conservation laws. These typically include spin conservation, and CP; they can also include CPT, momentum and energy conservation.

Let us conclude with a shortlist of challenging problems and interesting projects: It is important to get a better understanding of the *quantization* of field theories on noncommutative space-time. In the case of the lowest order noncommutative correction to ordinary field theories that we have discussed here, quantization is straightforward. Feynman rules can be obtained either in the canonical formalism or straight from the action in a path-integral approach. That is no longer true when one considers higher orders in θ or even sums the whole series: There are subtle issues related to time-ordering that can lead to apparent violation of unitarity, if one naively uses Feynman rules that have been directly read off the Lagrangian density, see[20]. A more careful canonical approach does, however, lead to a well-defined theory. The issue of *renormalizability* of the type of noncommutative gauge theories that we have discussed here has been investigated by several authors. We should recall that the primary goal of the models that we have studied is to provide an effective description of particle physics on a given noncommutative space. We should not a priori expect nor require such a theory to be renormalizable. The full θ-expanded action does in fact contain infinitely many power-counting non-renormalizable terms. Nevertheless these theories do appear to be almost renormalizable in the following sense: Noncommutative and commutative gauge invariance alone do not uniquely single out a specific action. There is in fact quite some freedom which includes also the freedom in the choice of Seiberg-Witten map. In quantum theory the constants that parametrize the freedom become running coupling constants. This issue has been carefully studied in the case of noncommutative QED and it has been found that at first order in θ at the one-loop level the quantum action has to contain (only) one additional term[12]. It is important to find more interesting processes that could serve as *experimental signatures* of space-time noncommutativity in accelerator physics, astrophysics and cosmology. Good candidates are processes that violate continuous and/or discrete space-time symmetries. Examples are the processes shown in figure 2: a $Z \rightarrow \gamma\gamma$

316

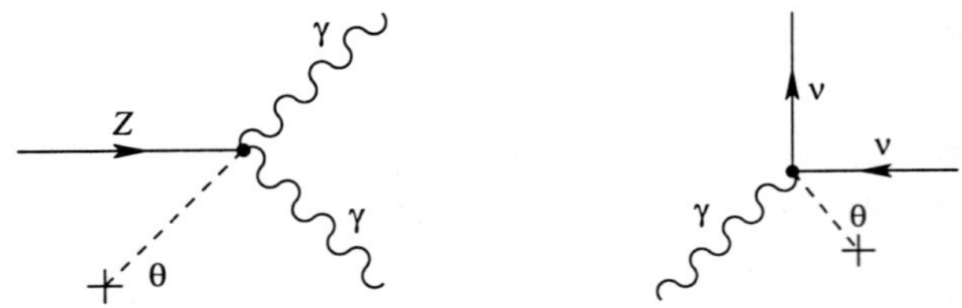

Figure 2. Standard Model forbidden Z and plasmon decays.

decay that violates spin conservation and a plasmon decay into a pair of neutrinos induced by space-time noncommutativity[21]. (In this context we would also like to refer to the review[2] of the phenomenology of noncommutative geometry and its extensive list of references.) On the theoretical side it would be interesting to study more complicated *noncommutative structures* as models of the microscopic structure of space-time including models where the noncommutative structures are dynamical and possibly include gravity. Here we have the intriguing possibility of processes that appear to violate energy and/or momentum conservation because four-momentum can be transfered to the noncommutative space-time structure.

References

1. H. S. Snyder, "Quantized Space-Time," Phys. Rev. **71**, 38 (1947); S. Doplicher, K. Fredenhagen and J. E. Roberts, "The Quantum structure of space-time at the Planck scale and quantum fields," Commun. Math. Phys. **172**, 187 (1995) [hep-th/0303037].
2. I. Hinchliffe and N. Kersting, "Review of the phenomenology of noncommutative geometry," hep-ph/0205040.
3. A. Connes, Non-commutative geometry (Academic Press, London 1994); J. Madore, An Introduction To Noncommutative Differential Geometry And Physical Applications (Cambridge Univ. Press 2000).
4. J. Madore, S. Schraml, P. Schupp and J. Wess, "Gauge theory on noncommutative spaces," Eur. Phys. J. C **16**, 161 (2000) [hep-th/0001203].
5. J. Madore, "The Fuzzy Sphere," Class. and Quant. Grav. **9**, 69 (1992).
6. B. L. Cerchiai and J. Wess, "q-Deformed Minkowski Space based on a q-Lorentz Algebra," Eur. Phys. J. C **5**, 553 (1998) [math.qa/9801104].
7. B. Jurco and P. Schupp, "Noncommutative Yang-Mills from equivalence of star products," Eur. Phys. J. C **14**, 367 (2000) [hep-th/0001032]; B. Jurco, P. Schupp and J. Wess, "Nonabelian noncommutative gauge theory via noncommutative extra dimensions," Nucl. Phys. B **604**, 148 (2001) [hep-th/0102129].
8. B. Jurco, S. Schraml, P. Schupp and J. Wess, "Enveloping algebra valued gauge transformations for non-Abelian gauge groups on non-commutative spaces," Eur. Phys. J. C **17**, 521 (2000) [hep-th/0006246].
9. B. Jurco, L. Moller, S. Schraml, P. Schupp and J. Wess, "Construction of non-Abelian gauge theories on noncommutative spaces," Eur. Phys. J. C **21**, 383

(2001) [hep-th/0104153].

10. N. Seiberg and E. Witten, "String theory and noncommutative geometry," JHEP **9909**, 032 (1999) [hep-th/9908142].

11. R. Stora, privat communication; D. Brace, B. L. Cerchiai, A. F. Pasqua, U. Varadarajan and B. Zumino, "A cohomological approach to the non-Abelian Seiberg-Witten map," JHEP **0106**, 047 (2001) [hep-th/0105192]; G. Barnich, F. Brandt and M. Grigoriev, "Seiberg-Witten maps and noncommutative Yang-Mills theories for arbitrary gauge groups," JHEP **0208**, 023 (2002) [hep-th/0206003].

12. A. Bichl, J. Grimstrup, H. Grosse, L. Popp, M. Schweda and R. Wulkenhaar, "Renormalization of the noncommutative photon self-energy to all orders via Seiberg-Witten map," JHEP **0106**, 013 (2001) [hep-th/0104097]; R. Wulkenhaar, JHEP **0203**, 024 (2002) [hep-th/0112248]; J. M. Grimstrup and R. Wulkenhaar, "Quantisation of θ-expanded non-commutative QED," Eur. Phys. J. C **26**, 139 (2002) [hep-th/0205153].

13. B. Jurco, P. Schupp and J. Wess, "Noncommutative line bundle and Morita equivalence," Lett. Math. Phys. **61**, 171 (2002) [hep-th/0106110]; P. Aschieri, I. Bakovic, B. Jurco and P. Schupp, "Noncommutative gerbes and deformation quantization," hep-th/0206101.

14. A. A. Bichl, J. M. Grimstrup, L. Popp, M. Schweda and R. Wulkenhaar, "Deformed QED via Seiberg-Witten map," hep-th/0102103.

15. M. Hayakawa, "Perturbative analysis on infrared and ultraviolet aspects of noncommutative QED on $\mathbb{R}^4$," hep-th/9912167.

16. P. Schupp, "Non-Abelian gauge theory on noncommutative spaces," PRHEP-hep2001/238 (2001) [hep-th/0111038].

17. X. Calmet, B. Jurco, P. Schupp, J. Wess and M. Wohlgenannt, "The standard model on non-commutative space-time," Eur. Phys. J. C **23**, 363 (2002) [hep-ph/0111115].

18. M. Chaichian, P. Presnajder, M. M. Sheikh-Jabbari and A. Tureanu, "Noncommutative standard model: Model building," Eur. Phys. J. C **29**, 413 (2003) [hep-th/0107055].

19. P. Aschieri, B. Jurco, P. Schupp and J. Wess, "Non-commutative GUTs, standard model and C, P, T," Nucl. Phys. B **651**, 45 (2003) [hep-th/0205214].

20. D. Bahns, S. Doplicher, K. Fredenhagen and G. Piacitelli, "On the unitarity problem in space/time noncommutative theories," Phys. Lett. B **533**, 178 (2002) [hep-th/0201222].

21. W. Behr, N. G. Deshpande, G. Duplancic, P. Schupp, J. Trampetic and J. Wess, "The Z $\to$ gamma gamma, g g decays in the noncommutative standard model," Eur. Phys. J. C **29**, 441 (2003) [hep-ph/0202121]; G. Duplancic, P. Schupp and J. Trampetic, "Comment on triple gauge boson interactions in the non-commutative electroweak sector," Eur. Phys. J. C **32**, 141 (2003) [hep-ph/0309138]; P. Schupp, J. Trampetic, J. Wess and G. Raffelt, "The photon neutrino interaction in non-commutative gauge field theory and astrophysical bounds," hep-ph/0212292; P. Minkowski, P. Schupp and J. Trampetic, "Non-commutative '*-charge radius' and '*-dipole moment' of the neutrino," hep-th/0302175.

LIST OF PARTICIPANTS

Dr. J. Garcia-Bellido
CERN
bellido@mail.cern.ch

Prof. Ikaros Bigi
Physics Dept., Univ. of
Notre Dame du Lac
bigi.1@nd.edu

Prof. Edward Blucher
The University of Chicago
e-blucher@uchicago.edu
Investigaciones Nucleares
acamacho@nuclear.inin.mx

Prof. Raymond Chiao
University of California, Berkeley
chiao@physics.berkeley.edu

M. Nicolas Dedek
LMU München
nicolas.dedek@cern.ch

Mrs. Dorothea Deeg
LS MukhanovTheoretische Physik,
LMU
dippa@deeg.net

Dr. Carl Dolby
Merton College, Oxford
dolby@thphys.ox.ac.uk

M. Mario Eichenseer
MPI für Quantenoptik
mre@mpq.mpg.de

Prof. Martin Faessler
Universität München
martin.faessler@physik.uni-muenchen.de

Dr. Christian Felix
University of Munich
felix@physik.uni-muenchen.de

M. Thomas Fraps
München
thomas.fraps@trick17.com

Dr. Eric Galapon
National Institute of Physics,
University of the Philippines
eric.galapon@up.edu.ph

Prof. T. W. Hänsch
Universität München
t.w.haensch@physik.uni-muenchen.de

Mrs. Nicole Helbig
Freie Universitaet Berlin
helbig@physik.fu-berlin.de

Prof. Frank Hinterberger
Univ. Bonn, H-ISKP
fh@iskp.uni-bonn.de

Prof. D. Hitlin
Caltech, Pasadena, USA
hitlin@hep.caltech.edu

Dr. C. Hull
Imperial College, London
c.hull@imperial.ac.uk

320

Prof. C. Jarlskog
CERN
Cecilia.Jarlskog@cern.ch

Dr. E. Joos
Schenefeld
ej@erichjoos.de

Dr. Alexander Kaganovich
Ben Gurion University
alexk@bgumail.bgu.ac.il

Dr. W. Kallies
JINR, Dubna
wkallies@off-serv.jinr.ru

Dr. Savely Karshenboim
D. I. Mendeleev Institute for
Metrology
sek@mpq.mpg.de

Prof. Claus Kiefer
University of Cologne, Theoretical
Physics
kiefer@thp.uni-koeln.de

Mrs. Christina Kraus
LMU München
*Christina.Kraus@physik.
uni-muenchen.de*

Dr. S. Krasnikov
St. Petersburg
redish@gao.spb.ru

M. Daniel Kremer
Tel-Aviv University
danikre@post.tau.ac.il

M. Danilo Mauro
Department of Theoretical Physics
University of Trieste
mauro@ts.infn.it

M. Eric Minassian
University of California
eminassi@landau.ucdavis.edu

Dr. Alexander Mück
Inst. f. Theor. Physik und
Astrophysik
mueck@physik.uni-wuerzburg.de

M. Thomas Müller
LMU München
mail@elfstone.de

Mrs. Christine Muschik
LMU München
cmuschik@aol.com

Prof. Renato Musto
Dip. Scienze Fisiche
Universita' di Napoli Federico II
musto@na.infn.it

Prof. Hermann Nicolai
Albert-Einstein-Institut, Golm
nicolai@aei.mpg.de

M. D. Oprisa
LMU München
dan-oprisa@yahoo.de

M. Matthias Ostermann
Theoretische Physik LS Wagner LMU
osterman@theorie.physik.uni-muenchen.de

Dr. Ekkehard Peik
PTB
ekkehard.peik@ptb.de

Prof. Huw Price
Univ. Edinburgh and Univ. Sydney
Huw.Price@ed.ac.uk

M. Denis Proskurin
Joint Institute for Nuclear Research
proskur@thsun1.jinr.ru

Dr. Irene Reinhard
GSI Darmstadt
i.reinhard@gsi.de

Dr. Alex Retzker
Tel Aviv University
retzker@post.tau.ac.il

M. Tobias Schneider
PTB Braunschweig
tobias.schneider@ptb.de

Prof. P. Schupp
Bremen
p.schupp@iu-bremen.de

Ms. Sabine Schwertel
Universität München
sabine.schwertel@ph.tum.de

Prof. F. Selleri
Bari

Dr. G. M. Shore
University of Wales, Swansea
g.m.shore@swansea.ac.uk

M. J. Simon
Regensburg
*johannes.simon@physik.
uni-regensburg.de*

Prof. L. Stodolsky
MPI München
les@mppmu.mpg.de

Prof. Leonard Susskind
Stanford University
susskind@stanford.edu

Prof. V. Telegdi
ETH Zürich
Valentine-Louis.Telegdi@cern.ch

Prof. G. 't Hooft
Spinoza Institute, Utrecht University
g.thooft@phys.uu.nl

Prof. G. Veneziano
CERN TH-Division
Gabriele.Veneziano@cern.ch

Prof. J. Walz
Mainz University
Jochen.Walz@uni-mainz.de

Dr. Lijun Wang
NEC Research Institute
lwan@research.nj.nec.com

M. Timo Wiegand
LMU München
Timo.Weigand@gmx.de

Prof. F. Zaccaria
Dipartimento di Scienze Fisiche,
Universit di Napoli 'Federico II'
zaccaria@na.infn.it

Prof. D. Zavrtanik
Nova Gorica Polytechnic
danilo.zavrtanik@ses-ng.si

Prof. W. Zurek
Los Alamos
whz@lanl.gov

Mr. Alexander Zvyagin
Universität München
alexander.zvyagin@cern.ch